Every Living Thing

BY JASON ROBERTS

A Sense of the World
Every Living Thing

Every Living Thing

THE GREAT AND DEADLY RACE
TO KNOW ALL LIFE

JASON ROBERTS

riverrun

First published in the United States in 2024 by Random House,
an imprint and division of Penguin Random House LLC.
This edition published in Great Britain in 2024 by

riverrun

An imprint of

Quercus Editions Ltd
Carmelite House
50 Victoria Embankment
London EC4Y 0DZ

An Hachette UK company

A CIP catalogue record for this book is available
from the British Library

Hardback ISBN 978 1 52940 046 5
Trade Paperback ISBN 978 1 52940 047 2
eBook ISBN 978 1 52940 045 8

10 9 8 7 6 5 4 3 2 1

Typeset by Jouve (UK), Milton Keynes

Printed and bound in Great Britain by Clays Ltd, Elcograf S.p.A.

Papers used by Riverrun are from well-managed forests and other responsible sources.

FOR
JESSE ELI

Contents

PART III ◎ GOD'S REGISTRAR

Carl Linnaeus and Georges-Louis de Buffon

Savants

Objects are distinguished and known by classifying them methodically and giving them appropriate names. Therefore, classification and name-giving will be the foundation of our science.

—LINNAEUS

The true and only science is the knowledge of facts.

—BUFFON

FOR MUCH OF THE EIGHTEENTH CENTURY, TWO MEN RACED each other to complete a comprehensive accounting of all life on Earth. At stake was not just scholarly immortality but the very nature of our relationship to nature—the concepts and principles we use to comprehend the living world. Their monumental works, entitled *Histoire Naturelle* and *Systema Naturae*, were far more than catalogues. They were embodiments of very different worldviews, attempting to substantiate those worldviews by fitting each piece of life's puzzle into a cohesive whole.

The task consumed their lifetimes. Both began believing that the Earth could not possibly hold more than a few thousand species, but as the decades passed, their grand compendiums remained vastly short of completion. The surprise of life's profusion, of its unexpected diversity and nuance, led them to develop even more starkly divergent views on the environment, on humanity's role in shaping the fate of our planet, and on humanity itself.

They were exact contemporaries, and polar opposites. Carl Linnaeus was a Swedish doctor with a diploma-mill medical degree and a flair for self-promotion, who trumpeted that "nobody has been a greater botanist or zoologist" while anonymously publishing rave reviews of his own work. Georges-Louis de Buffon, the gentleman keeper of France's royal garden, disdained contemporary glory as "a vain and deceitful phantom," despite being far more famous during his lifetime. Linnaeus, the anxious center of his own universe, preferred the company of student acolytes. Buffon, as renowned for his elegance as for his brilliance, cut a confident swath through the court of Versailles and the salons of Paris.

Linnaeus hewed closely to the Bible, while claiming divine inspiration for his own work. Buffon faced formal charges of blasphemy for suggesting the Earth might be older than Scripture indicated. Linnaeus thought his daughters unworthy of education, allowing them to learn only domestic skills. Buffon's closest friendship was with a woman, an accomplished intellectual he considered in many ways his superior.

Linnaeus blithely laid the groundwork for racial pseudoscience, not only creating the categories that would later be labeled "races," but assigning them fixed attributes (his *Homo sapiens europaeus* was inherently "governed by laws," while *Homo sapiens afer* was "governed by whim"). Buffon advocated passionately against the herding of humanity into rigid categories, emphasizing instead our vast, nuanced diversity and common origin.

Their rivalry ran deep. Buffon, with magnificent hauteur, publicly pitied Linnaeus as the "obsessed" victim of a mania, while Linnaeus took secret glee in naming a species *Buffonia* after his nemesis—a plant with slender leaves, as Buffon had "indeed very slender pretensions to a botanical honor." Yet both men had profoundly original minds. Both had astonishing capacities for work, pursuing their shared goal with unyielding discipline. As that goal continued to elude them, both persisted through chronic illness and bouts of staggering pain. Both attempted to raise their sons as successors, with tragic results. Both left behind legacies even more imposing than the sum total of their published works, legacies that contended for centuries. They still contend today.

Neither of them was a scientist—that term would not be coined until 1833, when the word *science* itself began to take on the meaning it has today. They were known in their time as *savants*, practitioners of a discipline that melded inquiry, philosophy, and a generous dollop of self-assumed authority. Yet many scientists of the nineteenth century eagerly claimed Linnaeus as a precursor, awarding him a posthumous fame that crowded the once-celebrated Buffon into obscurity. Gaining the upper hand during the chaos of the French Revolution and flourishing in the colonial expansionism of the Victorian era, pro-Linnaeus factions rushed not to disprove Buffon but to trivialize him, eviscerating his work while purporting to translate it. His *Histoire Naturelle* remained in print, but in abridged and adulterated versions that comprised near-parodies of the original.

Natural history, like human history, is written by the victors. For generations, adherents of the Linnean worldview were supremely confident of having won.

But flaws began to appear in that worldview as early as the 1860s, when Charles Darwin admitted that Buffon's theories were "laughably like mine." Attempting to reconcile Linnaeus's rigid hierarchies with the constant change of evolution, naturalists cobbled together increasingly elaborate taxonomies that were riddled with redundancies and errors. In the twentieth century, the development of genetics and the discovery of DNA further broadened our understanding of life, making even clearer the insufficiencies of existing systematics and pointing toward alternatives. In the twenty-first century, advances such as epigenetics and genome sequencing have prompted scientists to concede the limitations of the Linnean worldview, to debate plans for its replacement, and to consider Buffon's work and legacy anew.

This is the story of parallel lives lived on the grandest of scales, in pursuit of the core truths of our existence. The quest to know all life was an unwinnable impossibility, fueled by genius, hubris, the gleam of immortalizing fame, and the passion to simply understand—to know our world, and ourselves. There has never been a more human task.

—JASON ROBERTS

Every Living Thing

*The heroic statue of Buffon, commissioned
by King Louis XVI*

PRELUDE

The Mask and the Veil

AS JOURNALISTS SCRAMBLED TO ESTIMATE THEIR NUMBERS, some twenty thousand mourners lined the streets of Paris on the morning of April 18, 1788, jostling and craning to glimpse the funeral procession of one of the world's most famous men. What they witnessed was parade-like in its proportions, a solemn spectacle projecting "a luster rarely accorded to power, opulence, dignity," the *Paris Mercury* reported. "Such was the influence of this famous name."

First came a crier and six bailiffs, clearing the way for a convoy of nineteen liveried servants. Then a detachment of the Paris Guard marching in lockstep, followed by a contingent of schoolchildren, sixty clergymen, thirty-six choirboys chanting dirges to the accompaniment of four bass horns, and six guards bearing torches. At last the funeral carriage hove into view, drawn by fourteen horses clad in matching black silk with silver embroideries. In its wake passed a long, somber cortege of prominent mourners: aristocrats, academics, and artists walking shoulder to shoulder. One of them—Dr. Felix Vicq-d'Azyr, personal physician to Queen Marie Antoinette—was moved by the sight of ordinary people stepping from the crowd and marching alongside them. "You remember, gentlemen," he later reminisced,

> that innumerable retinue of people of all ranks, from all walks of life, who followed in mourning, in the midst of a huge and dismayed crowd. A murmur of praise and regret sometimes broke the silence of the assembly. The temple to which we walked could not contain this large family of a great man.

In life, Georges-Louis Leclerc de Buffon had held only the modest office of intendant, or keeper, of the Jardin du Roi—the King's Garden—which despite its name was far from any royal enclave, on Paris's working-class southern fringe. He'd begun inauspiciously, the son of a provincial tax collector in rural Burgundy, and earned no degree during his brief university studies, only a reputation for preferring dueling to studying. Yet he'd emerged as a leader of the intellectual revolution that would come to be known as the Enlightenment, a magisterial figure recognized throughout the world. Across the channel in London, *Gentleman's Magazine* was mourning Buffon as the last of the "four bright lamps" of France to be "totally extinguished," placing him in a pantheon alongside Montesquieu, Rousseau, and Voltaire. The historian Sainte-Beuve would go further, concluding that "Buffon, the last to vanish of the four great men of the eighteenth century, in a sense brought that century to a close on the day of his death."

The formal eulogy, delivered by the philosopher Nicholas de Condorcet, went further still. "The history of science presents only two men who by the nature of their works appear to come close to Monsieur Buffon—Aristotle and Pliny."

> Both indefatigable workers like him—amazing in the immensity of their knowledge and the plans that they have conceived and

executed, both respected during their lives and honored after their deaths by their countrymen—they have seen their glory survive the revolutions of opinions and of empires, survive the nations which have produced them and even the languages which they used, and they seem by their example to promise Monsieur Buffon a glory no less durable.

There was no talk of erecting a monument to the great man, as one already existed: a towering statue, secretly commissioned by a grateful King Louis XVI eleven years earlier and unveiled, much to its subject's embarrassment, as the Jardin du Roi's centerpiece. Bearing a fulsome inscription (*All nature bows to his genius*) and portraying a near-nude Buffon, the marble figure was by all accounts a remarkably accurate depiction of the man himself, who despite being seventy-one at the time of the unveiling appeared decades younger. The sculptor found little need to idealize, as Buffon had long struck observers as something of a heroic statue come to life. "An advantageous height, a noble air; an imposing face, a physiognomy at once pleasant and majestic," ran Condorcet's description in his eulogy.

"Monsieur Buffon has never spoken to me of the marvels of the earth," one longtime friend admitted, "without inspiring in me the thought that he is one of them."

Another monument, even more impressive, resided not in the Jardin but on bookshelves throughout the world. *Histoire Naturelle, Générale et Particulière*, Buffon's masterwork, stood at thirty-five volumes—three introductory ones on general subjects, twelve on mammals, nine on birds, five on minerals, and six supplemental volumes on miscellaneous subjects. It was, in one sense, a failure: His original plan had been to encompass "the whole extent of Nature, and the entire Kingdom of Creation," but despite forty-three years of single-minded dedication he'd fallen short of properly addressing plants, amphibians, fish, mollusks, or insects.

It was a staggering achievement nonetheless. Even incomplete, the *Histoire* dwarfed most encyclopedias, a task made even more impressive by his laborious writing process. Buffon famously read drafts aloud to friends, asked them to paraphrase what they heard, then rewrote anything muddled or misunderstood, a process he repeated up

to seventeen times for a single chapter. "Good writing is good thinking, good awareness, and good execution, all at once" was his dictum.
This painstaking emphasis on clarity, combined with his choice of
language—he wrote in contemporary French, not academic Latin—
had been a key to the work's enduring success.

To say that *Histoire Naturelle* was a bestseller is an understatement.
Hailed as both a landmark of science and a work of literature, it was a
publishing phenomenon that made Buffon the most popular nonfiction author in French history. Each new volume's release sparked a
flurry of sales and a round of public debate, and seemed poised to
continue to do so. Also walking in the funeral cortege was Buffon's
handpicked successor, a young French nobleman who'd carefully
trained in his mentor's methods and distinctive style. He'd already
published Volume 36 of the *Histoire*, on serpents and egg-bearing
quadrupeds, and was preparing Volume 37, on cetaceans. The great
man had passed away. The great work would go on.

That, at least, was the plan.

. . .

The death of Professor Carl Linnaeus, a decade before Buffon's passing, had not been an occasion for grand public mourning. The long-
retired academic was delivered to his rest on a winter evening in a
plain coffin, hewn from one of the yew trees on his property, his unwashed, unshaved body wrapped within a single winding sheet. The
torchbearers following the hearse were his tenants and servants. Most
in attendance had walked over from nearby Uppsala University, where
Linnaeus had taught medicine and botany for twenty-two years.

Many of them, students and faculty alike, knew him more by reputation than by experience. It had been fifteen years since the professor's progressive brain disorder compelled him to abandon his
appointment, five years since he had lost touch with reality, two years
since he could walk or talk. The ceremony was respectful, but tinged
with the awareness that the mind had failed well before the body. Linnaeus had spent his final years in a labyrinth of confusion, leafing
through his greatest work without comprehending that he himself
had written it.

No monuments had been erected, nor were any proposed.

Both Buffon and Linnaeus had been born in 1707. Both devoted themselves to compiling a massive work intended to capture the whole of nature, and neither had succeeded. But there the resemblance ended. Linnaeus was the foremost figure among the school of natural historians known as systematists, who prioritized naming and labeling above all other pursuits. Buffon was the leading practitioner of a more complex approach to nature—a perspective that, appropriately, he never saw the need to label. It may best be called complexism.

To Linnaeus's mind, nature was a noun. All species remained as created during Genesis, representing an unchanging tableau. To Buffon, nature was a verb, a swirl of constant change. To Linnaeus, classification *was* knowledge: How could life be understood without organizing it into tidy categories? Buffon believed that to classify was to oversimplify; that while useful for practical purposes, classification ran the risk of embedding fundamental misunderstandings. Linnaeus believed that a single specimen could exemplify the species it belonged to, displaying a distinguishing "essence." Buffon thought species were far more fluid, and that unknown forces connected them across an expanse of time.

Linnaeus trafficked in certainties, and congratulated himself heartily for doing so. As he described himself in one of his autobiographies (he wrote four),

> God himself has guided him. . . . He has let him look into His secrets and let him see more of his created works than any mortal man before him. God has given him the greatest insight into natural history, greater than anyone else has enjoyed. God has been with him, wherever he has gone, and has eradicated all his enemies for him and made him a great name, as great as those of the greatest men on earth.

Buffon, in contrast, had come to believe that the only way to study nature was in a state of permanent uncertainty—a willingness to compile observations and explore connections, but to maintain a sense of wonder, an expectation of surprise. Their worldviews were fundamentally different, as illustrated by Buffon's employment of a metaphorical mask and veil. "The greatest obstacles to the advancement of human knowledge lie less in things themselves than in our manner of considering them," he wrote. "Nature . . . carries only a veil, while we would put a mask over her face. We load her with our own prejudices, and suppose her to act and to conduct her operations even after the same fashion as ourselves."

To him, systems were masks imposed upon nature. They represented an urge to see nature as we wished it to be, not as it truly was. It was humbling to patiently observe, only occasionally catching glimpses beneath the veil, but to Buffon there was no other choice. Nature was profoundly, abundantly complex—perhaps beyond human understanding, as humans themselves were part of the equation. To sort into tidy categories was to deny life's inherent interrelation. "Nature, displayed in its full extent, presents us with an immense tableau," he wrote,

> a continuous series of objects, so close and so similar that their difference would be difficult to define. This chain is not a simple thread which is only extended in length, it is a large web or rather a network, which, from interval to interval, casts branches to the side in order to unite with the networks of another order.

If the crowds and tributes on that April morning were any measure, Buffon's complexist worldview had triumphed: Nature was a dynamic unity to be contemplated, not a static entity to be conquered. But within five years Buffon would be reviled as an enemy of progress, dismissed as a best-forgotten symbol of an irrelevant past. His earthly remains would be carelessly tumbled from the very coffin that had been solemnly borne through the streets, by scavengers stripping the coffin of scrap metal. A torch-bearing crowd would stream through the gates of his beloved Jardin, ignoring his monumental statue—its placement made it difficult to destroy—and clamoring to install an image of the man whose life's work stood in parallel and antithesis to his own: a plaster bust, hastily obtained and painted to resemble marble, of Carl Linnaeus.

PART I

The Great Chain of Being

There is nothing more difficult to plan,
more doubtful of success,
nor more dangerous to manage
than the creation of a new system.

—MACHIAVELLI

A nineteenth-century depiction of Linnaeus as child botanist

ONE

Of the Linden Tree

THE LANDSCAPE OF SEVENTEENTH-CENTURY SWEDEN WAS DOT-
ted with natural shrines. These were linden trees, viewed in a semi-
mystical light by Swedes who associated their heart-shaped leaves and
fragrant blossoms with Freyja, the Nordic goddess of love and fertility. Pregnant women visited them to gather leaves, hoping to invoke
Freyja's protection during childbirth by sleeping on pillows stuffed
with them. Travelers sought shelter under their branches during storms,
believing that Thor deferred to Freyja by aiming his lightning strikes
elsewhere. Their silver trunks were favored sites for both romantic
declarations and business negotiations, since Freyja was thought to
punish anyone speaking lies beneath their boughs.

One ancient linden tree was especially revered. It grew near the border of the southern Swedish province of Småland, in a field straddling Hvittaryds and Jonsboda parishes, rising in a massive triple trunk to spread a canopy shading most of an acre. Nestled near its roots lay a cairn of piled stones, dating to the Bronze Age and thought to mark the resting place of a Viking warrior. The natives of Småland had long since declared the tree and its grounds a *våarträd*, a treasured landmark believed to extend its protection to all of the surrounding countryside.

For the family that owned the land, that brought both honor and a special duty of care: Våartäds were customarily left as undisturbed as possible, even when they occupied what would otherwise be productive farmland. For generations the family tended to the tree. When twigs or branches fell, they carefully gathered them up (it was bad luck to break even a single one) and placed them in neat piles atop the roots, leaving them to weather in peace. They erected a perimeter fence to protect the tree from the bunting, rubbing, and ringbarking of grazing animals, but kept a path to the trunk accessible to visitors.

The family had no name. This was common in rural Sweden, where surnames were rarely necessary; one's lifelong presence on ancestral land served as identity enough. Most males, when required to provide one, adopted the simple patronymic of adding *-son* to their father's given name, a practice that would eventually populate Sweden with a healthy percentage of Johannsons, Petersons, and Svensons. But in the fall of 1690, a sixteen-year-old named Nils made a different choice. Instead of becoming Nils Ingemarsson, he decided to commemorate his family's tree by adopting the surname Linnaeus. It meant "man of the linden tree," not in Swedish but in Latin.

The choice betrayed ambition. Nils was leaving to attend the University of Lund, 150 miles away, where Latin was the language of academic discourse. Many professional scholars adopted Latin versions of their surnames, which is why Michel de Nostredame is better known as Nostradamus, and why Nikolaj Kopernik is remembered as Nicolaus Copernicus. Assuming a latinized name in advance was a jarring, presumptuous choice in the province of Småland. But provincial life was what Nils intended to leave behind.

. . .

Nils Linnaeus did not stray as far from the linden tree as he had hoped. His family's farming income was enough to send him to the university, but not enough to keep him there. He'd arrived with barely the money for a single year's tuition, hoping to fund the rest with a combination of scholarships and part-time work. When those failed to materialize, he abandoned his education and for most of the next decade wandered through Sweden and neighboring Denmark, establishing himself nowhere and in no profession. He was by all accounts a friendly presence, a genial, outgoing soul. He was also in no particular hurry to make his way home.

He was twenty-seven when he finally returned, to familial disappointment and the challenge of finding a livelihood. Lacking the resources to set himself up as a farmer in Småland, he began studying to join the Lutheran priesthood. Two years later, the freshly ordained Reverend Linnaeus found an entry-level posting as a comminister, or auxiliary curate, in the village of Stenbrohult, a farming community on the shores of Lake Möckeln. It was only twenty miles from his birthplace and his namesake tree. In 1706, he married Christina Brodersonia, the daughter of his immediate superior, who changed her name to Christina Linnea ("woman of the linden tree"). On May 23, 1707, she gave birth to their first child, a son. In honor of the king of Sweden, they named him Carl.

The birth and early childhood of Carl Linnaeus is densely ornamented, with family legend and later attempts to render him a kind of secular saint, destined for green paths of glory. Apocryphal accounts hold that the boy was born with a head of snowy white hair, the mark of a *skogsanda*, or forest spirit, which later darkened to brown. The infant was strangely fussy and inconsolable, soothed only by his mother placing bouquets of flowers above his crib. "Flowers became Carl's first and choicest plaything," an early biographer wrote. "The father took the little year-old son out with him sometimes into the garden, putting the child on the ground in the grass and leaving a little flower in his hand with which to amuse himself."

There were flowers in abundance. By now Reverend Linnaeus had succeeded his father-in-law as curate and moved his family to the rectory of Stenbrohult, where, free from the need to travel throughout the parish, he took up gardening as a hobby. The reverend's garden took on formal proportions, encompassing several hundred floral varieties and sporting as its centerpiece a "feast" of flowers—a round, raised soil bed in the shape of a table, set with plantings carefully tended to look like heavily laden dishes. Shrubs, planted at table's edge, were trimmed in topiary fashion to perform the roles of dinner guests. Young Carl is said to have spent hours in their imaginary company, and more hours still tending to a garden patch of his own nearby. According to one early biographer, the boy was "for ever searching fields and woodlands in quest of flowers. . . . His loving mother complained that no sooner had he got a new flower than he cruelly pulled it to pieces, for the little fellow loved to penetrate, as far as it was possible, into the secrets of nature." There is even a story of young Carl being caught surreptitiously pressing flowers between the pages of the family Bible. "The Bible is the Book of Life," he purportedly explained, "and surely if I put the flowers between its leaves they would retain their color, the Bible keeping them alive for ever."

Such accounts, if true, were irrelevant: The boy's profession had been fixed at birth. Just as Nils Linnaeus had replaced his father-in-law as rector of Stenbrohult, Carl was intended to succeed him, thus representing the fifth generation to occupy the hereditary pulpit that had been his mother's dowry. Tending a garden was a hobby for the father, a respite from tending to souls. It could be his son's hobby as well, but nothing more.

· · ·

Carl would later pinpoint the moment his fascination became an obsession. It was on a bright spring day in 1711, not long after his fourth birthday. The weather was so beautiful that many of the citizens of Stenbrohult put aside their chores to enjoy a picnic at Möklanäs, a meadow on a promontory jutting out into Lake Möckeln. Afterward, as the crowd relaxed in the lush grass, Reverend Linnaeus volunteered to entertain by delivering an impromptu lecture on botany. "The guests seated themselves on some flowery turf," Carl later recollected, as his father pulled up a few nearby specimens and "made various remarks on the names and properties of the plants, showing them the roots of the *Succisa, Tormetílla, Orchides*."

Even decades later, the Latin names of the random specimens rang clear in his mind. Resorting to the third person, he later described the moment:

The child paid the most uninterrupted attention to all he saw and heard, and from that hour never ceased harassing his father with questions about the name, qualities, and nature of every plant he met with; indeed, he very often asked more than his father was able to answer.

Carl's irrepressible new curiosity led to some tense moments in the family garden. As he later admitted (again in the third person), he would ask about a plant, "but like other children, he used immediately to forget what he had learned, and especially the *names* of plants" (italics his). Tired of repeating himself, the father gave his son an ultimatum: He would describe and name a plant, but only once. For the rest of his life, Carl Linnaeus would thank his father for two gifts: his introduction to botany, and "this harshness" of instruction, which honed his memory at an early age, "for I afterwards retained with ease whatever I heard."

In 1717, his father deposited ten-year-old Carl at the Trivial School of Växjö, thirty miles from Stenbrohult, arranging for his room and board and admonishing him to acquire "useful material for the ministry." The Trivial School was so called because it taught the three subjects of the classic academic trivium: grammar, rhetoric, and dialectic. It was a rarified curriculum: The grammar was Latin and Greek, the rhetoric grounded on Aristotle, and the dialectics drawn from Socrates. Trivial schools tended to overwhelm even the most intelligent provincial youths—which, as Carl quickly discovered, led to the schoolmasters "preferring stripes and punishments to admonitions and encouragements." By his second year at Växjö, Carl was slogging to his morning classes with a growing sense of dread.

He became, at best, a mediocre student. Although he would spend the bulk of his career writing in Latin, his command of that language would always be more workmanlike than elegant; even his Swedish would be ridiculed by more sophisticated colleagues. He spent the first five years mostly confined to the Trivial's small grounds, until his upperclassman status gave him the right to venture outside the school. Then he spent hours walking alone, in the forests and fields that skirted Växjö. If he made a friend, he does not mention it in any of his four autobiographies. He does, however, record that both students and teachers were now calling him *den lilla botanisten*, the Little Botanist, a reference to his stature (he would never grow past five feet tall), and his growing obsession. He would return from his solitary walks with clutches of flowers and leaves, which he pressed between the pages of his books. Desperately unhappy, homesick for a Stenbrohult where "Flora seems to have lavished all her beauties," as he now imagined, Carl retreated further and further into his private expertise. His father's garden seemed unattainably distant, a post-expulsion Eden. "Let the child enjoy his paradise," he would later write. "It will be driven from it by care soon enough."

The demands of the school only increased in his teenage years. While he did well enough in mathematics and physics, he performed abysmally in Hebrew, metaphysics, and theology, consistently ranking as one of the school's worst students. His mother and father, however, remained unaware that anything was wrong, and Carl did not confess his misery on his rare visits home. But by his seventh year at Växjö, the Little Botanist's dread was tinged with desperation. He was failing— *had* failed—to acquire useful material for the ministry. A reckoning would come.

· · ·

It came the following year, when Reverend Linnaeus contracted a minor but lingering illness. On his way to consult Johann Rothman, a doctor with a practice in Växjö, he decided to drop by the Trivial School for an impromptu visit. As Carl himself later recalled, his father was "hoping to hear from the preceptors a very flattering account of his beloved son's progress,"

> but things happened quite otherwise. . . . It was thought right to advise the father to put the youth an apprentice to some tailor or shoemaker, or some other manual employment, in preference to giving him a learned education for which he was evidently unfit.

The reverend could not contain his shock. *A tailor or a shoemaker.* It was a profound disappointment, to be sure, but also a major financial blow. The cost of boarding and schooling Carl for nine years had been a significant hardship. Now there would be more expenses to come, since Carl's younger brother Samuel would have to be trained in his stead. More pressing was the immediate question: What to do with Carl? The school's suggestion of apprenticing him to a trade had come too late. He was no longer a boy, but a young man of nearly twenty. It was difficult to imagine any manual tradesman taking the measure of the physically small, perpetually distracted young Carl and accepting him as an apprentice.

In the office of Dr. Rothman, an old friend, Reverend Linnaeus confessed his dismay. Rothman listened sympathetically, then confirmed that the harsh assessment was probably correct: He also served as the Trivial's part-time physics instructor, where he'd come to know the younger Linnaeus as a slogging, unmotivated pupil. But at the same time, he'd recognized Carl's clear intelligence and capacity for obsessive focus. Perhaps, he suggested, another scenario was possible.

At the time, botany was scarcely a profession in itself. It was the realm of the hobbyist, or the independently wealthy dilettante. There were professors who taught botany as a discipline, but as part of a medical curriculum, since knowledge of plants and their uses was a key aspect of medicine. Had Nils considered making a doctor of his son? Rothman offered to take Carl into his home and train him for one year. It would be an informal apprenticeship, but with any luck Carl would emerge prepared to attend medical school.

The proposal did not fill Nils Linnaeus's heart with joy. Medicine held far less social cachet in Sweden than being a member of the clergy, and Carl had never evinced an interest in the subject. He feared that his daydreaming son would find work only as a military surgeon, the least-respectable member of the medical class, treating wounds on the battlefield and syphilis in the barracks. But it seemed there were no better options. Nils assented and departed for Stenbrohult, still wondering how he was going to break the news to his wife.

Released from the tedium of the Trivial, Carl proved a willing apprentice, quitting the school and moving in with his temporary master. The villagers of Växjö grew used to the sight of the Little Botanist doing his best to transform himself into the Little Physician, shadowing the doctor as he went about his rounds. But when Rothman's year of instruction was up, Carl returned home to Stenbrohult and a strained reunion. The reverend had concealed from his wife the truth about their son's academic failure for as long as possible, and when she did find out, the news of a hastily arranged change of profession had been no consolation. Both angry and disappointed, Christina Linnea forbade any mention of gardening and botany within the house. Yet she gave her cold consent to Carl's attending medical school in the fall.

Carl left home on August 17, 1727. He carried with him a letter of recommendation from his former schoolmasters, technically required for registration at medical school but so disparagingly worded that he would never bother to submit it. He also carried a purse of silver Swedish dalers, a gift from a father who made it clear that, regrettably, no more funds would be forthcoming. It was enough for a year at best. After that, he'd have to improvise.

A statue of young Linnaeus, Uppsala Botanical Garden

TWO

A Course in Starvation

ON APRIL 19, 1729, IN THE SWEDISH CITY OF UPPSALA, PROFESsor Olof Celsius disappeared into the weeds and brambles of a neglected field. Plump, dressed in black clerical robes, his face framed by a powdered wig and a prodigious goatee, the middle-aged academic cut a dignified figure, which made all the more incongruous his sudden departures into an overgrown thicket fringed by grazing cows. To casual passersby the land seemed a vacant lot, but Celsius knew its secret. It held the ruins of a garden.

Nearly a century earlier, another Uppsala professor had planted a private teaching garden here, an open-air classroom designed to give student physicians hands-on experience in identifying medically useful plants. The collection flourished for generations, growing to nearly two thousand botanical varieties (including, in a first for Sweden, a curiosity known as the potato). What had happened to the Uppsala Botanical Garden? The same thing that had happened to Uppsala itself. For centuries, Uppsala had rivaled Stockholm for the status of

Sweden's foremost city. While Stockholm was the commercial hub and seat of government, Uppsala was its cultural and religious center, the headquarters of the Swedish church, home to northern Europe's oldest university and the coronation site of Swedish kings and queens (who traditionally maintained castles in both Stockholm and Uppsala). But the rivalry ended abruptly in 1702, when a fire of unknown origin swept through Uppsala, fed by strong winds blowing through narrow medieval streets like air through a bellows. The Great Fire reduced three-quarters of the city to smoldering ruins, bringing on an eclipse from which Uppsala would never quite recover. Instead of repairing Uppsala Castle, Sweden's king hauled away stones from the rubble to use in his Stockholm palace.

No longer a metropolis, Uppsala rebuilt on a smaller scale. The city was still rebuilding twenty-six years later, and the old teaching garden was no one's priority. Only a tenth of the plants had survived the fire. Now they grew mostly of their own accord, choking off pathways, twining across still-charred ground. Yet portions of the interior could be navigated, and Professor Celsius found it a fitting place to gather his thoughts. He was writing, or rather attempting to write, a book about plants.

It had seemed like a straightforward project. With a working title of *Hierobotanicum*, or *Priestly Plants*, it sought only to provide details on the 126 plants mentioned in the Old and New Testaments. But Celsius, despite being one of Sweden's foremost biblical scholars, was mired in uncertainty. Knowing the names of those 126 plants, he'd come to realize, was not the same as identifying them.

What, for instance, was the "hyssop" mentioned in Leviticus, Numbers, Exodus, and Psalms? The Bible cites such a plant twelve times, but in Celsius's time the name *hyssop* was attached to at least five plants: an herb, a nettle, an aquatic plant, a wildflower, and a variety of anise. Furthermore, *hyssop* was a transliteration from Greek; in Hebrew the plant is called *ezob* or *ezov*, which may or may not be another thing entirely. Why did it matter? The Bible specifies hyssop as a key ingredient in purifying a church, ridding a house of leprosy, preparing a corpse for burial, and properly sacrificing a red heifer. Celsius had no intention of performing these rituals, but an intriguing theme of sanitation ran through them all. Did the biblical hyssop hold disinfectant qualities, or safeguard against contagious disease? Such insights could only be determined by first deciding which, if any, of the five hyssops was the "true" plant in question. It was not a decision Celsius felt qualified to make.

In the ruined garden, Celsius was usually free to ponder such matters in uninterrupted solitude. But on this particular spring day, as he rounded the remnants of a path, he saw another visitor. A young man was sitting on a bench, intently scratching away in a notebook. He was small, not more than five feet tall, and so slightly built he seemed almost elfin. He wore no wig, which marked him as something less than a gentleman of means. His clothes were not only shabby but mismatched and ill-fitting, as if they'd been stolen or scrounged. A threadbare coat hung loosely on his thin shoulders. Scraps of newspaper showed through holes in his shoes.

Uppsala University was full of students of minuscule means, just barely scraping by, but this stranger seemed closer to a beggar than a scholar. Celsius drew closer and noticed the young man was not writing but drawing from life, capturing a nearby flower in crude, graceless strokes. The clear lack of artistic intent (and skill) meant the work in progress was not a still life but a field schematic. The flower was a specimen.

"What are you examining?" the professor asked.

The stranger answered politely. But instead of giving the plant's Swedish name he cited a little-known technical designation, one given by the French botanist Joseph Pitton de Tournefort. This was impressive. Celsius knew that the Tournefort system of plant identification was notoriously difficult to master, requiring the rote memorization of 698 distinct categories. "Do you know about plants? Have you studied botany?" Celsius asked. "What is your name?"

"Carl Linnaeus, sir."

"Where are you from?" he asked, although much of the answer was already apparent in the young man's manner of speech. He spoke with the airy, unemphatic lilt of the rural provinces, in an accent thick enough to mark him of peasant stock.

Celsius started to point. "Do you know the name of that plant? How about that one?" The Linnaeus fellow named them each in turn. Then, on his own accord, he began identifying the weeds as well.

More than just a diligent student, then. A serious, self-directed botanist in his own right. "How many plants have you collected and pressed?" Celsius asked.

"More than six hundred native wildflowers." That was three times the number of surviving plants in the Botanical Garden.

Celsius looked at the unusual young man, who spoke like a university don despite his shabby dress. "His eyes were full of plants, but his stomach was achingly empty most of the time," the famed naturalist John Muir would later write of this juncture in Linnaeus's life. "A course of starvation, it would seem, is a tremendous necessity in the training of Heaven's favorites."

Celsius made an impulsive decision. He and his wife kept a spirited, crowded household, with several children and a busy kitchen. "Come with me," he said, turning abruptly and walking in the direction of his home, three blocks away. He gave no reason why he should be followed, nor had he given the stranger his name.

· · ·

The Carl Linnaeus who sat at Celsius's table a few minutes later, wolfing down a meal, was a far leaner and more hardened figure than the youth who'd left Småland province two and a half years earlier. His pursuit of a medical education had taken him through two universities, and to the very brink of poverty.

His first stop had been the University of Lund, his father's alma mater near the Baltic Coast. In his father's day it had been a thriving university town, but a subsequent string of misfortunes—fires, an outbreak of the bubonic plague, alternating occupations by Swedish and Danish soldiers—had ravaged what was once the *Londinum Gothorum*— the London of the Goths—and the largest city in Scandinavia. By the time of Linnaeus's arrival, Lund was in a state of creeping ruin, its population down to fewer than twelve hundred. Entire neighborhoods had been abandoned, left to flocks of geese and foraging packs of feral pigs. The University of Lund's medical faculty had been reduced to a single professor, the elderly Johan von Döbeln, who mumbled through perfunctory lectures with only a few dozen students in attendance.

Linnaeus had no choice but to make the best of it. He rented an attic room in the house of Killian Stobaeus, a physician unattached to the university. Stobaeus had an impressive collection of texts on medicine, geology, and fossils, but the library was kept locked and off-limits to all but his assistant, a young German medical student named Koulas. Linnaeus struck a deal with Koulas: He would tutor him in exchange for illicitly borrowed books. The assistant smuggled the books down to Linnaeus, who read them through the night, then returned them in the morning before their landlord could discover them missing. The scheme only lasted until the insomniac Stobaeus, roaming the house with a candle at two o'clock in the morning, discovered his boarder asleep at a table, a number of forbidden books by his side.

Stobaeus was more impressed than angry. The doctor prodded him awake, quizzed him on his furtive studies, and grew more impressed still. To Koulas's dismay, Stobaeus was soon treating Linnaeus as his protégé, giving him the run of the house and free meals, and taking him along on his medical rounds. "He loved me not as a student but rather as his son," Linnaeus later recalled. Yet he could not quite bring himself to be grateful. Dr. Stobaeus's patronage did not change the fact that Lund was a dead end, a moribund medical program in a decaying city.

With no opportunities to supplement his limited funds, he took to once again spending his spare hours wandering the adjacent wilderness, seeking out specimens to add to his collection. In the spring of 1728, he was exploring the nearby marshes of Fagelsong when he felt a sting in his right arm. He ignored it at first, but within hours the flesh was inflamed and swollen. Soon he was bedridden, feverish, in increasing pain, and beyond the help of Dr. Stobaeus, who summoned a surgeon colleague named Schnell. The surgeon incised a deep cut, leaving Linnaeus with a scar from elbow to armpit and (in the patient's opinion) saving his life. The source of the sting was a mystery, but he came to believe he'd been attacked by a slender, airborne worm "the thickness of a human hair, grey with black extremities," a creature that darted down from the sky to inject itself deep within the victim's flesh. He invented a name for this creature, calling it the Shot, to convey its speed and capacity for injury. He would later assign it the scientific name *Furia infernalis*. The fury of hell.

Once he was sufficiently recovered, Linnaeus moved on to the much larger University of Uppsala. He immediately regretted the decision.

. . .

Uppsala, Linnaeus found, was only marginally better than Lund. While the university had more than a thousand students in residence, its school of medicine was a near-mirage. The program sounded impressive enough, offering instruction in anatomy, botany, zoology, theoretical and practical medicine, surgery, physiology, and chemistry. In reality, all these subjects were taught by two elderly professors, and only when they chose to do so. Abetted by a seniority system that made them nearly unaccountable to anyone, they scarcely showed up to teach, instead appointing assistants to read aloud from lecture notes compiled years, even decades, earlier.

Their languid proprietorship was now neglect. Neither anatomy nor chemistry had been taught for years. Instruction in practical medicine was given with only rare recourse to actual patients. The school's hospital was so little used that a portion had been converted into a tavern. The department's zoology collection consisted of little more than a stuffed "dragon" (likely a lizard) and a six-inch-long, two-headed snake. Botany was a subject so rarely taught that Linnaeus would never have the opportunity to take a single class on the subject. The botanical garden, as mentioned, was an overgrown field. The currency of an Uppsala medical education was so degraded that it no longer qualified its holder to practice medicine within Sweden. The program stopped short of actually awarding medical doctorates, obliging its students to earn their diplomas elsewhere. Most left Sweden to do so.

For Linnaeus, the challenge of paying for a third medical school was daunting, but a worry for the future. He began his Uppsala studies by attending the only lecture then on offer, a multi-week discourse on ducks, chickens, and the medicinal use of poultry. In December of 1728, he succeeded in earning a small scholarship, and he used it to travel to Stockholm, where a female convict was about to be hanged. To compensate for Uppsala's missing anatomy lessons, he paid sixteen dalers to view the dissection of her corpse.

That all but depleted his funds, and his options. By January of 1729, disappointment was being crowded out by hunger. Linnaeus, by his own account "obliged to trust to chance for a meal," was borrowing money and accepting donations of clothing from other students. He shuddered through the harsh Swedish winter in a cheap rented room. When his last pair of stockings wore out, he cut off the feet and wore the resulting sleeves around his calves, needing the warmth of the cloth. When holes appeared in the soles of his shoes, he plugged them with newspaper. Still enrolled as a student but unable to afford lectures, Linnaeus now spent hours in the medical library, poring over botanical texts. There were a few impressive rarities in the collection, chief among them the *Hortus Siccus*, or Dry Garden, which in lieu of illustrations had pressed plants attached directly to its pages. It was a massive work of more than three thousand pages in twenty-six volumes, and was so valuable that the Swedish army had extracted it from Denmark as a spoil of war. But Linnaeus had come a long way, and paid a dear price, just to look at dried leaves.

There was no clear way forward, and no turning back. The hereditary parsonage, once his birthright, had already passed on to his younger brother Frederick. Linnaeus dreamed of leaving Uppsala and retreating to Lund, where he might at least beg his old benefactor Dr. Stobaeus for a second chance. As the weather grew warmer, he began to augment his library sessions with solitary hours in the old botanical garden, sketching the blossoms as they emerged.

The garden, in all its ragged beauty, was beginning to look like the terminus of ambition, a last way station before he slipped into vagrancy—until Professor Celsius strolled by, glanced at his sketchbook, and demanded to know the names of plants.

Celsius's impulsive invitation led to further meals, and soon to an offer the ragtag scholar grasped at without hesitation: room and board, in exchange for helping the professor compile his book of biblical plants. Linnaeus's contributions to the *Hierobotanicum* would prove minor, since Celsius would continue to work on the manuscript for another eighteen years. But he would ultimately return Celsius's kindness by naming a plant *Hyssop officialis*, an intended resolution to the nagging question of the "true" hyssop.

A bitter herb, *Hyssop officialis* does indeed grow in the regions men-tioned in the Bible. But as Linnaeus never explained the logic behind his choice, its designation seems to arise from his self-proclaimed cer-tainty. As far as modern historical and linguistic analysis can deter-mine, the true "true" hyssop is *Capparis spinosa*, the caper bush.

Georges-Louis Leclerc, later styled de Buffon

THREE

The Salt-Keeper's Son

NO MYTHOLOGY SURROUNDS THE CHILDHOOD OF GEORGES-Louis Leclerc. Born in the rural Burgundy village of Montbard on September 7, 1707, only five months after Linnaeus, he was, by all accounts, an unremarkable youth. "One can cite of his childhood, and even of his adolescence," a relative would recall, "only those traits which are common to all children graced with some degree of native wit." As a student he was assessed as possessing "no pronounced superiority," neither failing nor excelling in his schoolwork. He was agreeable enough in appearance, most notably for a pair of brown eyes so dark they appeared nearly black. Yet the eyes held no particular spark, betrayed no particular ambition. This disappointed no one, as young

Georges-Louis had no expectations beyond assuming his father's slightly distasteful profession.

To the modest extent possible in seventeenth-century rural France, the Leclercs of Montbard exemplified upward mobility. The boy's great-great-grandfather had been a farmer who'd wandered from his fields into the village, where he set up shop as a barber. Each subsequent generation had climbed another rung on the ladder of social respectability. The barber's son became a doctor, the doctor's son became a judge, and the judge's son obtained an important federal office. After paying the customary fee to King Louis XIV (such offices were openly for sale), Benjamin-François Leclerc was appointed the regional administrator and enforcer of the *gabelle*, France's national tax on salt.

Few taxes have been more hated than the gabelle. Instituted in 1259 as a simple 1.66 percent sales tax on salt, it was grumbled against from the start as an inequity: The flat rate was a burden for the poor and a bargain for the rich. But salt was seen as one of life's necessities, essential not only as a condiment but as a food preservative in the days before refrigeration. This guaranteed the king a stream of income so steady that over subsequent centuries French monarchs kept enlarging it by piling compulsory purchases upon increasingly draconian rules. By the time Benjamin-François Leclerc signed on as gabelle-master of Montbard, the salt tax was the French Crown's largest source of income.

The difficulties of controlling consumption of a common mineral gave rise to drastic measures. Salt ponds, salt mines, and even the boiling of seawater were strictly regulated—you could, in theory, harvest your own salt, so long as you then submitted it to the government and arranged to buy it back. Shepherds were forbidden to let their flocks drink from salty water. Border guards inspected travelers' possessions for smuggled salt, driving spikes through luggage and inspecting the tips for white contraband. Especially resented were the *gabelous*, armed agents who roamed the countryside in search of illicit smuggling within France itself. The gabelle was collected at different rates in different districts: On one bank of the Loire River, for instance, the tax was eighteen times higher than on the other. Such disparities tempted black-marketeers and poor farmers alike, which led the gabelous, without the burden of warrants or probable cause, to conduct raids with impunity. More than three thousand men, women, and children were imprisoned or executed each year for salt tax–related crimes. If they had the audacity to die in custody prior to trial, their bodies were preserved in salt. The cost was charged to their families.

An arrest for violation of the gabelle

It was impossible to avoid the gabelle simply by consuming less salt. Most citizens over the age of eight were required by law to purchase an annual quota, sold directly from government warehouses at massively inflated prices. In Burgundy, that quota was slightly more than fifteen pounds of salt per person per year—a stroke-inducing if not fatal amount if actually consumed. As a further indignity, the salt was doled out in obligatory weekly portions, imposing a much-resented routine upon daily life. In Montbard, at least one member of each household was obliged to visit the salt-keeper Leclerc in his warehouse on their allotted day, wait patiently as he registered their payment, then cart home a rarely needed package. The tax on salt had become a ritual tinged with absurdity, and a tax on being alive.

Georges-Louis grew up knowing why no one wanted to visit his father's warehouse, and why everyone did. He grew up knowing why his parents, as they strolled through Montbard, were greeted with more civility than warmth. And he grew up knowing that his father's fate was almost certainly his own as well. The Leclercs had improved their station over five generations, but no further ascension could be reasonably expected. The boundary was no longer prosperity, but the hard stratification of class. In a village in provincial Burgundy, there was simply no room to climb higher. As the position was inheritable, the salt-keeper's son awaited his turn to become the next gabelle bureaucrat, sitting at a warehouse desk and handing out needless packages to resentful neighbors.

Then struck the extraordinary good fortune of his name.

. . .

Georges Blaisot lived a colorless life, spending most of his career as a little-noted civil servant on the staff of Victor Amadeus II, the sovereign Duke of Savoy. A small but independent ducal state, Savoy comprised a buffer zone between two nations, incorporating regions that would later be assimilated into Italy (the Piedmont, the city of Turin) and France (eastern Provence and Nice). It cannot have been easy for Monsieur Blaisot to gain the trust of Victor Amadeus, who resented France's attempts to convert Savoy into a client state. The duke, forced to marry one of Louis XIV's nieces during his mother's regency, shook off French influence as soon as he ascended to power, not hesitating to alienate his powerful neighbor when he saw fit. During both the Nine Years' War and the War of the Spanish Succession, Savoy entered hostilities as an ally of France, then switched sides when Victor Amadeus found it expedient to do so. As reward for his shifting loyalties, ensuing diplomatic carve-ups granted him sovereignty over farther realms. By 1713, the Duke of Savoy was, among a half-dozen other titles, also the King of Sicily.

The island kingdom interested him not at all. It was too far away, at the toe-end of Italy's boot (he would eventually trade it for the Kingdom of Sardinia), and at any rate he'd long been in the habit of delegating the actual governance of his domains to subordinates. Which is why the colorless, trustworthy Georges Blaisot came particularly in handy: The civil servant was experienced at imposing rule without the trappings of a ruler, enforcing laws and collecting taxes with a minimum of fuss. The duke wasted no time handing the keys to his new-found kingdom to Monsieur Blaisot, with the understanding that his compensation would include a percentage of any revenues he'd manage to collect.

Sicily is roughly the size of Massachusetts or Wales. Sicilians were notoriously uncooperative with authorities (the code of *omertà*, or silence, originates there), and they naturally resented being used as a participation prize in distant power struggles. Georges Blaisot's administration was brief, cut short by his death less than two years later, but it was also effective enough that Blaisot's widow was known to be comfortably well off. When she died in 1717, her will held two surprises, the first being the size of the estate. Madame Blaisot was *extremely* well off: The portion of Sicilian taxes, combined with previous revenue shares, would in modern currency be counted in tens of millions of dollars.

The second surprise was the beneficiary. The Blaisots were childless. Lacking any direct heirs, they bequeathed the bulk of their fortune to their great-nephew, a boy living in a rural village in central France. Their niece, Christine Marlin Leclerc, had flattered Georges Blaisot by making him the godparent and partial namesake of her firstborn child, Georges-Louis. The ten-year-old son of the salt-keeper was suddenly rich.

• • •

The inheritance rapidly severed the thread of the boy's middle-class Montbard existence. Under French law he would not assume full control of his fortune until the age of twenty-seven, which meant that his father would control the estate in the interim. Within months, Benjamin-François Leclerc had installed the family in a mansion in Dijon, Burgundy's provincial capital forty miles to the northwest. He enrolled Georges-Louis and his two younger siblings in the finest schools, and launched the boy's thorough indoctrination into the manners and social graces of the upper class.

Acceptance into that world was not a foregone conclusion. Despite his fortune, the boy could have been snubbed as a parvenu, a *nouveau riche* more suited to playing the squire back in rural Montbard. But the transformation was successful, and complete. Six years later, Georges-Louis was a graduate of Dijon's Jesuit academy, moving confidently within a circle of wellborn friends. He had grown into a young man of above-average height, with handsome features and a powerful physique, his arresting eyes now matched with a thick mane of black hair that would become a trademark. For the rest of his life he'd avoid the dominant fashion of wearing white wigs, instead powdering his own hair when occasion suited. Athletic in bearing and with well-honed social skills, the teenaged Georges-Louis Leclerc already displayed an impressive mix of presence and poise.

Still, no one was impressed by his intellect. His friends would remember him as more interested in sports than schoolbooks, although they were vaguely aware that he read other books on his own. He remained as undistinguished a student as he'd been in Montbard, progressing in academics but showing no particular enthusiasm for them. Having completed the equivalent of his high school education in 1723, Georges-Louis had then begun a similarly tepid course of studies at Dijon's School of Law, despite having no plans to become a lawyer. A legal diploma was a prerequisite to purchasing a lifetime appointment to the provincial parlement, the next and final step in solidifying his standing among the elite. Not a legislative body but a judicial one, the parlement had become the system in which Burgundy's ruling class had consolidated much of their power, holding sway over most taxation and regulation. Even the king's edicts were enforced only after the genteel magistrates, known as *noblesse de robe*, stamped their approval.

In 1726, the same year that nineteen-year-old Carl Linnaeus flunked out of Trivial School, nineteen-year-old Georges-Louis Leclerc collected his certification to the bar. He now stood poised to become a justice of parlement, to take at least a passing interest in the portfolio of vineyards, tenant farms, and other properties acquired in his name, and to cut a swath through Dijon's high society. To his parents' dismay, he did nothing of the sort. Instead, he withdrew into his mansion, reading books for months on end.

Burgundy's elite may have accepted Georges-Louis Leclerc, but he had not quite accepted them. It had all happened so quickly—quickly enough to create a lasting sense of detachment from his surroundings, of separation despite assimilation. His quiet demeanor masked a growing diffidence and restlessness. He did not mind being rich: A taste for fine wardrobes and quality furnishings would follow him throughout his life. What he did not like was being idly rich. Previous generations of Leclercs had at least worked hard, daring to rise above their stations. The life laid out before him, an endless stream of overseeing vintages and collecting rents from tenant farmers, appears to have struck him as only slightly less tedious than collecting salt taxes in Montbard.

After more than a year of lounging among his books, young Leclerc announced he was returning to academia, this time by attending the University of Angers, 350 miles to the west. "As for me, I shall do whatever lies in my power to keep myself away from Dijon as long as I can," he confided to a friend. "If there is anything at all that bring me back there with pleasure, that can only be the desire I feel to see again the very small number of those for whom I retain some feelings of esteem."

· · ·

The University of Angers was not a particularly distinguished institution, but it was as far away as he could get from Dijon and still matriculate in France. While there he sampled various disciplines, delving into mathematics and even attending lectures at the medical school. He had no illusions of becoming a practicing physician—that would be comically beneath his station—and it soon became clear that he had no ambitions to take a degree at all. The lecture halls competed for his attention with Angers's coffeehouses and taverns, and he divided his company between serious scholars and high-living companions. As one of his early biographers gingerly described him, Leclerc the university student "showed from the beginning a great disposition for work and pleasure."

> Nature had given him every advantage, stature, carriage, face, strength, and an ardor in every pursuit. . . . His youth appears to have been rather violent and impetuous; but, in whatever way he may have employed his evening, next morning he had himself called at a fixed hour to set to work again.

"Rather violent and impetuous" is a discreet way of putting it. Leclerc caroused wholeheartedly, engaging in flirtations and feuds alike, making not only friends but enemies. He began to fight duels.

Later generations of impetuous men would refine dueling to a rarely lethal, stylized act of honor. Britons and Americans dueled at paced-off distances with pistols more often than not mis-aimed or fired into the air. Germans and Austrians dueled with specialized rapiers and typically inflicted only facial scars, which were viewed as marks of pride. But duels in early-eighteenth-century France were pacts of mutual attempted murder: brutal clashes fought at close quarters, culminating in either blood or surrender. The most commonly used weapons were thick, curved-blade sabers, capable of hacking as well as stabbing, unsheathed at a two-foot distance by combatants stripped to the waist to show they were unarmored. The rampant mortality rate of duels had made them illegal in France since 1547, yet they continued with such regularity that no fewer than eight royal orders subsequently attempted to underscore that illegality, the most recent of which, in 1723, made it clear that "any gentleman who struck another should be degraded from his rank and forfeit his arms." Still the challenges were issued, the seconds recruited, the dueling grounds cleared, and the signal "Allez!" given.

Leclerc fought at least three duels in his three years at Angers. Accounts vary as to the pretexts that incited them—a woman's sullied honor, a suspect game of cards—but each ended in blood. The third, against an Englishman, was the bloodiest of all.

The bernicla, or barnacle tree, with fruit transforming into geese

FOUR

Vegetable Lambs and Barnacle Trees

IN LATE FALL OF 1729, TWO FRIENDS DIVIDED UP THE WORLD between themselves.

One of them was Carl Linnaeus, retrieved from the edge of destitution by his chance encounter with Professor Celsius. In addition to providing room and board, his new mentor had arranged to upgrade Linnaeus's scholarship, securing his studies for another year. No longer consumed by maintaining a precarious existence, Linnaeus was at last able to enjoy the non-academic aspects of student life: lingering in cafés, engaging in leisurely extracurricular discussions, and forging friendships.

Peter Artedi, two years older and from the northern Swedish province of Ångermanland, had a background remarkably similar to that of Linnaeus. He too was a pastor's son, with a Latin surname coined by his father. He'd also been raised to inherit the family pulpit, developed a fixation on natural history, and been shunted off to medical school instead. Linnaeus had noticed him months earlier in the university library, and silently registered that they seemed interested in the same books, but had not felt confident to approach Artedi until recently. Once he struck up a conversation, the floodgates opened. "We immediately started talking about stones, plants and animals," Linnaeus recalled. "I wanted his friendship; and not only did he give it to me, but he also promised me his help whenever I needed it."

Physically and temperamentally, they were markedly different. Linnaeus described Artedi as "tall, slow and serious," at the same time describing himself as "small, giddy, hasty and quick." Artedi was inclined to sleep all day and work at night, while Linnaeus was an early riser who kept regular hours. But they were fast friends, and determined not to become rivals. As a measure against future conflicts in their careers, they divided up the living world between themselves. Linnaeus would study insects and birds, and Artedi would take fishes (then a term for all aquatic creatures), reptiles, and amphibians. *Trichozoología* ("hairy animals") would be catalogued collaboratively: Both could study as many as they chose, so long as each informed the other first. Knowing that Linnaeus's chief interest lay in plants, Artedi deferentially chose only a few of them, chiefly carrots, parsley, and celery. They also agreed to safeguard each other's legacy: In the case of one's death, both vowed, the other would take possession of the deceased's research papers and carry on in his stead.

Yet as their amicable division of all life grew more detailed, Artedi and Linnaeus were confronted by the fact that no matter how neatly they drew their boundaries, some species refused to respect them. There was, for instance, the boramez, or vegetable lamb of Tartary. Reportedly native to parts of Asia bounded by the Caspian Sea, the boramez was an animal like an ordinary lamb, except it was also a plant. It emerged from the earth suspended on a stalk that served as a sort of rigid umbilical cord; the lamb would die if it was cut. It did not live long, as it could only graze on grass in its stem's perimeter. Its meat tasted like mutton, but its blood tasted like honey.

Then there was the bernicla, the barnacle goose tree. Supposedly native to a small island off the coast of Lancashire, the tree gave forth fruit in the form of barnacles, which dropped into the water and, after a few submerged months, emerged as geese. This was an especially tricky question for Linnaeus's and Artedi's respective specialties, as it

The boramez

was simultaneously a plant, a fish, and a bird. According to John Gerard, an English naturalist, these mussel-shaped shells would grow until they split open, revealing

> the legs of the Birde hanging out, til at length it is all come foorth. The bird would hang by its bill until fully mature, then would drop into the sea. Where it gathereth feathers, and groweth to a foule, bigger than a Mallard, and lesser than a Goose.

Were such fantastical species taken seriously in 1729? Barring a few skeptics, very much so. The vegetable lamb had its own entry in Ephraim Chamber's *Cyclopaedia, or a Universal Dictionary of Arts and Science*, published just the year previously. In Gerard's *Herball*, published in 1636 but used as a teaching text well into the nineteenth century, the barnacle tree is authoritatively catalogued, side by side with a description of a potato. Pope Innocent III had explicitly prohibited the eating of barnacle geese during Lent, deciding that despite their unusual reproduction, they lived and fed like conventional geese and so were of the same nature as other birds. In Jewish dietary law, Rabbeinu Tam had determined that they were kosher, and should be slaughtered following the normal prescriptions for birds.

To modern eyes, such creatures are patently impossible. The vegetable lamb likely arose from a misreading of Herodotus, who wrote of a plant whose "fruit whereof is a wool exceeding in beauty and goodness that of sheep." He was describing cotton. Until closed by Innocent III and Rabbeinu Tam, the barnacle tree was probably a fictional loophole for those who wished to eat fowl but pretend it was fish. To understand why even experienced naturalists documented their existence without blinking, it's helpful to understand the near-universal acceptance of a master plan for organizing all life, a pattern commonly acknowledged as existing in nature.

It was a straight, ascending line.

First popularized by Aristotle in his *History of Animals* as the concept of *scala naturae*, or Ladder of Life, the schema was simple: a single line of increasing sophistication—commonly known as "perfection"—rising from simple plant life at the bottom step to humanity at the topmost. Further elaboration over the centuries transformed the rungs of the ladder into the links of a chain, refining the metaphor into a Great Chain of Being. At the bottom, the chain descended to mineral life. At the top it stretched up past humanity, to angels and finally to God Himself. Creatures like the bernicla and boramez struck no one as a violation of categories, as they constituted their own categories, their own links on the chain. If anything, intermediary organisms like vegetable lambs and geese-fruiting trees seemed necessary linkages from one level of perfection to another.

The Great Chain of Being was more than a metaphor. It was an instrument of temporal power. The particulars varied from version to version, but many depictions of the Great Chain awarded kings and nobility their own links, directly above ordinary people, thereby giving sanction to a ruling class as an integral aspect of the natural order. This attitude would be codified into the eighteenth-century hymn "All Things Bright and Beautiful":

> The rich man in his castle
> The poor man at his gate
> God made them high or lowly,
> And ordered their estate.

The Great Chain reinforced a monarchical perspective in lesser realms as well. The eagle was elevated to the status of the "king" of

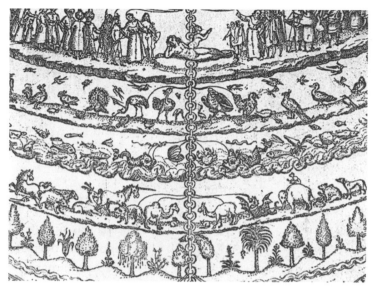

The Great Chain of Being (detail), from Rhetorica Christiana, *1579*

birds. Typically the elephant or the lion was heralded as the king of beasts, and the whale the king of fishes. The oak tree was the king of plants. Extending beyond living things, the Great Chain declared gold the king of metals, diamonds the king of gems, and marble the king of stones. By the late Renaissance, the Great Chain of Being was displaying even more precise calibrations. Wild animals held higher place than domesticated ones, since their untamed nature was evidence of larger souls. Worm-eating birds were higher than seed-eating ones. Celestial beings were introduced toward the top of the hierarchy, with seraphim declared the highest order of angels—since a seraphim was king, or "primate," of the angels.

The schema grew in both complexity and acceptance, to the extent that by 1667 the scholarly British Royal Society was defining its mission in direct relation to the chain:

> This is the highest pitch of humane reason: to follow all the links of this chain, till all their secrets are opened to our minds; and their works advance'd and imitated by our hands. This is truly to command the world; to rank all the varieties and degrees of things so orderly upon one another . . . we make a second advantage of this rising grouynd, thereby to look the nearer into heaven.

Yet for all the praise it had garnered over the centuries, the Great Chain raised a host of questions. If the lion was king of the animals, were other cats higher up on the chain than dogs? Were nutritious turnips more "perfected" than ornamental rosebushes? Such questions were debated by students like Linnaeus and Artedi, but there were no clear answers.

• • •

The respectability conferred by Celsius's patronage had opened up another vista of opportunity for Linnaeus, namely a lucrative trade in tutoring his fellow medical students. They were turning to him in increasing numbers, drawn by his real-world experience—unlike most of them, he'd spent a year assisting a practicing physician. His time with Dr. Rothman in the village of Växjö would, ultimately, be the only meaningful medical education Linnaeus would ever receive, and at the moment it made him a valuable if somewhat circumspect resource.

Rothman had begun Linnaeus's medical apprenticeship by instructing him in two subjects. The first was physiology, the mechanics of how the body works. During the course of his village rounds, he'd prompted Linnaeus to study the articulation of a patient's limbs, to see and feel how muscle and bone connect and coordinate, and to notice how illness or injury impeded that motion. A gash or a broken bone was a rare opportunity to witness flashes of the interior body itself, glimpsed between pulsings and blood. That was why Linnaeus had paid most of his scholarship money to witness the dissection of the executed woman in Stockholm: to confirm his mental picture of the world beneath the skin.

The second subject was *materia medica*, the identification and preparation of substances used in medicine. Aside from a few items such as tincture of opium, drugs in the modern sense of the term did not exist. In their place was an arsenal of salves, poultices, elixirs, and other concoctions collectively known as *physicks*, a term that gave rise to calling the doctors who applied them "physicians." A precursor of pharmacology, materia medica was essentially a stock of recipes for physicks, accompanied with instructions on how to obtain the necessary therapeutic ingredients. Some treatments were readily available:

Patients suffering from lethargy, diarrhea, or postpartum pain were frequently prescribed generous amounts of wine, and asthmatics were treated with red sugar candy. Some physicks involved animals: The ague, for instance, was treated by wrapping the patient in the skin of a freshly killed lamb. Others were mineral in nature: Paralysis, bad breath, and melancholy were treated with *Aurum potabile*, a drinkable suspension of flecks of gold. But most physicks drew their primary ingredients from plants. Carl learned to identify a plant and harvest its medicinal components, quickly and with confidence.

Unusually for his era, Rothman had cautioned Linnaeus against use of the Doctrine of Signatures, a philosophy of materia medica commonly accepted at the time. Rooted in prehistory but refined in the sixteenth and seventeenth centuries, this was the belief that each plant had been designed by God to serve a specific human purpose, and that clues to that purpose were conveniently incorporated into the plant's appearance. Hence a blood-red leaf was the sign that a plant helped strengthen the blood. The walnut, because it resembled a human brain, treated mental illness. Strong-smelling plants, because they excited the nose, would excite a patient's nerves when ingested. The notion was patently flawed and even dangerous: Birthwort, a plant commonly administered to pregnant women because its flowers resembled birth canals, is now linked to kidney disease and cancer. But this idea would hold a place in mainstream medicine for generations to come.

While Linnaeus's rejection of signatures reduced the amount of minutiae to memorize, even his version of materia medica could not be gleaned solely from books. Most texts on medicinal plants, if illustrated at all, had only moderately detailed woodcuts or engravings of the plants themselves, insufficiently detailed for the student to accurately locate them in the field. Rothman had given Linnaeus access to a copy of the standard work on the subject, Theophrastus's *Historia Plantarum*, but it was more confusing than useful. In addition to being over two thousand years old (Theophrastus had been a pupil of Aristotle), the text only described about five hundred kinds of plants, few of which were found in northern Europe. Linnaeus attempted to reconcile Theophrastus with the plants of southern Sweden, but found that "there were many which had not at the time been examined with sufficient botanical accuracy, and which, not being reducible to the rules of that system, involved our young botanist [himself] in great perplexity."

Even well-known plants were difficult to recognize in the pages of Theophrastus, who could only describe by painting word pictures, using comparisons now obscured by two millennia. For instance, his description of the sacred lotus (*Nolumbo nucifera*) compares the stalk to the thickness of a man's finger, the flower bud to a wasp's nest, and the blades of its leaves to a Thessalian hat. How large was a Thessalian hat? The question was as mysterious to Linnaeus as it was to Rothman, who admonished his apprentice not to put too much stock in Theophrastus, or for that matter formal classical identification in general. "To know a flighty Latin word or the name of a plant was nothing," he'd informed his pupil, urging him to rely on his own senses and field experience.

Yet ancient names were an essential aspect of European medicine. They transcended regional differences, as illustrated by Linnaeus's namesake tree. In Sweden it was called *lind*. In Germany it was *linden*, in Romania it was *tei*, and in England it was either *basswood* or *lime* (the latter being a further confusion, as it bore no relation to the citrus tree of the same name). But a Swede, a German, a Romanian, and an Englishman could all discuss the same tree by referring to it as *tilia*, the term used in Latin translations of *Historia Plantarum*. The use of Latin names was more than a tribute to antiquity. It was a tool for contemporary clarity.

Such practice required going beyond the ancients. It was necessary to coin Latin names for plants that Theophrastus and others of his era never mentioned. The challenge of this was embodied by a book Rothman had made available to Linnaeus: *Elements of Botany, or A Method for Recognizing Plants*, by the French botanist Joseph Pitton de Tournefort, laboriously translated into academic Latin as *Institutiones Rei Herbariae*. The translation had taken over five years to produce, as it required tracking down or inventing Latin names for nearly seven thousand plants.

The materia medica Linnaeus had learned—and was now trying to teach his fellow students—was subtle and contextual, informed by the observation that many plants' appearances and even medical proper-

ties change throughout the year. In ripened form the *fläderbär*, or el-
derberry, is a staple of Swedish cuisine; in immature form it looks less
like a berry than a kind of pea, and is poisonous. Different parts of a
plant yielded different treatments as well, as in the case of the linden,
Linnaeus's own namesake tree. Leaving aside the dubious Doctrine of
Signatures (which held that its heart-shaped leaves were good for irregular heartbeats), practical experience showed that a linden tea—
brewed from blossoms, not leaves—could combat anxiety. However,
the same tea might exacerbate the symptoms of someone feeling dizzy
or light-headed, in which case a better treatment might be a decoc-
tion of bark from the same tree. Linnaeus could teach these nuances
in the abstract, but the efficacy of such medicine still hinged on deriving treatments from the correct plant, not a similar-appearing one.
How to be certain? In the absence of extensive time in the field,
Tournefort provided an alternative.

Institutiones Rei Herbariae was not strictly part of the medical curriculum, since it encompassed plants with no known medicinal purpose.
Yet Linnaeus had been enthralled, not only by the book's massive scope
but also by its attempt to organize the subject into an overarching
whole. Tournefort's system separated trees from herbs, then classified
the latter chiefly on the characteristics of their petals. Since some
plants had no petals (these were classified as "apetalous"), the schema
was not a model of clarity, and even this pared-down approach quickly
bogged down in complexity. After dividing plants into 22 distinct
petal-shape groupings, Tournefort further subdivided them into 698
genera, broad categories based on other physical resemblances. There
Tournefort halted. The book's subtitle was "a method for recognizing
plants," but sorting nearly seven thousand plants into nearly seven
hundred categories only brought the reader partway down the path of
recognizing individual species. For field identification, one would either need to tote along a very large book or memorize all 698 genera.

Linnaeus the medical apprentice had memorized them, but Linnaeus the tutor found few of his fellow students interested in putting
forth a similar effort. Or, for that matter, much effort at all. Why go to
the trouble? Their professors administered no tests, engaged in no
classroom discussions, and only evaluated students based on submitted written work. Not overly bound by scruples, Linnaeus let out dis-

creet word: For the proper fee, he'd not only edit their papers but write them himself. This dubious trade kept him busy and well compensated until December, when another writing project loomed, one that he dreaded. It was time to write a poem.

Uppsala students under a professor's mentorship were traditionally expected to present their patrons with an original poem of praise on New Year's Day. Linnaeus did not feel remotely up to the task of composing verse, but ghostwriting and the tedium of trying to pound Tournefort through his fellow students' heads had started him thinking about a streamlined, easier-to-grasp approach to plant identification. In the waning days of 1729, he began work on an alternative gift for Professor Celsius. *I am no poet but something, however, of a botanist,* he wrote in Swedish, taking care not to blot lines or waste paper. *I therefore offer you this fruit from the little crop that God has granted me.*

The "vegetable letters" of Linnaeus's sexual system

FIVE

Several Bridegrooms, Several Brides

ON THE FIRST DAY OF 1730, OLOF CELSIUS SETTLED DOWN TO read his New Year's gift from his student boarder and assistant. Linnaeus's tribute may have seemed a disappointment at first—no verse, no praise, a less-than-florid dedication—but he soon found himself engrossed in its startling contents. The handwritten twenty-two-page booklet, entitled *Praeludia Sponsaliorum Plantarum*, or *Prelude to the Betrothal of Plants*, was thoroughly besotted with sex.

It opened placidly enough, with Linnaeus celebrating the inevitability of spring's arrival. "See how every bird, all the long winter silent, bursts into song!" he began. "See how all the insects come forth from their hiding places where they have lain half-dead, how all the plants push through the soil. . . . Words cannot express the joy that the sun brings to all living things." But then the bucolic scene takes a turn toward the carnal.

> Now the black cock and the wood grouse begin to frolic, the fish to sport. Every animal feels the sexual urge. Yes, Love comes even to the plants. Males and females, even the hermaphrodites, hold their nuptials (which is the subject that I now propose to discuss).

Linnaeus was not writing erotica, but proposing a new system of plant identification. Just as Tournefort had organized his work around the physical characteristics of petals, previous botanical systematists (as they were called) had devised their own approaches to classification. One of the earliest European systematists was Charles de L'Ecluse (1526–1609), who grouped plants according to the geographic regions in which they were found. Andrea Cesalpino (1519–1603) attempted to organize them by their pattern of fruits and seeds. If they bore neither external fruit nor seed, Cesalpino used the shapes of their ovaries—a tricky business, considering that some plants appeared to have two sets of ovaries. John Ray (1628–1705) indexed approximately eighteen thousand species in his massive *Methodus Plantarum* according to what he termed "gross morphology," the overall size and shape of the mature plant, bush, or tree.

All these systems demonstrated their deficiencies. First, they required an intimate knowledge of each specimen plant: The observer needed to examine minute interior details, or large-scale growth patterns, or both at the same time. Second, these systems inevitably broke down into a morass of arbitrary rules, becoming so complex that they scarcely improved over randomness (as already demonstrated by Tournefort's 698 categories). The efficient recognition of plants seemed to require a new schema, one capable of being easily grasped and applied.

Hence Linnaeus's emphasis on copulation. Sex, as he put it, was "the great analogy that exists between Plants and Animals, and their similarities in propagating their species." He began by describing how spring sunshine renders "plants to be smitten by love," commencing a sort of vegetable romance. "The flowers' leaves . . . serve as bridal beds which the Creator has so gloriously arranged," he wrote, "adorned with such noble bed curtains, and perfumed with so many soft scents that the bridegroom with his bride might there celebrate their nuptials."

Linnaeus was not the first to notice the resemblance between flowers and genitals. It was commonly acknowledged that floral stamens were masculine in appearance, and pistils feminine, yet this was generally accepted as mere resemblance, since most flowering plants had both pistils and stamens in various combinations. Linnaeus, however, had recently read of Sebastian Vaillant, a botanist at the Jardin du Roi in Paris, who twelve years earlier had theorized that the pollination patterns of flowering plants were indeed a form of sexual reproduction—a controversial idea, to say the least, and the mechanisms of plant sexuality were far from clear. Ingeniously, Linnaeus did not propose to discover those mechanisms, only to use their visible externals as signposts for zeroing in on a plant's identity.

That plants had no neat divide of genders was, to Linnaeus, not a bar but an opportunity. "Yes, love seizes the very plant, among them both males and females," he continued, "and the very hermaphrodites celebrate their nuptials." He postulated that plants did not have strict, singular male and female parts but various combinations of both, and engaged in a spectrum of sexualities. These he compiled into a sort of botanical Kama Sutra. Each plant category in *Praeludia Sponsaliorum Plantarum* reflected a different reproductive configuration, which he described as a "marriage," although they quickly proceeded past conventional conceptions of the term. These ranged from the prim *monoclinous* ("husband and wife have the same bed") to the positively orgiastic *polyandria* ("twenty males or more in the same marriage"). Interim variations on the unions of several bridegrooms and brides included *monoecia* ("husbands live with their wives in the same house") and *polygamia* ("husbands live with wives and concubines"). Even plants without flowers were described as having *cryptogamia*, or "hidden marriages," where "nuptials are celebrated privately." Drilling down into the bedroom arrangements still further, he created the category of *didynamia*, consisting of "four husbands, two taller than the other two" (two long and two short stamens), while his *tetradynamia* had "more than four husbands, two shorter than the others." If the stamens rose in a single bundle, they connoted *monadelphia* ("husbands like brothers").

In all, *Praeludia Sponsaliorum* recorded a total of twenty-six floral categories—a number that, as Linnaeus noted with pride, corresponded to the letters of the alphabet. These "vegetable letters," as he referred to them, provided the most convenient shorthand of all: A field botanist would not need to write down, say, *polygamia* ("husbands live with wives and concubines"), but merely the letter Y. As far as he was concerned, an impeccable logic lay revealed beneath the choreography of coupling.

· · ·

After the initial shock of the language (and the images it summoned) wore off, Professor Celsius could not help but be impressed; even as an amateur botanist, he recognized the immediate utility of such a method. The vividness of its central analogy made it easy to grasp. Mastering 26 categories was certainly an easier task than memorizing Tournefort's 698, and it would no longer be necessary to minutely examine several components of a plant. One simply counted the pistils and stamens participating in a "marriage," noted their positions, and arrived at the proper vegetable letter. Was this system overly broad, or too sweeping in its assumptions? Perhaps. But its learning curve was far more appealing than that of any previous system. Celsius, pleased with his protégé, did not mind sharing the booklet with others.

Suddenly, Linnaeus's life began to change. Within weeks, handwritten copies of *Praeludia Sponsaliorum* were circulating throughout Uppsala, and soon afterward in Stockholm. In April, one of them was anonymously submitted to a Swedish learned society, which endorsed it for publication. In May, he was offered employment as a seasonal "demonstrator" at the botanical garden, where he would deliver impromptu lectures to his fellow students. In June, he was moving out of Celsius's house with his mentor's blessing, shifting lodgings into the house of Olof Rudbeck, Uppsala's elderly professor of anatomy and botany. Linnaeus's "sexual system," as it had come to be known, was catching on, and even the conservative Rudbeck was eager to take its author under his wing.

Audiences numbering in the hundreds were soon crowding into the overgrown garden, straining to hear Linnaeus apply his system of vegetable letters to cultivated plants and weeds alike. After Linnaeus had taken up residence in Rudbeck's home, the next step seemed clear: The professor, retired in all but name, had long been in the habit of appointing recent graduates to deliver lectures in his stead. He made it known that when classes resumed in the fall, Linnaeus would perform that function, making him the de facto professor of botany. It was an astonishing elevation for a second-year student (much less one who had never attended a single class on botany), and Linnaeus was justifiably brimming with pride. What had begun as a New Year's gift had, in a few months, transformed him from a marginal figure on campus to a respected voice at its very center.

The headlong year of 1730, with its sudden onset of attention, would do much to shape Linnaeus's professional personality. He cultivated an air of authority, a public-facing confidence that would come to strike at least some observers as a near-impermeable certainty about his conclusions. In his defense, there was little call for self-doubt: His sexual system had moved directly from concept to curriculum, without additions or refinements. In fact, even he could not seem to improve upon it, which persuaded him that his best work consisted not of slow, iterative discovery but of flashes of insight. The stodgy, staid world of botany, seemingly immune to innovation, now had a one-person avant-garde. Never notably humble, Linnaeus now began to view himself as a man of destiny. That self-conception would soon be tested.

. . .

Linnaeus's rise did not go unchallenged. In March of the following year, his classroom lectures were interrupted by the return of Nils Rosén, a graduate of Uppsala and an earlier favorite of Rudbeck. Just one year older than Linnaeus, Rosén had begun his medical studies at the age of sixteen, earning a reputation as a brilliant student. Rudbeck had tacitly anointed him as an eventual successor, but university rules dictated that Rosén first needed the *potestatem docendi*, or right to teach, a status only imparted by an accredited medical degree. Since Uppsala itself was not accredited, Rosén had left for an extended academic tour of Europe, studying with some of the best medical minds of the era and eventually earning a degree in Holland. When he returned, fully credentialed at last, he found a second-year student at the podium in his place.

Rudbeck attempted to smooth matters over by awarding him the lectureship in anatomy, but Rosén was far from content with the arrangement. Petitioning the academic senate to supersede Linnaeus's appointment, he reminded them of the *potestatem docendi*, arguing that a teacher should actually be qualified to teach. The senate agreed, forcing Linnaeus to step down and return to mere student status.

The demotion devastated Linnaeus, who saw himself as not just losing a job but being stripped of the legitimacy his ideas had seemed to impart. Humiliation quickly turned to consuming rage. Linnaeus publicly vowed to stab Rosén, and began carrying a short sword for exactly that purpose. Word of his threats reached the academic senate, where a motion to ban him permanently from the campus was staved off by Olof Celsius, who was able to talk down sanctions against his former protégé into a formal reprimand. Linnaeus, unchastened, confided to friends that he still planned to kill Rosén at the earliest opportunity.

Evelyn Pierrepont, the Duke of Kingston

SIX

The Greater Gift of Patience

NO ONE WAS QUITE SURE WHAT TO DO WITH EVELYN PIERRE-
pont. Born in 1711 to a family of wealth and peerage in Nottingham-
shire, England, the boy acquired estates and titles the way other
children accumulate toys. By the age of fourteen, he was the Marquis
of Dorchester, Viscount Newark, Baron Pierrepont, and heir to miscellaneous fortunes. He was also an orphan. He'd never known his
father, who died when he was two. When his mother died in 1722,
Evelyn was old enough to begin the typical education for a boy of his
station—namely, being packed off to boarding school—but none of
his relations stepped forward to attend to such matters. Instead, the
boy did more or less as he pleased for the next three years, until someone got around to enrolling him in the elite Eton School. His studies
there lasted less than a year, ending when the death of his grandfather
made him the Duke of Kingston. He was fifteen years old.

What to make of this indifferently raised teenage boy, now elevated to enormous wealth and the highest non-royal title in the United Kingdom? "The Duke of Kingston has hitherto so ill an education 'tis hard to make any judgement," his aunt Lady Montagu observed at the time. She encouraged him to embark on a modified version of a British upper-class tradition, the Grand Tour of the Continent. Usually undertaken around the age of twenty or twenty-one, the Grand Tour was a leisurely, luxurious circuit through Europe's ancient capitals and cultural sites, intended to broaden one's horizons and impart a patina of sophistication prior to formally taking up social station.

"According to the law of custom, and perhaps of reason, foreign travel completes the education of an English gentleman," the historian Edward Gibbon wrote of his own Grand Tour. "At home we are content to move in the daily round of pleasure and business . . . but in a foreign country, curiosity is our business and our pleasure; and the traveller, conscious of his ignorance and covetous of his time, is diligent in the search and the view of every object that can deserve his attention."

Yet the young Duke of Kingston clearly needed to continue his education, not just cap it off. For this he acquired a full-time tutor as his traveling companion, a thirty-one-year-old gentleman scholar named Nathan Hickman. An Oxford graduate and experienced naturalist, Hickman had solid credentials as a savant: His work in entomology had recently earned him election to the Royal Society for the Propagation of Useful Knowledge, making him one of the youngest members of Britain's oldest and foremost learned societies. Just as important, Hickman was heir to a considerable fortune himself. While he was not a member of the nobility, his wealth rendered him socially acceptable to travel in the same circles as the duke, allowing him to be received as a genteel mentor rather than a servant. Kingston and Hickman left England in July of 1726, with the understanding that they'd remain abroad for a year or two, perhaps three. They would not return for over a decade, trailing clouds of scandal in their wake.

Unlike other upscale Britons touring the Continent at the time, the young duke was not intent on making his Grand Tour particularly grand. The duo traveled in high style, of course, with a retinue of servants and staff. But they were in no hurry to visit the great cultural capitals, instead proceeding at a leisurely pace to wend through the countryside and smaller cities. In 1730, four years into their ongoing idyll, they landed in Dijon.

As with most stops on their tour, their arrival created a local sensation. The duke, now nineteen, was not only titled and wealthy but extraordinarily handsome—the writer Horace Walpole would describe him as "of the greatest beauty and finest person in England"— and Dijon's social elite scurried in welcome. Amid the rounds of galas and receptions, Kingston and Hickman took notice of one figure who held himself slightly apart from the festivities, as if he wished to be elsewhere. It was Georges-Louis Leclerc. The young man who'd vowed "to keep myself away from Dijon as long as I can" had been forced to retreat there, his university days abruptly ended. He was waiting for an Englishman to die.

Leclerc's taste for dueling had gone too far. In June of that year, he'd wounded his English opponent so severely that the man's survival was in doubt. While not exactly fleeing the scene of the crime (all parties involved knew who he was and where he might be found), he'd hastily departed Angers for his manor in Dijon, to await word from a prudent distance. Arrest and prosecution hinged on whether his opponent lived or died, and whatever the outcome the ensuing scandal already guaranteed that the University of Angers would not welcome him back as a student. Having ignominiously returned to the very place he'd hoped to escape, he reclaimed his quarters within the manor, resumed reading books, and began seriously pondering the question of what, if anything, he was going to do with his life.

In the wake of their chance meeting, the duke and Leclerc found much to admire in each other. Both were uninterested in taking up the roles demanded by their station (Kingston would never get around to taking his seat in the House of Lords). Both, while socially adept, had reputations as rakes—Leclerc for his dueling, the duke for gambling and womanizing. Both were rich, attractive, and temperamentally unconventional.

The Englishman did not die. Leclerc was free to move on. When Kingston and Hickman left Dijon to continue their tour on November 3, 1730, their new friend came along as well.

. . .

For the next year and a half, the three young men traced a leisurely path through France and Italy. Leclerc cut a resplendent figure alongside the duke, acquiring the habit of silks and fine brocades that would become a trademark. He also acquired a more refined name. Traveling with a friend who was simultaneously a duke, a marquis, a viscount, and a baron had one downside: The people they met were easily confused as to the role a man named "Leclerc" played in the entourage. To subtly highlight his own elevated station, he turned to the portfolio of real estate his father had purchased in his name. His holdings were extensive, but what interested him the most was the village of Buffon, four miles northeast of his birthplace in Montbard.

Buffon was a minuscule settlement, little more than a patchwork of farms and a dozen or so buildings alongside a stream. But he liked the name, and thanks to his acquisitive father he owned the entire village lock, stock, and barrel. That was an important distinction. Since the place was entirely his, it was in his rights to style himself as "de Buffon"—not as good as a title, but effectively conveying an aristocratic aura. At the duke's side in the reception rooms of southern Europe, he began bowing to introductions as Monsieur de Buffon. He would not be formally ennobled until late in life, when a grateful Louis XVI made him a count, but by then the gesture would seem the mere correction of a technicality.

The freshly dubbed Buffon was now aware that while his companions did little to dispel their reputations as sybarites and rakes, they were also men of considerable intellect. Hickman, the Oxford scholar, learned on the road that he'd been awarded the university's prestigious Radcliffe fellowship in absentia. He'd also taken his role as Kingston's tutor seriously, helping to nourish a nimble, active mind. "He is an extremely likable man whose intelligence is astonishing," one acquaintance observed of Kingston at the time. "No matter what subject is put before him, he reasons with it as if he were the most accomplished master of the art."

Buffon had struck no one as especially intelligent, but then again there'd been no reason to parade his intellect: He'd had other means of impressing the bourgeoisie of Dijon and the demimonde of Angers. But now, in the company of two peers who valued elegance and intelligence alike, he began to acknowledge an interior life in which his interests already bounded well beyond the ordinary.

One of the books he'd read during his Dijon self-exile was *Analysis of the Infinitely Small*, an early French treatise on the new mathematics of calculus. It was his first encounter with Sir Isaac Newton, a figure he would come to idolize, and the occasion for a startling self-realization. While poring over Newton's advanced and still-esoteric work, he'd recognized one of the equations, a generalized binomial theorem, as familiar. It was one that he himself had worked out, as an idle thought exercise.

This was, Buffon realized, an impressive feat: a twenty-year-old French provincial anticipating Newton. We have only his word on this casual achievement, but when later recounting the story he would do so reluctantly, fully aware that it sounded like improbable boasting and almost painfully embarrassed to draw comparisons with the profound genius of his idol. But the discovery gave him pause, and allowed him to think he might have a place among an intellectual movement dominated by Newton, an emerging new aristocracy of the mind that included young Frenchmen like Montesquieu and Voltaire. At the time, it was largely referred to as "the new science," but it would later be recognized as ushering in *le siècle des Lumière*. The Age of Enlightenment.

· · ·

In the 1730s, the word *science* did not have its modern meaning. Derived from *scientia*, the Latin word for knowledge, it was freely used as a catch-all term for any kind of acquired expertise. In the opening scene of Shakespeare's *Measure for Measure*, Duke Vincentio praises Escalus's political savvy by saying,

> Since I am put to know that your own science
> Exceeds, in that, the lists of all advice
> My strength can give you: then no more remains

In the Europe of 1731, higher education remained rooted in the practical. Buffon's former professors at Angers had addressed abstract concepts, but only long enough to harness them to real-world applications. The professor of mathematics, for instance, taught the subject of geometry through the task of building fortresses, then through an analysis of artillery trajectories needed to destroy them. Learning not only had purpose at its core, it was also demarcated differently. Today we make distinctions between sciences and humanities, but in Buffon's time the great division was between history and philosophy. Both words had very different meanings than they have today. In the modern sense, *history* means "things that happened in the past," while *philosophy* means "concepts and ideas relevant to existence." At the time, history meant all that was tangible, occurring in observable reality—things as well as events. Conversely, philosophy concerned itself with the intangible—ideas and principles, which might manifest themselves in the physical world but exist on their own as abstractions. History and philosophy were in turn bisected by a single criterion: whether or not they pertained to humans. The word *natural* was applied to all non-human phenomena, producing tidy quadrants:

Natural History	Natural Philosophy
things & events	*ideas & principles*
History	Philosophy

This reflected an important divide in the perceived nature of knowledge. Botany and zoology, both studies of living things, came under the rubric of "natural history," a label that also encompassed astronomy (the sky above) and mineralogy (the earth below). Yet theorizing—seeking to understand *why* life manifested and behaved as it did—belonged to the entirely different discipline of "natural philosophy," where it kept company with mathematics and what we know today as physics. Natural historians surveyed and catalogued, leaving principles to be discerned by natural philosophers. The quantitative and the qualitative were separate realms.

The New Science respected no such boundaries. Figures such as Spinoza, Newton, and Leibniz were ranging freely across subjects, making connections and synthesizing knowledge in new and unexpected ways. Fueling these interdisciplinary approaches were not universities but learned societies, small but influential groups that delved into knowledge for knowledge's sake. Membership in these societies conveyed legitimacy, and their journals and presses were an important platform for disseminating ideas—Newton's groundbreaking *Principia Mathematica* had been published not by his employer, the University of Cambridge, but by the British Royal Society for Improving Natural Knowledge.

In France, the foremost institution of savantry was the Académie des Sciences, established sixty-two years earlier (the *science* in its name reflected the fact that it had been initially created to consult with the king on patent disputes). No academic credentials were required, only acceptance as a peer by the intelligentsia of the day. As Buffon had come to realize, a savant's life was a respectable pursuit for a person of means. Newton, born into a wealthy family, had served both in Parliament and as master of the Royal Mint. Presently, the Académie's influential permanent secretary was the well-to-do Bernard de Fontenelle, who'd abandoned a law career (after a single case) in favor of a life of the mind.

"Rome is at this hour in all its glory," wrote the young man who used to be Georges-Louis Leclerc in January of 1732. "The carnival began two weeks ago; four magnificent operas and as many comedies, this without counting the minor theaters, are here 'ordinary' pleasures, and I admit that for me they are most extraordinary." But between the operas and carnivals Buffon was continuing to read advanced texts, and striking up correspondences with several of the more renowned mathematicians of the day. He was unsure of the role he'd play in the future, but he recognized himself as someone with an original mind, in original circumstances. He resolved to live an original life.

. . .

Buffon briefly left the tour in the spring of 1732, when he learned that his mother, Christine, was mortally ill. He rushed to Dijon to be with her in her final days, then mourned the loss of the parent he was closest to—the one he considered the source of his intelligence and social acumen, and the one whose prudent choice of namesake had transformed his life. There had never been much warmth between him and his father. The distaste of being the salt-keeper's son persisted, and Buffon had grown to resent the control Benjamin-François still maintained on what was, after all, primarily his fortune. He rejoined Kingston and Hickman in time to complete what would become the final leg of their protracted tour. That August the trio landed in Paris, intending to stay indefinitely.

It was here that Buffon began to subtly distance himself from his friends. While the duke and doctor took up common quarters as usual, Buffon sought out an address of his own. If he wished to enter Paris society on his own terms, perceived as a friend of the duke rather than a member of his entourage, the choice was an excellent one: The now-stationary Kingston lapsed into dissolution, racking up gambling debts and conducting indiscreet affairs. (He would depart suddenly for England four years later, taking a friend's wife with him.) Buffon, in contrast, rented rooms in the respectable household of Gilles-François Boulduc, who in addition to being the King's apothecary was a prominent savant and a senior member of the Académie des Sciences.

But Paris would have to wait. Buffon was just settling in when he learned that his father was planning to remarry, a union that could cloud a number of legal claims to his estate. He hurried back to Dijon and negotiated a settlement with his father, accepting cash reserves of eighty thousand livres—approximately thirty million dollars today—and title to all the lands purchased in his name. It was a considerable fortune, and assuming control of it quickly became a full-time occupation.

Without a backward glance, Buffon left Dijon to his father and new stepmother and returned to his birthplace, the village of Montbard. Examining his real estate portfolio, he found he was technically

The Parc Buffon in Montbard

no longer *de Buffon*, as his father had sold the namesake hamlet four miles away; he quietly repurchased it. He also purchased the properties surrounding the old family townhouse off the village square, demolished them along with his childhood home, and constructed an imposing three-story mansion in the latest neoclassical style. With its steep mansard roofs and distinctive line of chimneys capped with pointed spires, it would come to be known as the Castle.

Impressive as it was, Buffon's new residence would not dominate the landscape of Montbard. That honor belonged to the ancient stoneworks looming over the village, the source of its name (from *mont qui barre*, a defended mount), and whose foundations dated back to at least the fifth century. Since 1189, a succession of dukes of Burgundy had constructed numerous structures on the elevated site— a hunting lodge, a chateau, a series of towers and ramparts—but at present they were unoccupied and falling into ruin. Buffon did not purchase the seventeen-acre hilltop from the current duke, but years earlier his father had acquired in his name the antique office of Castilian, which gave him near-absolute proprietorship over the grounds. Once construction on his mansion was underway, he began transforming the mount into what would become famous as the Parc Buffon. More than two hundred workers cleared land, carried earth away in baskets, and dug irrigation lines, but mostly they planted trees.

Hundreds of rows of trees. The cost was enormous. "If you were to cover my gardens with six-franc coins," he later observed, "that would still not be the price that they cost me."

He was populating the park with experiments he knew would not bear results for years, if not decades. He planted the same variety of sapling in different soils, giving instructions that they be equally shaded and watered. On mature trees slated for harvesting, he first carefully peeled back the bark to determine if months of exposure might toughen the wood when sawn. He devised a standardized test to determine the breaking point of different kinds of wood, and meticulously catalogued the results. Reading up on other research in the field, Buffon was particularly inspired by a British savant named Stephen Hales, who had conducted a series of experiments in how plants "breathed" and circulated sap. Hales's work pleased Buffon so much that he translated it into French and praised the methodology in a preface. "Through these keen, reasoned and sustained experiments . . . Nature is forced to show her secrets," he wrote. "All other methods have never succeeded."

There was little precedent in France for such experiments as Buffon's orderly rows of trees, planted solely to yield a harvest of whatever insights might arise. These would not be his only experiments, but Buffon intended them to be his first and last. He would tend to his groves for the next forty years, transgressing the old borders between natural history and natural philosophy, observing both how trees grew and theorizing about what his observations implied. As the French historian Jacques Roger notes, "Buffon's originality here is considerable."

> He was the first to consider a forest not as a collection of trees but rather as an entity in itself, a whole in which the individuals maintained specific relationships and acted upon one another; in short, something which was a precursor of what today is called an ecosystem. He noted the relationships between the different types of trees according to the way they were grouped, and how the copse played a role in the growth of trees for timber. He even noticed the role of birds in the scattering of seeds and that of field mice making their winter provisions.

It was the work of a rare individual, young, unconstrained financially, and above all persistent. "Genius," he quipped, "is the greater gift of patience." To Buffon, this meant two kinds of patience: the careful acquisition of information, and the careful maintenance of a contemplative state in which to process that information. For these pursuits he was building a controlled environment, tailored for personal occupation. He expected to live a long, productive life exploring numerous fields, including mathematics, optics, metallurgy, microscopy, and any other disciplines that struck his fancy. In anticipation of a lifetime lost in thought, Buffon contrived not to be lost at all. The seventeen-acre park would become both a living laboratory and a private theater of inspiration. The flashes of insight would come when they would come. Meanwhile, he would wander among his trees.

• • •

Buffon continued to maintain his residence in Paris and make strategic appearances there, although his mind strayed increasingly to the renovations underway in Montbard. "I sigh for the tranquility of the country," he wrote to a friend. "I would rather spend my time making water flow and planting hops than to waste it here with useless errands, and more uselessly paying court." But the rounds of introductions and receptions led him to an important acquaintance with Jean-Frédéric Phélypeaux, the Count of Maurepas. Only six years older than Buffon, Maurepas was minister of the royal household for King Louis XIV, a position he'd inherited from his father at the age of seventeen. Born in Versailles and raised to comfortably walk the corridors of state, Maurepas kept a low political profile but wielded enormous power. Among his portfolio, as Buffon well knew, was supervision of the Académie des Sciences.

Maurepas was also the king's marine minister, which meant he was in charge of the French navy, all French colonies, and all French imports and exports proceeding by ship. In this role one of his greatest challenges was not at sea but on land: Large wooden ships required large timbers. The French shipbuilding industry was routinely scouring the countryside for massive old-growth trees suitable for masts, spars, and hulls, but the native supply had thinned considerably. Finding alternatives would be a tremendous boon to both the French navy and maritime trade. As Maurepas was delighted to learn, there was now one person in France systematically studying the growth of trees from a practical perspective, among other things experimenting with making them grow faster, straighter, and stronger. At the minister's behest, Buffon was invited to the French Academy to deliver a guest lecture. Not on trees (a subject unlikely to interest the members) but on geometry.

Buffon was more than up to the occasion. In his lecture, he began by considering basic questions of probability. What are the odds of a flipped coin coming up heads? That was elementary: Since the desired outcome (heads) is one out of two possibilities, the odds are one in two. And if one randomly draws a card from a deck of playing cards, what are the odds of it being the jack of spades? One in fifty-two, as there are only fifty-two possible outcomes and one jack of spades. Calculating probability seemed to be a simple matter of counting all possible outcomes, then dividing them by the outcome in question. But what happens when you can't count all possible outcomes? Buffon invited his audience to imagine a needle tossed onto a square-tiled floor. What are the odds of it landing so that it intersects a tile's edge? Calculating all the permutations, every single possible site and angle of a dropped needle, becomes a task that quickly approaches infinity.

Does that mean the odds are unknowable? Not at all, Buffon asserted. Take an entire box of needles and dump them on the floor, he directed, so that they scatter in a single layer. The pattern would be random but still provide useful information. Count the number of needles crossing a tile, then gather them up. Toss and count them again. Repeat a dozen times, or a hundred. Average up the results of each toss and count, and you begin to arrive at an approximation. If x number of needles are strewn across y number of tiles, your multiple random samples provide the z factor: the number that can be expected to land crossing a tile.

The full power of Buffon's insight, however, became clear when he pointed out the process could be reversed. Want to count all the needles in a very large box? Instead of laboriously tallying them all, it's a much quicker matter to scatter them all on the tiles, then count *only* the intersecting ones. If the odds are, say, one in seven thousand that a random needle crosses a boundary, then 152 boundary-crossers indicates the probability of 1,064,000 needles.

An experimental application of Buffon's Needle

Probability is not certainty, of course. But if you increase the sample size by gathering and casting the needles several times, then averaging the results, you end up with a number that, in practical terms, is likely to be *more* accurate than one produced by rote counting—when millions of objects are involved, human error has a way of creeping in. It was, in short, a way to quantify large, even vast numbers with a minimum of tedious tallying. It was also a powerful means of encapsulating chaos, since it allowed for unpredictable behavior within a framework of predictability.

Buffon's Needle, as the proof is still taught, is recognized as one of the seminal mathematic theses of its century. It marks the beginning of the field of geometric probability, the mathematics of accurately estimating large numbers of physical objects by analyzing a bounded area of the space they occupy. Scientists today use techniques descended from Buffon's Needle to tally the number of cells in a tissue sample, to calculate the internal surface area of a lung, and to quantify the number of neurons in a human brain. Meteorologists use them to predict the growth of storms. Financial firms use them to calculate investment risks, and governments use them to set economic policy. Physicists used them to develop the first atomic bomb.

Such implications were unimaginable in 1734, but the proof was impressive nonetheless. On January 9, when Maurepas submitted the

names of two Académie candidates to the king, there were no objections when he added Buffon's name as a third. For most members of
the Académie, election was an honor that came at either the height or
sunset of a brilliant career. For Buffon, it was an opening act.

"I have long been predicting that he would give evidence one day of
a good head," wrote one of his closest friends. It was not a tongue-in-
cheek statement. Buffon was in the habit of keeping his work to himself, revealing only the results and only in the context of his choosing.
Having achieved lasting renown in mathematics, he now moved on,
not even bothering to publish his treatise on Buffon's Needle except
as an afterthought, thirty years after the fact. The twenty-six-year-old
sat through an initiation ceremony, then returned to the business of
planting trees.

. . .

In Montbard, Buffon designed a life of maximum efficiency, calculated to keep himself physically healthy and mentally alert for as many
years as possible. Unlike his idol Isaac Newton, a solitary man who
died a self-admitted virgin (he had "neither passion nor weakness,"
according to Voltaire), Buffon recognized himself as hedonistic by
nature, given to indolence and dissolution. In order to combat his
natural tendencies, he adopted a strict schedule. The day began at
5:00 A.M. with the entrance of his valet Joseph, who was under specific instructions to rouse him no matter what. Joseph took this task
seriously, at times dragging him from bed. One morning when no
amount of coaxing would do, Joseph flung a basin of ice-cold water at
his slumbering form, then ran into the next room, fearing he had gone
too far. His employer's bell recalled him. "Give me some linen," said
Buffon, without anger. "But, in the future, let us try not to scramble
ourselves, we will both win." He gave the valet three francs as a reward.

There was another hour or so of preparation before Buffon was
ready to leave his bedchamber. He insisted on dressing each day in full
formal attire—not for guests but for himself. He believed it helped
him focus, bringing a consistent sense of gravity and occasion, re-
minding him of his purpose even when alone. After a hairstylist finished curling and powdering his hair, he walked by first light through
the streets of Montbard to the Parc Buffon.

Making his way up the slopes of the ancient fortress, now fenced, tiered, and surrounded by a series of groves, he entered one of two isolation chambers. The first, his primary workshop, was a single-room structure perched at the edge of the park, so modestly proportioned that it was easily mistaken for a garden shed. The second, a circular stone chamber inside one of the remaining medieval towers, was a well-insulated, naturally cool retreat for the hottest summer months. Each space contained only a writing table, a fireplace, and a portrait of Sir Isaac Newton. Aside from a view of the distant hills, there were no other distractions. As one rare visitor remarked:

> There, in a bare room, before a wooden secretaire, he meditated, he wrote. No papers in front of him, no accumulation of books; all that erudite lumber only impeded Buffon. A subject deeply thought out, contemplation, silence and solitude, those were his matter and his instruments.

He worked from memory of prior reading and his own ruminations, consulting no texts or notes. But he was not a claustrophile: The chambers were small because in effect he used his entire park as a workshop, darting in and out as inspiration suited. "He walked around thinking long and deeply about what he wanted to write," another visitor observed. "He filled himself up, he nourished himself, so to speak, with the wonders of nature, his ideas rose to the point of enthusiasm; then he took up his pen, wrote a few lines, and promptly returned to his walk and his meditations."

At 9:00 A.M. he paused for a quick breakfast, which always consisted of a roll and two glasses of wine. Then back to work until 2:00 P.M., when he enjoyed an unhurried meal with family, friends, and guests. Afterward, a brief nap, then another walk in his gardens alone.

This afternoon amble was so important to him that if any guests or workers accidentally encountered him on a garden path, they knew better than to acknowledge his existence. He returned to the isolation chamber at 5:00 P.M. and ended his workday exactly two hours later. To keep his workspace equally sparse for the next morning, he handed his day's writing off to a secretary, who transcribed a clean copy and added it to any manuscript in progress. The original pages were consigned to the fireplace.

To the frustration of future chroniclers, Buffon habitually inciner-
ated almost all of his papers once he was through with them, considering them as extraneous to his process as sawdust is to a carpenter. "I
burn everything," he explained to a visitor. "Not a single paper will be
found when I die. I have taken this course of action realizing that
otherwise I would never make it. One would be buried in one's papers."

Beginning at 7:00, he conducted his salon hours, the most social
time of his day. These consisted of more wine and conversation with
invited guests, most of whom knew well enough to dine beforehand—
Buffon did not eat supper, choosing to fast until the following day. At
9:00 P.M. he excused himself and went to bed.

He kept to this schedule for the next fifty years. While his Paris days
could not be quite so regimented, he maintained a contemplation-
centric schedule as much as possible. Far from seeming like depriva-
tion, this rigorous cultivation of solitary focus "was my whole pleasure,"
as he later reminisced.

> It is necessary to look at one's subject for a long time. Then little
> by little it just opens out and devélops. You feel as if a small charge
> of electricity struck you in the head, and at the same time reached
> your heart. That is the moment of genius; that is when one experi-
> ences the pleasure of working. . . . I have given myself over more to
> such delight than I have been concerned with glory. Glory comes
> afterwards, if it will. And it comes almost always.

Linnaeus in indigenous Laplander clothing,
holding a twinflower

SEVEN

Now in Blame, Now in Honor

"NEVER HAVE I KNOWN A WORSE ROAD," THE MAN IN THE green cap and pigtailed wig wrote in his journal. It was May 27, 1732, and he was trudging northward, toward the point where the purported road ("a mass of boulders, with great twisted tree roots and between them potholes of water") disappeared completely. He rode an unsaddled and unbridled horse, struggling to control the creature by tugging on a rope tied to its jaw. A short sword dangled at his side, but his urge to use it had long since faded.

Nils Rosén, the Academic Senate, and the rest of Uppsala University were hundreds of miles at his heels. His immediate concern was finding food and respite from the rain.

Abandoning hopes of either reemployment or revenge, Linnaeus had retreated from Uppsala to his parents' house in Småland in the summer of 1731, where he grasped at one last chance to restore his reputation. Appealing to Anders Celsius, the nephew of his old patron and head of an Uppsala learned society (and incidentally the inventor of the Celsius thermometer), Linnaeus applied for a modest grant of 675 copper dalers to launch a natural history expedition to a remote corner of Sweden. He was awarded 450 copper dalers—barely enough to provision a single expeditionary, traveling on horseback and on foot. On May 12 he left Uppsala, riding the first of a series of rented horses. In a small leather bag he carried one shirt, an inkstand, a magnifying glass, a spyglass, a comb, and an ample supply of gauze. He did not carry foul-weather gear, a surprising omission considering that his destination was Lapland, above the arctic circle.

Although Lapland was a province of Sweden at the time (as was Finland), its remoteness had kept the region little known to outsiders beyond the occasional missionary or explorer. As far as Linnaeus knew, the last natural historian to pass through had done so thirty years earlier. Linnaeus began his northward journey optimistically, filling his notebook with bucolic prose ("the whole land laughs and sings"), but after two weeks on the road his entries grew as dark as the prevailing weather. "All the elements were against me," he wrote of the day's travels.

> The branches of the trees hung down before my eyes, loaded with rain-drops, in every direction. Wherever any young birch trees appeared, they were bent down to the earth, so that they could not be passed without the greatest difficulty. The aged pines, which for so many seasons had raised their proud tops above the rest of the forest, overthrown by the wrath of Juno, lay prostrate in my way. The rivulets which traversed the country in various directions were very deep, and the bridges over them so decayed and ruinous, that it was at the peril of one's neck to pass them on a stumbling horse. It seemed beyond the power of man to make the road tolerable.

More maddening still were the thick, ubiquitous clouds of insects. Despite the gauze he wrapped around his face, "the gnats kept inflicting their stings. I had now my fill of traveling," he recorded, just three weeks into the expedition. "A divine could never describe a place of future punishment more horrible than this country, nor could the Styx of the poets exceed it. I may therefore boast of having visited the Stygian territories."

As Linnaeus himself was realizing, he was temperamentally ill-suited to be a field researcher, a fact made clear not only by his misery but by his methodology, which swung wildly between minutiae and the cursory. He acquired a Laplander shaman's drum and several items of clothing for his own wardrobe, while failing to note that the conical hat he obtained was traditionally worn by women. He sketched precise floor plans of the contents of a Laplander tent, right down to the racks of cheese, but he also concluded that "like all people addicted to fishing, the Laplanders are very fond of brandy," and "no Laplander can sing, but instead of singing utters a noise resembling the barking of dogs." If he learned the name of a single Laplander, he did not commit it to the page.

He collected several saddlebags of specimens, including so many minerals that he stopped to ship them back to Uppsala. His most significant botanical acquisition was samples of *Campanula serpyllifolia*, an arctic plant known commonly as the twinflower for its tendency to bloom in pairs. Linnaeus enjoyed it so much he adopted it as a personal emblem, posing for portraits with a twinflower sprig in his hand and later incorporating it into his family coat of arms. Did it symbolize his friendship and planned collaboration with Peter Artedi? He never explained its personal significance. Linnaeus would in time allow it to be renamed *Linnea borealis*—the only species he ever allowed to be named in his honor. He described it as "lowly, insignificant, disregarded, flowering for a brief space—from Linnaeus, who resembles it."

More than anything else, Linnaeus's arduous journey gave him ample time to think. His earlier thought exercises with Artedi had centered on quantifying life, on setting a reasonable number of how many species could be catalogued. But a comprehensive study of nature would require more than an inventory—it would require a master plan for organizing it. Developing that master plan would be a signal achievement, far outstripping his vegetable letters and inevitably vaulting him back into the academic forefront, forever keeping the Roséns of the world at bay.

• • •

Could someone create a master plan for all of life on Earth, a logical arrangement that accommodated every single species? Linnaeus weighed the matter, and decided it was distinctly possible. To understand why, it's necessary to examine some of the central assumptions Linnaeus and his contemporaries shared about nature—assumptions so ubiquitous in European culture at the time that they scarcely called for acknowledgment, like the presence of air in a room. A first step is unlearning what we presently mean by *species*. Our current popular understanding is that it means a particular kind of lifeform that exists as a distinct population, one capable of propagating itself over time. To Linnaeus's generation, the concept was far more fluid.

The word is simply Latin for "appearance." *Fallaces sunt rerum species et hominum spes fallunt*, wrote the philosopher Seneca in the first century A.D. *Appearances are deceptive, and betray the hopes of men.* When a botanist used the term *species* in 1729, it was a shorthand for "plants that have this appearance." Equally serviceable were the Latin words *exemplum* and *descriptionem*, but *species* had come to the fore; *exemplum* had a negative connotation (as in "make an example out of you"), while *descriptionem* tended to convey more of a verbal than a visual depiction. The working definition of *species* depended on the context of the work itself. A botanical guide to ornamental gardening might not make a distinction between a calla lily and a lily of the valley, as both are white and serve similar decorative purposes. A medical text might take great pains to differentiate the two, since the lily of the valley is poisonous and the calla lily is not. For two daydreaming Uppsala students, the term could be defined as loosely or as strictly as they wished.

Other contemporary assumptions: We take as a given that each species has a representative population, and that this population is subject to change. If meaningful differences arise in that population over a long span of time—hundreds if not thousands of generations—we call that process evolution. If evolution produces profound enough changes, we declare that population a new and different species. If a population no longer maintains itself, we say the species ends. This process we call extinction.

But these concepts belong to our age, not to the mid-eighteenth century. *Extinction* and *evolution* would not be used in a biological sense for decades; for that matter, even the concept of *biology* itself belongs to the subsequent century. When Linnaeus set out to assess the living world, one of his lodestones was the assumption that it represented a static target. This concept, so omnipresent in its day as to be nameless, represents a stark contrast from contemporary conceptions. As a term of differentiation, we shall refer to it as a sort of lens. The lens of fixity.

The lens of fixity was a legacy of the Old Testament. In fact, generating such a lens was the ancient text's first priority: Genesis, the first book of the Bible, establishes a strict sequence for the arrival of life. It attests that after spending the initial day separating darkness from light, God devoted the second day to raising up dry land, naming it Earth, and willing into existence the very first lifeforms: "plants yielding seed, and fruit trees of every kind on earth that bear fruit with the seed in it." Summoning "two great lights" (the sun and moon), adding stars, and fixing them in the appropriate places occupied the third day. On the fourth day, to populate the newly wrought sea and sky, "God created the great sea monsters and every living creature that moves, of every kind, with which the waters swarm, and every winged bird of every kind." Land-based creatures were the sixth day's agenda, "the wild animals of the earth of every kind, and the cattle of every kind, and everything that creeps upon the ground of every kind." Later that day, the Lord capped off His work by creating humans, then bidding them to "fill the earth and subdue it; and have dominion over the fish of the sea and over the birds of the air and over every living thing that moves upon the earth."

Plants and trees on the second day. Birds, fish, and sea monsters on the fourth. All other organisms on the sixth. By the dawn of the seventh day, Creation was complete. Thereafter the tableau of existence was static.

When one viewed life through this lens of fixity, it was against faith to envision new species coming into existence, or existing ones fading into extinction. Nature did not change—could *not* change, for that matter, since the Maker had long since put away His tools and closed up His workshop. To suggest otherwise, to imply that the perfection of Genesis was subject to revision, would be blasphemous. It also held an undeniable logic, once you considered the interaction of species: crops and cultivators, parasites, predators, and prey. A watchmaker does not add parts after the watch is up and ticking. The parts and purposes of Nature were, necessarily, the same today as in the beginning.

It seems natural for us to organize species as "related to" one another, and therefore similar. But through the lens of fixity, this was as artificial a system as any: All life was equally related, having been created at the same time. To Linnaeus, a minister's son who did not dispute the biblical account, the lens of fixity provided a finite scope of ambition. While any given species might wax and wane in population, the overall number of species themselves remained unchanging.

Later passages of Genesis confine the scope of life still further. When urging preparations for the Great Flood ("everything that is on the earth shall die"), the Lord provided instructions specific enough to make mathematical projections possible. Noah was to construct an ark exactly 442 feet long, 44 feet wide, and 73 feet high, containing three decks to accommodate

> seven pairs of all clean animals, the male and its mate; and a pair of the animals that are not clean, the male and its mate; and seven pairs of the birds of the air also, male and female, to keep their kind alive on the face of all the earth.

Given these parameters, and a knowledge of animal husbandry, numerous theologians and scholars produced informed estimates on the sum total of passengers, ranging from approximately three hundred to more than two thousand species. The uncertainty arose from how much feed was required, which animals were considered "clean" (requiring seven pairs of specimens) or "not clean" (only a single pair), and whether or not some of the aquatic birds could have paddled alongside. Still, the parameters were specific enough that in 1768 the *Encyclopedia Britannica* would declare the ark "abundantly sufficient for all the animals supposed to be lodged in it." In fact, the fit was so proper "that the capacity of the ark, which had been made an objection against scripture, ought to be esteemed a confirmation of its divine authority."

Surely, an energetic young man could seek out and study three hundred to two thousand species of animals. The problem was not number but dispersion. In the days since the Flood, land-based animals had ranged far across the Earth, and some would be harder to track down than others. What of marine life, the creatures "with which the waters swarm"? Since those had occupied only a half-day of Creation, they were reasonably assumed to be half as many, at most. Vegetable species were presumably limited to the descendants of the plants that Noah's family used to feed the animals and themselves, as well as those whose seeds had survived being submerged for the 371 days of the Flood (the olive tree, from which Noah's dove had retrieved an encouraging leaf, being a prime example). Thankfully, generations of botanists had already devoted themselves to enumerating these. The most recent comprehensive survey, John Ray's massive *Methodus Plantarum* of 1703, indexed approximately 18,000 species, a number that even decades later Linnaeus would estimate as 90 percent of the total plant population.

The total was somewhere between 20,450 and 23,000 species, the vast majority of which had already been identified and described to some degree. A complete catalogue of all life was an achievable goal. The real challenge would be organizing it.

• • •

In referring to his sexual system as a vegetable alphabet, Linnaeus was acknowledging that, like the alphabet, it too was an artificial system. Counting stamens and pistils was a convenient guide to field identification, but not truly illustrative of some deeper structure of life. It also lumped together wildly disparate plants, requiring one to ignore some of the stranger pairings (in Linnaeus's sexual system, elm trees and carrots shared a category). There was nothing wrong with artificial systems—in fact, they could be more useful than natural systems. But they were not the timeless universal truths Linnaeus was looking for.

What constitutes the difference between a natural system and an artificial one? Consider alphabetical order. The A-to-Z sequence we assign letters is a foundational standardization, drilled into our head through songs and picture-book primers from the beginning of our education. Yet there is nothing inherent in letters (or other language symbols, for that matter) that requires lining them up in one particular order—even the idea of a single, agreed-upon "order" to written characters is a social construct. Japanese has two different traditional equivalents of alphabetical order, gojüon and iroha. Chinese has at least fourteen, some of which were zealously kept family secrets for generations of hereditary bureaucrats.

We develop and rely upon such systems because their very artificiality is a strength: They only need to be memorized, not researched. One indisputably "natural" system for organizing books on a library shelf would be arranging them according to date of publication, chronology conveying a clear inherent order. The same goes for books organized by height, by the number of their pages, or by the color of their cover design. But none of those would make individual books particularly easy to find. Despite having no reality beyond a social construct, alphabetical order admirably performs that task.

While Linnaeus's sexual system was artificial, it was also eminently practical. So why wasn't he content with it? Because what is artificial is also mutable. That *J* is the tenth letter in the alphabet is an arbitrary choice dating back to 1524, when Italian linguists began to differentiate it from the letter *I*. The tradition of listing the letter *Z* last is an even more recent invention. Up until the early twentieth century, English-speaking schoolchildren were routinely taught that the final letter in the alphabet was the ampersand (&), a symbol now exiled from the alphabet into the realm of punctuation. Linnaeus's vegetable alphabet could be subjected to a similar process of revision, or for that matter scrapped entirely—he was betting that every single plant species perpetuated itself by means of some arrangement of stamens and pistils, even if those parts were currently undetected. If the plants he shunted into the category of cryptogamia ("hidden marriages") could be proven to lack stamens and pistils altogether, the entire system could come crashing down.

There was a second, still more crucial limitation to Linnaeus's present system: It couldn't be extended from the realm of botany into that of zoology, since the factor used for differentiation was sex itself. While plants had developed a multitude of approaches to reproduction, so far as Linnaeus knew animals seemed to have settled on only one. There were variations in the mechanisms of gestation, such as live-bearing versus egg-laying, but conception appeared to be the province of one female and one male. In other words, all animals fit into the sexual system's monoclinous category, which meant to zoologists it was no system at all. The challenge before Linnaeus was twofold: to find a natural system for plants that could replace his vegetable letters, and to find another, equally natural system to apply to the animal kingdom.

How to begin? Linnaeus found his answer to the latter challenge on the Lapland road, when he noticed a curious object left to weather on a fencepost. He stopped to examine.

> Close to the road hung the under jaw of a horse, having six fore teeth, much worn and blunted, two canine teeth and at a distance from the latter twelve grinders, six on each side. If I know how many teeth and of what peculiar form, as well as how many udders, and where situated, each animal has, I should perhaps be able to contrive a most natural methodical arrangement of quadrupeds.

He'd had no trouble recognizing it as the jawbone of a horse, he realized, because of its particular arrangement of teeth. Animals ate, and conveyed nourishment to their young, but the means by which they did so required distinctive features of their anatomy. If he could quantify and inventory those features, would not such an accounting be definitive of the animals themselves? He trudged on, beginning to sort the possibilities of such a method in his head.

· · ·

Returning to Uppsala in October, Linnaeus found his academic status only slightly improved. The university allowed him back but as a mere academic assistant, teaching only when Rosén found it inconvenient to do so. In the long weeks between his occasional lectures he entertained friends, conducted private tutoring sessions, and discreetly ghost-wrote dissertations for a price. He decorated his rooms with field souvenirs and specimens, and had his portrait painted dressed in his Lapp clothing (complete with woman's hat, jauntily worn). He worked on the official account of his Lapland journey, polishing and improving on his field notes to the point of fabrication.

He began with simple embroideries on the truth, painting a water transit as an arduous voyage when in fact he was ferried across in a rowboat. He progressed to vast exaggeration, estimating mountain passes as "more than a land mile high"—a "land mile" being a now-archaic unit of measurement equal to thirty-six thousand feet, which would make them taller than Everest. Then he entered the realm of sheer fiction. "Oh, how many weary steps we took up that mountainside, how we sweated," he wrote of his trek to the summit of Mount Caitumbyn. "Sometimes we were enveloped in clouds which half-blinded us. Sometimes we had to make a detour to avoid a stream, sometimes to strip and wade through icy water. Had we not had this cold snow water to revive us in the heat we could never have survived." Evocative imagery, considering that the excursion was never mentioned in his travel journal and would have broken the bounds of human endurance. Mount Caitumbyn was an 840-mile detour from his itinerary, a side trip that would have demanded an average marching pace of sixty miles a day. Since he was describing incompletely surveyed terrain, such discrepancies were not exposed for the better part of a century. But they inevitably would be, a fact that speaks to Linnaeus's near-desperate desire to impress his sponsors and earn at least a short-term return to relevance.

The tactic failed. The Swedish Royal Society published only a small portion of his report and slow-tracked the rest (it would not see print for another six years), citing a limited budget that also held no funds for further expeditions by Herr Linnaeus. The latter did not particularly discomfit him: He would never attempt field research again. But the lack of a forum for reigniting his career convinced him that his fortune awaited elsewhere. At eight in the morning of December 19, Linnaeus left Uppsala again, this time with no plans to return. "I said good-bye to Uppsala Academy, to which Almighty God so marvelously conducted me," he wrote in his diary. "Living now in difficulty, now in enjoyment, now in poverty, now in abundance; now in blame, now in honor." He carried with him a thick sheaf of papers. These were notes and draft pages, comprising the beginnings of a vastly ambitious project.

An eyewitness portrait of the Hydra

EIGHT

The Seven-Headed Hydra of Hamburg

IN MAY OF 1735, IN HAMBURG, GERMANY, A WANDERING EN-
tertainer encountered a monster.

The entertainer was Linnaeus, now twenty-eight years old. Dressed
in his exotic Laplander costume and beating his shaman's drum, he'd
just finished performing in a private reception for the learned gentlemen of Hamburg, regaling them with his (exaggerated) arctic adven-
tures, showing his portable collection of over a thousand Swedish
insects, and holding forth on his ideas about the organization of
plants. "All that this skillful man thinks and writes is methodical," the
newspaper *Hamburgische Beriche* had reported prior to his arrival. "His
diligence, patience and industriousness are extraordinary." The notice
(which Linnaeus had written himself) had earned him the reception,
and now, as a reward, it had earned him access to the monster. The

learned gentlemen opened a cabinet to reveal one of the most valuable specimens in the world, pridefully called the Hydra of Hamburg.

It bore sharp claws, a dense armor of leathery brown scales, and a menacing writhe of seven angry heads—all poised to attack, rising up on sinuous necks and baring fourteen vicious pairs of fangs. The specimen had remained frozen in this posture for eighty-seven years, ever since it mysteriously appeared upon the altar of a church, yet even in stasis its ferocity was unsettling. The fact of its existence was more unsettling still: The beast seemed to have slithered directly out of chapter 12, verse 3 of the Old Testament's apocalyptic book of Revelation, making it not only a terrifying proof of monsters but evidence of biblical end times drawing nigh.

> And there appeared another wonder in heaven; and behold a great red dragon, having seven heads and ten horns, and seven crowns upon his heads.

The creature before him was more brownish than red—perhaps the color had faded over time—and while its horns had apparently gone missing, it certainly sported seven heads. "Many people said it was the only one of its kind in the world," Linnaeus observed, "and thanked God that it had not multiplied."

The Hydra first appears in historic records in 1648, when a German general acquired it as a prize of war at the end of the Thirty Years' War. Passing through many hands in the ensuing decades, it was now the prized centerpiece in the collection of Johann Anderson, the wealthy former mayor of the Hamburg city-state. Anderson did not keep it on public display but rather guarded it zealously, in keeping with its perceived enormous value. The King of Denmark had once offered thirty thousand thalers for the creature, roughly equivalent to four million dollars today, but Anderson had turned down the offer—a choice he likely regretted, since he was now negotiating to sell it to a private buyer for the smaller but still tidy sum of two thousand thalers.

Its authenticity seemed beyond dispute. For generations scholars from throughout Europe had conducted pilgrimages to Hamburg, to the cabinet of wonders where the Hydra now resided. All had pronounced the creature genuine. One eminent naturalist, the Bishop of Bergen, had authenticated the specimen as "a natural truth." The natural historian Albert Seba had recently included the Hydra in his *Thesaurus*, the most authoritative zoological text of the day, commissioning an artist to sketch it in the scaly flesh and solemnly pronouncing it "in no way a work of art, but truly one of nature."

As they gathered behind him, their faces a mixture of expectation and pride, Linnaeus scrutinized the body's transitions, from the forest of necks to the club-like tail. He leaned in close, the better to take the measure of all 322 jagged teeth.

Then Linnaeus began to laugh.

"O Great God," he said, "who never set more than one clear thought in a body which Thou has shaped!"

Who never set more than one clear thought in a body. He was snickering at the idea that the Hydra had ever been alive. Since seven heads presumably held seven brains, how could they independently move their necks, while simultaneously cooperating on moving the body as a whole? This of course was an objection that Linnaeus could have raised before walking into the presence of the Hydra, but he did not elaborate further. Instead he moved in closer, pointing to the surfeit of jaws. Clearly, he explained, these once belonged to seven unfortunate weasels. The skin stretched over them? Taken from snakes, and just as patently repurposed. The construction was a taxidermic hoax, likely perpetrated by Prague clergymen to put the fear of end times into their parishioners. "I was the first to see this was no wonder of nature but of art," Linnaeus later wrote, relishing the neat reversal of Dr. Seba's authentication: *not a work of art, but of nature.*

It was like the lifting of a veil, the dispelling of an illusion. With a few deft observations, Linnaeus had shown the learned gentlemen something they could not now unsee. In the ensuing silence one of the party, a Dr. Jaenisch, stepped forward to diplomatically offer a suggestion. It would be a good idea for the Swedish medical student to leave Hamburg as soon as possible.

Linnaeus gathered up his insect box, costume, and drum, and took his leave. The Seven-Headed Hydra of Hamburg disappeared soon after, disposed of as a worthless hoax.

. . .

Two days later, Linnaeus recorded the encounter in his usual third person ("Linnaeus must hasten his departure to escape the Burgo-master's vengeance"), struggled with seasickness on a stormy passage to the Netherlands, and approached his ultimate destination. This was Harderwijk, a Dutch town thirty-five miles from Amsterdam, notorious for selling (as a song of the day went) "bloaters, bilberries, and degrees." Bloaters were cold-smoked herring, bilberries were a local fruit, and the degrees were courtesy of the University of Harder-wijk, specializing in catering to medical students in a hurry. No one needed to find long-term student housing in Harderwijk, not with a pro forma course of study designed to be completed in less than a week.

Linnaeus's time on campus lasted all of two days. On the first day, he enrolled, submitted a pre-written thesis on the cause of intermit-tent fevers (he blamed clay soil), sat through an oral examination, wrote a brief essay on the *Aphorisms* of Hippocrates, and visited an examination room long enough to diagnose and treat jaundice in a sin-gle patient. After an enforced pause of three days—time for a printer to strike off copies of his thesis—he mounted a podium to defend his conclusions on intermittent fevers, in the form of a brief debate with Harderwijk professor Jan de Gorter, after which de Gorter awarded him a signed diploma, a silken hat, a gold ring, and his con-gratulations. On June 23, 1735, the twenty-eight-year-old Linnaeus was officially a doctor, certified not only to practice medicine but to teach it.

This was the *potestatem docendi* that Nils Rosén had used to displace him as lecturer—in fact, Rosén had obtained his own degree at Hard-erwijk, in a similarly hurried manner. But Linnaeus had no immediate plans to return to Sweden, much less to renew the dispute with his Uppsala nemesis. He knew Rosén had followed his visit to the di-ploma mill with a multi-year extended tour of European schools and clinics, studying with and working alongside some of the best medical minds of the day. Despite the fact that Linnaeus was rapidly running out of money, he too needed practical experience.

As a first step toward finding employment he'd made sure that the *Hamburgische Beriche* (the same newspaper that had won him the invita-tion to inspect the Hydra) carried a notice that he intended to stay in Holland for several years, "in order to frequent the famous men there, and in particular Herr Boerhaave with whom he has already carried on a scholarly correspondence." Herr Hermann Boerhaave, a retired professor at the University of Leyden, was one of the best-known physicians in the world. He was also perfectly aware that he'd never corresponded with a Swedish medical student named Linnaeus, and was less than pleased after reading the false self-promotion. When the student arrived at his doorstep bearing a letter of introduction, Boerhaave kept him waiting a week before begrudging a frosty reception. The great man would eventually warm to him, but for now Linnaeus had alienated a key source of referrals and connections.

The one bright spot in Linnaeus's visit to Leyden came a few days later, when he spotted a familiar face in a tavern. It was Peter Artedi, the friend and fellow student with whom he'd divided up all earthly life. "Our tears showed what joy we felt," Linnaeus wrote of the chance reunion. Artedi had traveled to England to continue his studies of fish, but dwindling funds had forced him back to the Continent. "He had been made penniless by the expenses of his journey," his old friend observed, "and was worried about how to get clothes, food, and all the books he needed for his work." While almost penniless himself, Linnaeus didn't hesitate to put his resourcefulness to work. To render Artedi presentable he scrounged up three shirts, then took him to Amsterdam and finessed a meeting with seventy-year-old Albert Seba, the apothecary and naturalist who had certified the Hydra of Hamburg as "in no way a work of art, but truly one of nature."

Bad blood might have been expected, but Linnaeus knew Seba had forgiven him for debunking the specimen. He also knew Seba was continuing to compile his multi-volume *Thesaurus*, and was presently busy on a volume dedicated to fishes. As he anticipated, the wealthy Seba eagerly offered to hire Artedi as an assistant. Linnaeus received the grateful thanks of both Artedi and Seba, left them to their project, and moved on to the pressing matter of his own survival.

In Amsterdam, the only suitable employment he could find was temporary work as live-in assistant to Johannes Burman, a botanist writing a book on the subject of tropical plants. It was a humbling arrangement, because it paid little more than room and board and because Burman, a full professor at the University of Amsterdam, was only one year older than Linnaeus himself. But it did provide access to interesting dried specimens (Burman was currently studying the flora of Ceylon), and on one momentous occasion brought him into contact with the living plants themselves. On August 13, Burman permitted Linnaeus to accompany him on a day's excursion to a place called Hartekamp, sixteen miles outside the city. It was the private estate of George Clifford, the grandson of an Englishman who'd moved to Holland and prospered as a banker. Clifford maintained the family business but also held a seat on the board of the Dutch East India Company, the government-sanctioned shipping and import monopoly that dominated the international spice trade. Now one of the wealthiest private businessmen in Europe, Clifford was pouring copious amounts of money into making Hartekamp a showplace of one of the most rarified status symbols a rich man could display: his collection of exotic life.

In an era when tropical fruits are commonplace fixtures of local supermarkets, it's easy to forget how even a single pineapple or banana was once a sight that Europeans clamored to see. What impressed them was not only the item itself but the long, expensive chain of effort required to transport it to their climate. Live plants and cuttings needed to be rushed across the planet in fast-sailing ships, tended to in transit by experts, painstakingly transplanted in specially built structures called adonis houses (precursors of our modern greenhouses), and maintained around the clock to mimic tropical conditions. Even successful importations often failed to thrive, making it necessary to restart what was often a years-long process. No wonder Charles II of England, when presented a pineapple by his royal gardener in the 1670s, saw fit to commission a portrait of the occasion. By the 1730s, commoners were paying the equivalent of $8,000 in contemporary currency for a single pineapple.

No uncrowned head in Europe had more living prestige items than George Clifford. Drawing on specimens delivered by Dutch East India Company ships, he'd transformed Hartekamp into what an awestruck Linnaeus called "the miniature of Paradise," with extensive grounds that held not one but four adonis houses, filled with "such a wealth of plants that a son of the North could only stand as though bewitched, unable to say to what strange quarter of the earth he had been transplanted." It also encompassed the most diverse private zoo in Europe, a well-stocked menagerie and aviary filled with tigers, apes, wild dogs, Indian deer and goats, and South American and African boars. Linnaeus tore himself away reluctantly, dreaming of living and working there.

He was bound to his room-and-board arrangement with Burman at least through the winter, but that did not stop him from positioning himself for a job at Hartekamp, adopting a strategy based on two observations. An extensive staff of gardeners, groundskeepers, and animal handlers tended to the complex, but no naturalist stood at their head. And the lord of the manor was notable not only for his collection but also for his hypochondriac tendencies. Soon Clifford was receiving letters from personages like Dr. Gronovius, suggesting how marvelously convenient it might be to hire a supervising botanist who could also serve as a resident physician, and how fortunate it was that a young man with such capabilities had recently arrived from Sweden.

There were substantial marks against Linnaeus. His practical medical experience consisted of treating only one patient. He had never previously seen, much less worked with, the vast majority of the specimens at Hartekamp, and he spoke not a word of Dutch. The idea nevertheless appealed to Clifford, who made overtures of employment only to run up against Burman's unwillingness to release his contracted assistant. The banker called upon the professor, and when the latter mentioned a book he'd never seen but particularly coveted (Sloane's *Natural History of Jamaica, Volume II*), Clifford saw his chance. "I happen to have two copies," he told Burman. "I will give you one if you will let me have Linnaeus." The bargain was struck. Linnaeus shifted his few belongings to Hartekamp, blissfully happy to have been traded for a book. Less than a week later he was rushing back to the city, steeling himself for one of the grimmest moments of his life.

Amsterdam was a city of canals, but not of streetlights. To a late-night pedestrian, inexperienced, exhausted, inebriated, or a combination of all three, the demarcation between walkway and waterway was not always easy to discern. On September 28, Albert Seba invited his new assistant Peter Artedi to a dinner party. When Artedi emerged from the Seba mansion late that evening, he wandered unaccompanied and lost his way on the dark city streets. His corpse was spotted in the early light of the following day, floating in a canal.

"When I saw the lifeless, stiffened body, and the froth upon the pale blue lips . . . I burst into tears," wrote Linnaeus, who was called to identify the drowned man in the City Hospital's morgue. The death was an accident, but Artedi would not have been at Seba's house, or in Amsterdam at all, without Linnaeus's intercession. The corpse was wearing one of Linnaeus's loaned shirts.

"I felt that the love I cherished for my friend compelled me to fulfill my promise," Linnaeus concluded, referring to their pledge from student days—that if one of them died, the other would continue their collaboration. Linnaeus began quarreling with Artedi's former landlord, demanding possession of his late friend's notebooks. As soon as he obtained them he edited and adapted their contents, incorporating them into a manuscript already being readied for the press.

. . .

On December 13, 1735, a tall, thin volume written in Latin went on sale at a bookstore in Leyden. It was expensively priced at two and a half guilders, especially considering it was more an oversized pamphlet than a book: just fifteen pages of densely set type, printed on folio sheets fifteen inches high. Its author was billed as Caroli Linnaei, Doctoris Medicinae, with a last-minute addition of Sveci ("Swede") in smaller type, inserted between name and qualification. It bore the title *Systema Naturae*, The System of Nature.

"The first step in wisdom is to know the things themselves," Linnaeus wrote in his introduction. "This notion consists in having a true idea of the objects; objects are distinguished and known by classifying them methodically and giving them appropriate names. Therefore, classification and name-giving will be the foundation of our science."

In *Systema Naturae*, Linnaeus began by breaking the Great Chain of Being. First, he severed the link connecting humans and Heaven's celestial legions, choosing to let theologians decide if nephalim were ranked higher than seraphim, or vice versa. Now firmly earthbound, he divided his subject into three parts. "*Minerals* grow; *Plants* grow and live; *Animals* grow, live and have feeling," he wrote. "Thus the limits between these kingdoms is constituted." The use of *kingdoms* still retained the tone of the Great Chain's feudalistic schema, as did much of his subsequent terminology. Linnaeus was carefully choosing terms that seemed familiar, while freighting them with new meaning.

Although the volume's title announced it as a singular truth—*the* system of nature—*Systema Naturae* employed three very different systems. For the plant kingdom Linnaeus simply grafted on his extant sexual system, knowing fully its limitations. "No natural botanical system has yet been constructed," he wrote. "I may, on a future occasion, propose some fragments of such a one. Meanwhile, until that is discovered, artificial systems are indispensable."

The shortest section of *Systema Naturae*'s first edition comprises its treatment of the mineral kingdom. Linnaeus divided it into three categories: *Petrae*, or rocks and simple stones, *Minerae* ("stones composed of constant rock particles impregnated with foreign particles"), and *Fossilia* ("stones of aggregate rocky particles, or a mineral mixture"). The fact that some *Fossilia* "show an impressed image of an animal or vegetable" was, to him, only an amusing coincidence; he did not consider them anything that had formerly been alive. He postulated there were two "primary soils," sand and clay, produced in the beginning, "from which by the working of the Elements, we assume the whole Kingdom of minerals to have been produced. From them the remaining Stones originated in the time since Creation." He believed that soil came first, only gradually hardening into rocks: Just as sandstone was hardened sand, marble was hardened clay, and slate the hardened decayed vegetable matter of boggy soil. Minerals were rocks "impregnated with some substance foreign to its simple component." Quartz "is a parasitic stone; as it is produced in cavities of other stones and grows out from there." Gemstones he classified "not as true species but as varieties, as they differ distinctly by the color only."

These were vague delineations, involving beliefs that had little traction in 1735, and even less so today. Quartz is not a parasite, marble is not hardened clay, and fossils are far too numerous to coincidentally resemble life. Linnaeus's work on the mineral kingdom would have no impact whatsoever on geology, and it would be roundly ignored by even his most ardent followers. But he would continue to maintain the category in subsequent editions, updating it for the rest of his professional life.

That left the animal kingdom. Bypassing the distinction between wild and domesticated, he broke animals into what he considered obvious categories—fish, birds, amphibians, insects. For these, he decided to use the familiar category of *classes*—another borrowing from several versions of the Great Chain, in which human social classes occupied separate and distinct links. Drilling down further, he made divisions based on a prominent physical characteristic—the size and shape of birds' beaks, for instance—which he called *orders*. It was another comfortably familiar word. Monks and priests, for instance, belonged to orders, wearing distinctive robes to tell themselves apart.

He further clustered these into groupings based on physical similarities, which he termed *genera* (plural of *genus*, a Latin word meaning "family" or "type"). Finally he arrived at the specific: the category he labeled *species*, meaning that which could be represented by a single individual organism. Differences below the level of species—varieties of dogs or breeds of horses, for instance—Linnaeus considered incidental but not essential. He did not include them in his system.

Kingdoms, classes, orders, families, species. He was not the first to use such terminology. *Genus* dated back to Aristotle, Tournefort had designated floral classes in 1694, and *species* was commonly used in a

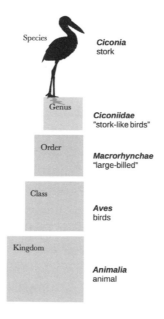

Species — *Ciconia* stork

Genus — *Ciconiidae* "stork-like birds"

Order — *Macrorhynchae* "large-billed"

Class — *Aves* birds

Kingdom — *Animalia* animal

The nested boxes of Systema Naturae's *"Animal Kingdom"*

variety of contexts. But Linnaeus was the first to fix their meaning to what they connote today, and to arrange them in a series of five nested boxes.

Linnaeus firmly believed that these five boxes sufficed to contain all animal life. He would later defend his plan by drawing parallels with what he saw as similar five-nested hierarchies. In geography: realm, province, district, parish, hamlet (or alternately, kingdoms, cantons, provinces, territories, and districts). In the military: regiment, battalion, company, platoon, soldier. It is easy, of course, to find more than five such divisions in each instance, but Linnaeus considered fivefold synchronicities a validation of his work. Whether one accepted his premise or not, on the face of it the taxonomy appeared to make the zoologist's task more difficult, not easier. Why make five decisions about a species—each dependent on the other—as opposed to describing the species itself? Could life truly be fixed with such certainty?

Linnaeus believed it could. *Systema Naturae* is credited as the beginning of our contemporary system of "scientific" names, but this is something of a misnomer: Linnaeus's innovation was not the awarding of names but the giving of addresses, specific locations on the map of creation. Viewing life through the lens of fixity, he considered such locations permanent. "There are no new species," he wrote. "Like always gives birth to like." As such, the species level (the name) was permanently fused to the remainder of the hierarchy (the address). Pinning down the address with fivefold accuracy, that was not the concern of other zoologists—Linnaeus assigned the task to himself alone and was confident in his work. The discovery of a bird similar to the stork, for instance, might cause him to expand the genus *Ciconiidae* (literally "stork-like birds"), but the stork itself wasn't going anywhere. It would remain *Ciconia,* genus *Ciconiidae,* order *Macrorhynchae* ("large-billed"), class *Aves,* kingdom *Animalia.* These conceptual containers might expand, but he did not expect them to collapse.

Despite its title, *The System of Nature* contains no fully natural system. The plant kingdom's organization is admittedly artificial, and the mineral kingdom is sorted by criteria so inchoate as to render it effectively arbitrary. It is only in the animal kingdom that Linnaeus attempted a detailed naturalistic logic, and then at a crucial remove.

His approach to classifying animals was neither artificial nor truly natural, but a hybrid that could best be called multi-logical.

. . .

If you've ever walked the aisles of a grocery store, you've immersed yourself in a multi-logical system. Grocery stores do not shelve their items alphabetically (an artificial order), nor by date of manufacture (a natural order)—either would strike the shopper as a useless jumble. Instead they're organized in an order that switches criteria from aisle to aisle, grouping some products by their contents, some by origin, some by typical manner of use, and some by method of preservation. We're used to shopping for vegetables by the latter criterion, seeking fresh ones in the produce section, canned ones in another aisle, and frozen ones in a third location. We're used to locating milk, cream, yogurt, and butter by criterion of origin, finding them all in a "dairy" section. Many larger stores have a "breakfast" or "dessert" section, which identifies products by customary use.

A multi-logical system employs choices that, in context, seem perfectly natural. Yet the context shifts within the overall plan—one set of organizational logic is applied in one area, and a different set in another. So long as one experiences each region individually (and doesn't dwell upon the imperfect seams between the switch from one logic to another), it appears easy enough to navigate. In grocery stores, shifts in organizational logic are ubiquitous enough to frequently go unnoticed, even when they create cognitive dissonance. Dairy aisles now regularly include non-dairy products, such as soy milk and margarine, and one type of product might be found in multiple locations (one of America's largest grocery chains, for instance, displays salsa in four separate areas). The so-called breakfast aisle usually contains only a subset of what one might eat for breakfast—muffins are over by the bread, and yogurt's back in that dairy section.

In *Systema Naturae*, Linnaeus employed similarly shifting criteria. For the kingdom of *Animalia* he designated six classes, beginning with *Quadrupedia* (Latin for "four-footed"), which he defined as "Body hairy. Four feet. Females give live birth, and lactate." Then *Aves* ("birds" in Latin), "Body feathered. Two wings. Two feet. Bony beak. Females lay eggs." This was followed by *Amphibia* (Greek for "having two ways of

being"), described as "Body naked or scaly. No molar teeth. Perpetually resting. No feathers." This encompassed not only animals later
classified as amphibians (frogs, toads, and salamanders) but also alligators, crocodiles, and other semi-aquatic creatures. The criteria of
"two ways of being" applied as well to animals not associated with the
water, such as the winged lizard (which leaps from ground to air) and
the skink, a lizardlike denizen of the African Sahara (which undulates
through sand dunes in a motion similar to swimming). This was a
realignment of the term *amphibian*, which had been used since the
1630s but in reference to any animal that regularly inhabited both
land and sea: crocodiles, walruses, and beavers. Then came *Pisces*
(Latin for "fish"): "Body footless, propelled by wings (fins), naked or
scaly." As gills were not a criteria, these included whales, dolphins,
seacows, and narwhals. Following were *Insecta*: "Body shell of bony
skin covering it (entirely). Head equipped with antennae." Despite its
definition, this class encompassed numerous species without antennae, including scorpions and aquatic crabs.

A sixth class was entirely invented. He named it *Vermes*, a term
seemingly drawn both from the classical Latin *vermis* ("worm") and
the later Latin *verminium* ("troublesome creature," which survives in
English as *vermin*). Bearing the shortest description of all ("Muscles of
the body each attached to a solid base"), it constitutes the first instance of what biologists have come to call a "wastebasket" category, a
dumping ground for species that don't seem to fit elsewhere. In it
Linnaeus placed a motley miscellany: reptiles, mollusks, sponges, and
squids. The earthworm, *lumbricus*, is also present, but classified as a
reptile and described as "one and the same species" as the intestinal
tapeworm.

Linnaeus included a final class, which unlike the others was unnumbered and unorganized. *Paradoxa* was a gallery of species that de-
fied his new categories, but that readers would expect to be mentioned.
Here he placed the barnacle tree ("believed by the Ancients to be
born from decaying wood thrown in the sea") without expressing an
opinion on its existence. But he did attempt to disavow other creatures he considered fanciful. He listed, then dismissed, the unicorn
("a painters' invention"), the satyr ("a species of monkey, if ever one
has been seen"), and the phoenix (which he speculated was not a bird

but a kind of palm tree). He noted his personal debunking of the hydra ("by most people it is considered a real animal species, but wrongly so").

Bizarrely, *Paradoxa* included the pelican, an abundantly documented creature even in 1730s Europe; Linnaeus appears to have included it to discredit the legend that they stab their own breast to feed their young. Also included is the "frog-fish" or lungfish, which he pronounced "very paradoxical" (it is quite real). The lungfish could not exist, he concluded, as "Nature would not admit the change of one genus [*frog*] into another of a different class [*fish*]." An audacious statement, considering he'd just invented said genus and class.

It was on the next level of hierarchy, the orders, that Linnaeus's multi-logical approach became more obvious, as he chose different sets of sorting characteristics for each order. Quadrupeds were grouped by the arrangement of their teeth (inspired by his Lapland encounter with a horse's skull), fish by the shape of their fins, birds by the shape of their bills. Frogs, lizards, turtles, and snakes were shunted into the order *Serpentia* with no explanation at all. The order *Anthropomorpha* ("manlike"), encompassing humans, apes, monkeys, and sloths, was defined not by a feature but by the lack of it: the absence of tails.

On the fifth and final hierarchy—the level of species—Linnaeus applied only single-word names. The sloth was *Bradypus*, or "slow foot." The baboon was *Cynocephalus*, or "dog-headed." For humans, Linnaeus did not choose the Latin word *Humanus* but instead *Homo*, a word the Romans had cribbed from the Greeks, literally meaning "of the same" but understood in Latin as "man."

Of the 549 species described in the first edition of *Systema Naturae*, Linnaeus knew *Homo* was the most controversial listing of all. It was treading in dangerous waters to place humans in the animal kingdom, much less in the same genus as sloths and baboons. Doing his best to avoid ruffling religious sensibilities, he placed a biblical verse on the title page (*O Lord, how manifold are thy works!*). He added a preamble, stating "it is necessary to attribute this progenitorial unity to an Omnipotent and Omniscient Being, namely God, whose work is called *Creation*" (italics his). And when it came to the description of species *Homo*, he used no physical characteristics at all. *Nosce te Ipsum*, he wrote, borrowing a well-known Latin aphorism. *Know yourself.*

...

Today, *Systema Naturae* is viewed as a foundational text, but at the time of its publication in 1735 it barely registered as a minor curiosity. Its treatment of the vegetable kingdom was a rehash of Linnaeus's sexual system, already well known in botanical circles, and its ordering of the other two kingdoms was primarily received as a superfluous exercise. While ambitious on its face, it had little practical application—almost no one, except the owners of extensive cabinets of curiosities, needed an organizational system for animals, vegetables, and minerals. Linnaeus's contemporaries took far more notice of his work at Hartekamp, the banker Clifford's miniature of paradise, where his latest achievement was beginning to draw crowds.

It was a *pisang*, or banana tree, painstakingly transplanted from the tropics. Despite the previous supervisor's best efforts, it had never flowered, much less given fruit. Conventional wisdom held that the air of Europe, even when properly heated in an adonis house, was insufficient for blooming, but Linnaeus suspected there was nothing detrimental in the local atmosphere—it was simply that banana plants were adapted to the intense cycles of dryness and water brought on by tropical monsoons. He instructed the attendants to create an artificial monsoon season, alternating dry soil with copious amounts of water. Just four months later, he triumphantly recorded that the banana plant "showed the first signs of flowering." After three more weeks, it was blooming so vigorously that a steady stream of admirers arrived at Hartekamp. Both Clifford and Linnaeus greeted them as befitted their social station, conducted them through the greenhouse, and basked in their praise.

The feat was not unprecedented. Banana plants had been brought to bloom three times before in Europe, but the attention pleased Clifford so much that he commissioned the private printing of a short book about the plant in question, to be given as a gift to future visitors. Linnaeus wrote the forty-six-page work rapidly, naming both the species and the book *Musa Cliffortiana*. In the latter, he flattered his patron in fulsome language ("Nature has never granted, to anyone below the rank of a Prince, so as to produce flowers in the European parts of the world") while apportioning no small amount of praise for

himself. For *Hortus Cliffortiana*, a catalogue of his patron's entire botanical collection published the following year, Linnaeus commissioned a frontispiece that jumbled allegory with self-promotion.
Beneath a bust of Clifford and a banana plant in flower, a female per-
sonification of Europe receives the floral tribute of figures representing Africa, Asia, and the Americas, the three known parts of the world.
Enlightening the proceedings with a torch is a near-nude Apollo,
treading on a serpent reminiscent of the Hydra of Hamburg. He bears
the face of Linnaeus himself.

Hoping to capitalize on the tide of attention, Linnaeus also self-
published *Fundamenta Botanica* and *Critica Botanica*, two pamphlets in
which he urged a new, standardized "science of names" and proposed
certain rules. The right of naming a new genus or species, he argued,

Frontispiece of Hortus Cliffortiana, *with Linnaeus as
torch-bearing Apollo*

should go to its discoverer, by which of course he meant the first European to report it to the European botanical community. Furthermore, all chosen names should be limited to either Greek or Latin, for conformity's sake. Yet in the case of *Musa Cliffortiana*, Linnaeus realized he was violating two of the rules he'd just proposed. First of all, the genus name *Musa* (already used by previous generations of botanists) was derived from *moaz*, the Arabic word for banana. Not wanting to contradict his decree that all names be either Greek or Latin, he conveniently decided that *Musa* could retroactively pay tribute to Muse, "that goddess of the ancients," an attribution employing a vague gloss on mythology—the Muses, daughters of Zeus and Mnemosyne, goddess of memory, have aways been plural. To shore up the tenuousness of the connection, he added yet another interpretation: The genus name could also honor Antonius Musa, an obscure physician to the Emperor Augustus. Linnaeus was unable to finesse the second violation, namely that he was far from the "discoverer" of the banana plant, and thus had no right to name it *Musa cliffortiana*. He chose to let it stand, although years later he would deprive Clifford of his immortality by changing the name to *Musa sapentia*. It meant "muse of the wise."

· · ·

In May of 1738, Linnaeus closed Hartekamp's adonis-house doors one last time and climbed into his loaned carriage. He'd spent two years on Clifford's estate, and would remember his time there as among the happiest periods of his life. But it was time to leave.

His official reason? He'd acquired a fiancée before leaving Sweden, and it was unreasonable to make her wait much longer. She was Sara-Lisa Moraeus, the daughter of a well-to-do physician in the rural province of Darcarlia. Linnaeus had gotten to know the family while visiting the mines there, gathering information for the *Kingdom Mineralia* portion of his *Systema Naturae*. In fact, his future father-in-law had been the impetus for his sojourn in the Netherlands: In bargaining for the betrothal, Dr. Moraeus had offered a generous dowry, but only upon the condition that Linnaeus first acquire his medical degree. Word had reached Hartekamp that other suitors had begun to circle Sara, waiting for the wealthy doctor to finally lose his patience with the arrangement. Linnaeus did not need to hurry back, but it was prudent to at least send earnest proof that he was on his way.

There were other reasons. Linnaeus found himself abysmally inept at acquiring foreign languages: After three years in Holland, his command of Dutch remained limited to a few dozen words and phrases, which meant his management of the gardening staff entailed a mix of pointing, pantomime, and frustration. More important, Clifford was an employer, not a patron. Linnaeus had tallied every variety of plant and animal on the grounds (he counted 1,251) and compiled copious notes for a possible second edition of *Systema Naturae*, but while Clifford was willing to pay for publications specific to Hartekamp, he had shown no inclination to pay for independent works.

Linnaeus set off for Sweden, with one important stop to make along the way. Matrimony, and the task of addressing an uncertain future, awaited in his homeland. But first he had a pilgrimage to make.

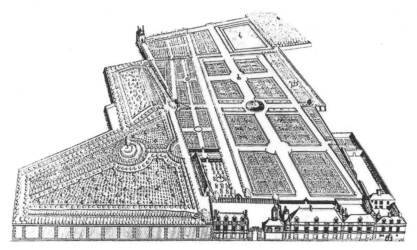

The Jardin du Roi, Paris

NINE

An Abridgment of the World Entire

IF LINNAEUS TOOK NOTE OF THE TUILERIES PALACE, THE Louvre, the cathedral of Notre Dame, or any of the other landmarks of Paris when he arrived in June of 1738, it was only while striding purposefully past them on his way toward the semi-rural eastern edge of the city. There, hemmed in by a monastery and the municipal market for horses and pigs, stood a patch of green: the Jardin du Roi, a royal garden in a decidedly unroyal setting. To Linnaeus's eyes, it was one of the marvels of the world.

The Jardin was a unique institution, in large part because it pretended to be no institution at all. Created in the previous century as a stealth project to subvert the academic status quo, it had survived and even flourished. Yet its very existence remained controversial.

As both law and long-standing tradition dictated, the teaching of medicine in Paris was the sole purview of the University of Paris, colloquially known as the Sorbonne. Since its founding in 1150, the Sorbonne had created and formalized many of the core concepts of higher education: strict qualifications for professors, the system of chancellors and deans, and the awarding of doctorate degrees. But five centuries had ossified several of the Sorbonne's departments into hyper-conservative, static bodies. Particularly hidebound was the college of medicine, which conducted no research, held no classes other than lectures, and stuck to a dusty curriculum that did not extend past the writings of Galen, the Greek physician born in 129 A.D. By 1626, King Louis XIII had so little confidence in Sorbonne-trained doctors that he appointed as his personal physician one Guy de la Brosse, a man whose primary qualification was that he had studied in the provinces, not Paris.

De la Brosse wished to break the Sorbonne's ancient monopoly on medical instruction. He found a sympathetic ear in the king's chief minister Cardinal Richelieu, who ultimately counseled an oblique approach. At Richelieu's prompting, the king commissioned de la Brosse to plant a garden in the Faubourg Saint-Victor, a sparsely settled area then technically outside the city. The Jardin Royal des Plantes Médicinales, as it was then called, would serve as the monarch's personal medical chest, a ready source of the ingredients for physicks, poultices, and decoctions. Nearly two thousand botanical varieties took root there over the next five years.

Since the king's physician would naturally need to compound such treatments at a moment's notice, the Jardin acquired buildings to house apothecaries and their requisite equipment. The king then commanded that the royal medicine chest be expanded to a stockpile of every drug known to exist. This called for the construction of a warehouse, the addition of a staff to attend it, and an ongoing research program to ensure the collection stayed up to date. As this new concentration of natural knowledge could benefit kingdom as well as king, Louis XIII magnanimously allowed de la Brosse to share the Jardin with other physicians. For that matter, even medical students were welcome.

The Sorbonne howled. This was clearly an attempt to teach medicine, in violation of their sanctioned exclusivity. The king sidestepped their protests, maintaining that the Jardin was not a school but simply a garden: It held no entrance examinations, charged no tuition, and awarded no degrees. Yes, the grounds now employed several learned gentlemen to hold forth on various topics in botany, chemistry, and even anatomy, but they were "demonstrators," not professors, and what appeared to be lectures were only demonstrations. The stockpile of drugs expanded into a burgeoning collection of "all rare things found in nature," designated the Cabinet du Roi (natural history collections were commonly referred to as "cabinets," no matter how vast they might be). As a final proof that the Jardin did not encroach upon the Sorbonne's monopoly, in 1640 the gates were thrown open to the public. Now anyone who wished could stroll the grounds, wander into demonstrations, and enjoy a convivial exchange of knowledge that was most certainly in no way a formal education.

The Sorbonne continued to lodge numerous complaints with the French parliament. Rumors spread that de la Brosse was an atheist and a libertine, taking advantage of the isolated location to conduct orgies. The Sorbonne faculty grew especially enraged in 1673, when the Jardin's "demonstrators" were audacious enough to demonstrate the circulation of blood, a principle discovered by Sir William Harvey in England a scant forty-five years earlier. Yet the Jardin Royal des Plantes Médicinales, eventually renamed the Jardin du Roi, endured as a technically non-academic educational institution, as four generations of intendants presided over what remained an egalitarian anomaly.

For Linnaeus it was hallowed ground. It was here that Tournefort had created his botanical system with 643 categories and where Valliant had put forth the idea that plants had sexual organs. The gates now opened to a lush campus of nearly twenty acres, with an amphitheater for chemistry demonstrations and (as was customary due to noxious odors) a separate structure for demonstrating anatomy. When it proved impractical to level a small hill in the eastern corner, Guy de la Brosse had incorporated it as a vista, landscaping a winding circular path along its height and planting a single tree at the summit, where it sheltered visitors enjoying views of the Seine and the surrounding countryside. Most of the public were content to enjoy that view, to stroll the geometrically tidy planting beds, and to tour the Cabinet of the King. But others lingered, learned, and marveled. Here, "one has found the means of shortening and flattening the surface of the earth," the philosopher Denis Diderot wrote of the Jardin. "Here one sees productions from all the countries in the world, and so to speak an abridgment of the world entire."

Linnaeus was eager to test the premise that anyone could wander into a class in session. When he saw a group gathering around one of the garden's demonstrators, he quietly joined them. They were listening to thirty-seven-year-old Bernard de Jussieu, a Lyonnaise apothecary's son who had joined the staff at the age of nineteen.

Bernard de Jussieu

The informal class was a refreshing departure for Linnaeus, who was used to elderly professors droning on from prepared texts. The genial, good-natured Jussieu engaged directly with his audience, asking questions of them and tailoring his explanations accordingly. When Jussieu pointed out a plant and asked if anyone present could identify its origin, a voice confidently rang out in Latin.

"Facies Americana." *The plant is native to the Americas.*

"Tu es Diabolis aut Linnaeus," Jussieu replied. *You are either the Devil or Linnaeus.*

It was a dramatic meeting, but not a startling one. Linnaeus would later cite the encounter as proof of his fame preceding him, but in truth he'd already paid a social call to the Jussieu household, leaving behind a memorably immodest letter of introduction ("Behold Charles Linnaeus, the prince of botany, if ever one existed"). Jussieu was familiar with the young man's recent publications—Linnaeus had sent copies from Hartekamp—and while the Jardin had no intention of replacing its own venerable Tournefort system with Linnaeus's vegetable letters, Jussieu was at least intrigued enough to extend a warm welcome.

Since Linnaeus understood French no better than Dutch, he and Jussieu continued to communicate in Latin, establishing a cordial friendship that would continue via correspondence for the rest of their lives. Linnaeus spent every day of the next month with Jussieu, who not only gave him the run of the Jardin but organized (and paid for) excursions to the countryside surrounding Paris. These included trips to the royal gardens of Versailles and Fontainebleau, off-limits to the public, where Jussieu called in favors to give him access to some of the rarest plants in France. This cordial treatment came as a great relief to Linnaeus, who knew that his sexual system had received a mixed reception in different countries. English botanists, initially wary of the system—one called Linnaeus "the man who has thrown all botany into confusion"—were beginning to embrace it. Most German botanists, on the other hand, remained openly hostile, rejecting it as uselessly artificial. "I have never pretended that the method was natural," Linnaeus wrote in appeasement to one German professor. "Become yourself a creator of a similar system, and I will immediately acknowledge you. . . . PEACE BE WITH US!"

Through Jussieu, Linnaeus came to understand that the current reception of his sexual system in France lay somewhere in the middle. The general consensus was that while it should not replace the native Tournefort system, the ingenuity of its author should be applauded and encouraged. In their years of correspondence that followed, Jussieu would prompt Linnaeus not to rest on his laurels, but to continue seeking a truly natural botanical system.

After overstaying his planned visit by several weeks, the "prince of botany" reluctantly prepared to depart Paris. As a parting gift Jussieu brought him to the Louvre meeting rooms of the Académie des Sciences, where to his delight Linnaeus was inducted as a foreign correspondent. He would later vastly inflate this polite mark of professional courtesy, claiming that the academy's president begged him to stay and "become a Frenchman," with full membership and a generous stipend, an offer he said he turned down because it would have entailed learning French. In truth he had a fiancée waiting for him in Sweden, and her dowry was the only foreseeable source of income for the near future. Still, he found it hard to part with Jussieu, and particularly hard to leave the Jardin du Roi. It was a brilliant institution, he believed. If only a similar one could arise in Sweden.

. . .

When Jussieu introduced Linnaeus to the savants of the academy, Buffon was not among them. Regular attendance at the academy's meetings was a requirement of membership, but Buffon insisted—and would insist, for the rest of his life—on departing Paris for Montbard every August, returning only in April the following year. He was there at that moment, preparing the first French translation of Isaac Newton's *Method of Fluxions* and drafting a preface that addressed one of the most quarrelsome questions of the day: Who had invented calculus, Newton or Gottfried Wilhelm Leibniz?

While Leibniz had been the first to publish on the subject in 1684, Newton's private papers showed his own discoveries in the field dating back as early as 1666: The work Buffon was translating dated back to 1671. Leibniz and Newton had corresponded with each other, and there was ample evidence that the German, with the help of mutual friends, had read at least some of the Englishman's unpublished notes and manuscripts. But their methodologies were slightly different, which seemed to indicate something more nuanced than mere plagiarism. The dispute raged long after the death of both men, becoming both an investigational matter and a metaphysical one: Did the right of "discovery" belong to the first expression of a concept, no matter how crude, or to the expression that, through elegant clarity, best revealed an underlying truth of the universe?

Buffon, who idolized Newton, argued passionately in his favor. His argument was lopsided at best—he knew very little about Leibniz's goals and professional context—and would in time be supplanted by a general consensus that both men had discovered calculus independently. But the effort underscored his reputation as a rising star of French savantry. Voltaire, a fellow pro-Newtonian, was particularly impressed. "I am the lost child of a party of which Monsieur de Buffon is the head," he wrote. "He pleases me so much that I would like to please him."

Linnaeus's portrait in the second edition
of Systema Naturae

TEN

Loathsome Harlotry

LINNAEUS AVOIDED UPPSALA UPON HIS RETURN TO SWEDEN IN
September of 1738, aware that even his growing reputation in natural
history gave him no prospects there. The university still held only two
slots for professors of medicine, and both were filled. Lars Roberg,
the professor of practical medicine, was aging but showed no sign of
retiring. Olof Rudbeck still technically clung to the chair of theoretical medicine, but since Linnaeus's old nemesis Nils Rosén had assumed Rudbeck's teaching duties, that succession seemed ordained.
Instead Linnaeus settled in Stockholm, where he hoped to resume his
hybrid profession of independent naturalist and private-practice physician. Barring the acquisition of another fabulously wealthy patron
like Clifford, this meant setting out his shingle and soliciting a medical clientele.

He was soon miserable. The city "received me as a stranger," he would recollect. "There was no one who would put even a servant under my care. I was obliged to live as best I could, in virtuous poverty." He'd expected that an entrance into Stockholm society would present challenges, but not that he'd be summarily ignored as the months passed without a single patient. Had he a less dramatic turn of mind, he might have realized how thin his credentials were: Aside from the practical portion of his medical exam, he had never treated a patient. He had also neglected to apply for, much less pass, the oral examination required for all Swedes returning with foreign medical degrees, rendering him technically disqualified to practice. But Linnaeus attributed darker reasons. "I was the laughingstock of everybody on account of my botany," he wrote. "No one cared how many sleepless nights and weary hours I had passed, as all declared in one voice that Siegesbeck had annihilated me."

Siegesbeck was Johann Georg Siegesbeck, professor of botany at the Russian Academy of Sciences and director of the St. Petersburg botanical garden. He and Linnaeus had struck up a cordial professional relationship when the latter was at Hartekamp, exchanging specimens and compliments in a series of letters. It shocked Linnaeus, then, to learn that Siegesbeck had published a pamphlet entitled *A Critical Analysis of the Well-Known Linnaeus Sexual System of Plants*, which excoriated his work on both theological and moral grounds. In it, Siegesbeck pointed out that the book of Genesis specifically stated that when God created plants on the third day of Creation, "the earth brought forth grass, and herb yielding seed after his kind, and the tree yielding fruit, whose seed was in itself, after his kind." He read this to mean they came into being fully formed and mature—the first apple trees materialized already bearing apples, and the first rosebushes arrived awash in blooms. It was therefore ridiculous to attempt to discern the purpose of flowers and seeds, as their purpose was fulfilled by existing. Siegesbeck renounced Linnaeus's sexual system as extraneous to the Bible, but moreover a moral affront so repellent he called it "loathsome harlotry." To describe a vegetable kingdom as naturally welcoming "eight, ten, twelve, twenty or more husbands in the same bed with one woman" was to warp the sensibilities of young students, who might be "corrupted by the immorality that had broken out among the lilies and onions."

Did the bourgeoisie of Stockholm read and agree with the pam-

phlet of a Prussian expatriate working in Russia? Linnaeus believed
they did. "What a fool have I been, to waste so much time, to spend so
many days and nights in a study which yields no better fruits, and
makes me the laughingstock of all the world," he lamented, before
deciding to abandon botany altogether.

> Aha! said I, Esculapius [the god of medicine] is the giver of all
> good things; Flora bestows nothing upon me but Siegesbecks! I
> took my leave of Flora; condemned my too numerous observa-
> tions, a thousand times over, to eternal oblivion; and swore never
> to give any answer to Siegesbeck.

In fact he would eventually answer, pointedly awarding the name
Siegesbeckia to a small, foul-smelling weed. But at the moment he was
pondering the best way to destroy his notes and manuscripts, rejecting the catharsis of a bonfire and deciding to bury them instead.

Linnaeus may have earnestly planned a grave for his life's work, but
soon he was too busy to dig one. Compelled to find some means of
plying a trade, he stopped waiting for patients to arrive and began
haunting the more popular coffee shops of Stockholm, making a daily
circuit that occupied his waking hours. He took a great interest in his
fellow patrons, particularly men his age or younger, preferably dressed
in uniforms or sporting a worldly air. Circumspectly but intently, he
gazed at their wardrobes, their heads, and their hands. If he saw what
he was looking for he approached them, suggested a glass of Rhine
wine, and discreetly struck up a conversation.

He was looking for neither friendship nor companionship, but for
loose-fitting clothing, a sign the wearer had recently lost weight. He
was looking for irregular patches of thinning hair on wigless heads,
and for reddish-brown spots on the palms of hands. All were signs of
what Linnaeus called a wound *in castris Veneris*—in the camp of Venus,
the goddess of love and sex, a euphemism that gave rise to the term
venereal disease.

At the time, syphilis and gonorrhea were thought to be one disease
(the distinction would not be made until 1761), and the standard
treatment was almost as horrific as the malady itself. It involved substantial quantities of toxic mercury rubbed into the skin under high

heat conditions and/or inhaled as fumes. It was painful, and if patients did not die of mercury poisoning they regularly suffered kidney failure, loss of teeth, facial ulcers, and mental impairment. Its effectiveness against the disease, if any, was often short-lived, requiring sufferers to undergo a prolonged course of treatment. "A night with Venus," one saying went, "a lifetime with Mercury."

In Holland, Linnaeus had made the acquaintance of a physician named Gerhard van Swieten, who'd had promising results with a new substance he dubbed *liquor swietenii*. The primary ingredient was corrosive sublimate, a compound of mercury and chlorine that was just as toxic but more efficient, producing faster results with fewer applications. This new treatment would come to be called salivation, so named because a side effect was an increased production of saliva. But that was infinitely preferable to disfiguration, dementia, and death. With his stock of *liquor swietenii* and his discreet manner, Linnaeus quickly became the busiest salivationist in Sweden, administering topical applications to upward of sixty patients a day. "My adverse fate took a sudden turn, and after so long a succession of cloudy prospects, the sun broke out upon me," he reported to a friend in September of 1739. "I emerged from my obscurity, obtained access to the great, and every unfavorable presage vanished."

Access to the great. Linnaeus's booming practice operated beneath the acknowledgment of polite society, but it nevertheless gained the attention of Marshal Carl Gustaf Tessin, at the time head of the Swedish parliament and the nation's most powerful politician. Even more so than Linnaeus, who was beginning to marvel that "alas, almost all the young men . . . had been infected," Tessin grasped that Sweden's wounds in the field of Venus were reaching crisis proportions, and nowhere more so than among the sailors of the Swedish fleet. He moved quickly to scale up the new doctor's services by co-opting him into the government, appointing him chief medical officer of the Admiralty's naval base in Stockholm. Just seven months after his return, Linnaeus was overseeing a lucrative private practice as well as the nation's largest naval hospital, filled with up to two hundred patients at a time.

What proportion of those patients was suffering from what the Swedish navy euphemistically called "the French disease"? That was a closely guarded number, as Marshal Tessin and his administrators sought to minimize the stigma of seeking treatment by surrounding it with a milieu of respectable medicine. To blunt the opprobrium of putting a pox-doctor in charge, Tessin took rapid steps to bolster the man's respectability as well. Within days Linnaeus was living as a guest in Tessin's private mansion, administering cough drops and other innocuous medicine to his aristocratic friends, and giving public lectures on botany and mineralogy sponsored by the national Council of Mines. The crowning touch came a few weeks later, when Tessin facilitated the launch of a Swedish Academy of Sciences, organized along the lines of the French institution. Linnaeus was not only welcomed as a founding member, but chosen as its first president. The post was not quite the honor he would make it out to be for the rest of his life—the Academy consisted of six members, the presidency had been decided by a random drawing, and his term lasted three months. Yet it nonetheless capped a dizzyingly rapid ascent. Under Tessin's patronage the thirty-two-year-old Linnaeus had moved from the furtive margins of Stockholm savantry to its very center.

At last positioned to follow through on his engagement, he married his longtime fiancée Sara-Lisa Moraeus on June 26, 1739, collected his dowry, and procured a house large enough to begin a family. He and Bernard de Jussieu had kept up a regular correspondence since his visit to the Jardin. In one letter, written during his honeymoon, he could not help but brag.

> I have succeeded in obtaining quickly the largest medical clientele of this town and I have also been appointed titular physician to the admiralty. I have just married the woman whom I wanted for years to marry and who, if I am permitted to speak this way between ourselves, is quite wealthy; I am therefore at last leading a quiet and satisfactory life.

· · ·

In July of 1739, forty-one-year-old Charles de Cisternay Du Fay, the current intendant of the Jardin du Roi, died suddenly of smallpox. "All the medical world and all the Academy fought for that position," wrote a contemporary observer. "It is worth a thousand crowns in salary, one of the most beautiful residences in Paris, and the right to make nominations for all the positions which depend upon it." Among the few savants not coveting the job was Bernard de Jussieu, despite his reputation as the Jardin's presiding eminence. Modest to a fault, Jussieu was famously reluctant to even publish his research, much less assume a higher public profile.

If anyone considered Buffon a candidate, it was only as a distant long shot. He'd published nothing on natural history; his reputation at the time was as a mathematician, albeit with a side interest in forestry. He was sequestered 170 miles away in Montbard, uninterested in returning to Paris to campaign for the job in person. Still, he hoped to be considered. "The Intendancy of the Jardin du Roi needs a young, active man who can brave the sun, who knows plants and the way to multiply them," he wrote to a friend named Hellot. "I am what they are looking for; but so far I do not have any great expectations, and consequently I will not be greatly grieved to see this position filled by another."

Buffon knew Hellot had also been a friend to Du Fay, and had tended to him in his final hours. He did not know that Hellot was the executor of Du Fay's estate and had already helped compile the late intendant's last will and testament. It included an extraordinary codicil: a signed endorsement of Buffon as his successor. Since the person responsible for conveying this recommendation to King Louis XV was none other than his patron the Count de Maurepas, the matter was sealed. On July 26, 1739, Georges-Louis Leclerc de Buffon was appointed the fourth intendant of the Jardin du Roi.

Du Fay, even in his last days, was no fool. He knew better than anyone that the Jardin remained a controversial institution. The Sorbonne's long-standing antagonism was being increasingly matched with scrutiny from Catholic Church officials, wary of the unorthodox ideas (such as sexuality of plants) emerging from the grounds, and disdain from aristocrats who found the Jardin's egalitarian learning model distasteful on principle. A truly effective leader would have to be as much a courtier as a savant, appeasing rival factions at Versailles while retaining the king's favor and funding. Even before Buffon had written his letter, Hellot had interrupted Du Fay's dying to discuss a young man far removed from Parisian intrigue, too wealthy to be bribed, whose ruthlessly efficient ascension into the academy revealed a strong strategic intellect. The choice puzzled many and infuriated others. But Du Fay's last wish, guided or otherwise, comprised a fierce defense of his legacy.

. . .

Exactly one month after Linnaeus's wedding, a letter from Jussieu arrived with news of Buffon's ascension to the head of the Jardin. Linnaeus immediately began regretting his career choices. "I now grew fond again of plants," he wrote, at the same time complaining about how the life of an in-demand doctor was beginning to wear him down: "From seven in the morning until eight at night I hardly have a moment to snatch even a hasty meal." Even though he could now afford to take up natural history again as a side interest, he could not spare the time. Besides, without the backing and support of an institution, even his most earnest efforts would consign him to the ranks of amateurs. "Botany is very difficult," he'd written in *Musa Cliffortianus*,

> but it is very expensive also, because the earth does not produce everything everywhere, and the various families of plants are distributed all over the world. To hurry to countries far away, to hit one's head against the borders of the world, view the never-setting sun, this is not for the life or even the purse of a single botanist, and his vigor will fail in these endeavors. The botanist needs global commerce, a library with all the books published on plants, gardens, hot-houses, and gardeners.

With the income from his practice and the dowry money, Linnaeus nevertheless felt flush enough to underwrite a second edition of *Systema Naturae*, published in Stockholm and dedicated to his political patron Tessin. Although he insisted in the introduction that "I was obliged to publish at the instigation of the audience" (i.e., by popular demand), he also seemed to be wrapping up the project. "I have added many things in the entire kingdom of Nature, especially the species of Quadrupeds and the names of Swedish Animals," he wrote. "Farewell."

Physically half the size of the original but larger in scope, the seventy-eight-page second edition included anteaters alongside the sloths, monkeys, apes, and humans in the order *Anthropomorpha*. Far more significantly, the description of the species *Homo* now included the phrase *homo variat*: Man varies. What followed were four fateful entries:

Europaeus albus (White European)
Americanus rubescens (Red American)
Asiaticus fuscus (Tawny Asian)
Africanus niger (Black African)

A slightly different version of these terms had appeared in the first edition, but in highly abbreviated form and type so small it was difficult to discern what the author meant by them. Now the import was clear: Humans had exactly four variations, correlated to skin color and geographic origin.

Why four? It's been argued that Linnaeus was straining for symmetry, attempting to add gravitas to his work by aligning it with the then-accepted medical doctrine originating with Hippocrates, namely that the body held four liquids, or "humors"—white phlegm, red blood, yellow bile, and black bile. Physicians regularly diagnosed illness as an imbalance of these humors. A patient lacking energy was "phlegmatic" from an excess of phlegm, a depressed one "melancholic" from an abundance of melanic (black) bile. But Linnaeus himself drew no direct parallels, and at any rate this particular color-coding was jarringly different from other widespread conceptions of human ethnicity. Several languages had adopted versions of the Portuguese word *negro* to signify African origin or descent, but that usage only began to emerge in the mid-1500s; prior to that a common descriptor was *ethiop*, from a Greek word meaning both "fiery-looking" and "sunburned." Documents dating back to 1387 employ the Middle English word *blewmane*, alternately spelled *bloman* or *bleuman*, as well as references to Africa as *blewmen londe*. Contemporaries of Chaucer had perceived darkened coloration not as black but as shades of blue. Linnaeus's audience in 1740 was familiar with the usage of *niger* (one of several Latin words for "dark" or "black"), but the idea that all Europeans were uniformly *albus*, white, was far from given. In 1684, for instance, the French *Journal of Savants* had noted that Egyptians and Indians "are not blacker than many Spaniards," and that skin color therefore "did not seem sufficient to comprise a particular species."

More readers still were likely puzzled by Linnaeus's corralling of Asian ancestry into the category of *fuscus*—"tawny," or dark yellow. In his *Travels*, first published in 1302, Marco Polo described both Chinese and Japanese as *bianca*, or white. In the 1330s, Odoric of Pordenone, a Franciscan missionary in southern China, characterized his hosts as "beautiful in body" (*di corpo belli*) with pale (*pallidi*) skin. Tomé Pires, a Portuguese apothecary who ventured as far south as Malaysia in 1512, was emphatic that the Chinese were "white like us," comparing the males to Germans and the females to Spaniards. When the first Japanese diplomatic delegation arrived in Europe in 1585, observers described them as variously pale, olive, and "the color of Africans." That same year, Juan González de Mendoza, another missionary to China, reported that while "they of the most inward provinces are white people" and "some [were] more white than others, as they draw into the cold countrie," other Chinese manifested a range of skin tones from fair to "Moorish" dark and even *verdinegroes*, a darkish green. Despite the broadness of this spectrum ("it is a strange thing to see, the strange and great difference betwixt the colors of the dwellers of this kingdom"), yellow was not included—except in an early English translation, which in one instance mistakenly rendered Mendoza's *rubio* (blonde, fair) as "more yealow . . . like unto the Almans [Germans]." Even that "yellow" was in the context of reassuring Europeans that Asian skin looked much like theirs.

"Man varies," Linnaeus had written, at the same time providing almost no variety at all. *Europaeus albus, Americanus rubescens, Asiaticus fuscus,* and *Africanus niger* were four segregations of humanity, drawn with broad permanent strokes that would later be encoded as *race*.

. . .

While Linnaeus was putting the finishing touches on his second edition, news arrived of the death of eighty-year-old Olof Rudbeck, Uppsala University's long-serving professor of botany, whose patronage Linnaeus had briefly enjoyed before being displaced by Rosén. The successorship was all but guaranteed to his old nemesis, but Linnaeus

decided to pose a challenge by applying anyway. He lost, and decisively: Rosén emerged not only as the new professor but as head of the
medical department. "Rosén, who cannot even recognize a nettle, has
received Rudbeck's post," Linnaeus complained to a friend. "That's
the way things are going here."

In his new role of department head, Rosén appears to have greased
the wheels for the exit of the remaining elderly professor Lars Roberg, encouraging him to retire—something he was understandably reluctant to do, as Uppsala medical professors had no pension while
alive. (After their death, their widows received a small stipend paid
entirely in corn.) Rosén's motives were less than benign: He wanted
to claim some of Roberg's duties as his own, particularly the teaching
of anatomy.

Linnaeus put himself forward for this second professorship, even
though he had no real interest in it—his heart was in botany, not
anatomy—and he knew Rosén would do his best to block the candi-
dacy, which he promptly did. Rosén demanded that his rival publicly
demonstrate his command of Latin, even though Linnaeus had been
publishing in Latin for years. And when Wallerius, another candidate
for the job, decided to mount a public attack on Linnaeus by publishing a thesis, Rosén unabashedly sought to legitimize him by presiding
over a reading of the thesis on campus.

Wallerius's thesis, entitled *Twice Ten Medical Theses* because it
launched twenty separate attacks against Linnaeus, backfired spec-
tacularly. Even those previously neutral on the appointment were dis-
mayed by the overkill of Wallerius's tirades, and by Rosén having
allowed the spectacle to happen in the first place. Students sympathetic to Linnaeus brought the event to a raucous end by shouting,
standing on chairs, and rushing the lectern to tear the offending text
into pieces. Back in Stockholm, Linnaeus's political patron Tessin
took note of the uproar and seized the opportunity. On May 5, after
parliamentarians in the Rikstag had denounced and censured university leadership for permitting the event, Sweden's King Frederick I
signed a certificate of appointment. The job belonged to Linnaeus.

It was not, however, a job he particularly wanted. On November 3,
Linnaeus arrived at the podium in Uppsala's largest auditorium to
teach his first class as professor of practical medicine. It would also be

his last. Later that same day, he and Rosén jointly petitioned the university chancellor to reallocate their duties. During a series of wary negotiations, the two adversaries had worked out what amounted to a swapping of jobs.

Rosén would take charge of the university hospital and teach anatomy, etiology (the origin of diseases), physiology, and pharmaceutical chemistry. Linnaeus agreed to teach materia medica, dietetics, semiotics (the study of how disease visibly manifested, such as the yellow skin of jaundice), nosology (the categorization of diseases), and botany. "By God's Grace I am now released from the wretched drudgery of a medical practitioner in Stockholm," Linnaeus wrote to Jussieu. "If life and health are granted to me, you will, I hope, see me accomplish something in botany."

In a return letter Jussieu congratulated Linnaeus, at the same time admonishing him that his primary task should be the replacement of his admittedly artificial sexual system. "You may now devote yourself entirely to the service of Flora," he wrote, "and lay open more completely the path you have pointed out, so as at length to bring to perfection a natural method of classification, which is what all lovers of botany wish and expect."

· · ·

Professor Linnaeus relocated his wife and child (a son, Carl Junior, born the previous year) to Uppsala in March of 1742, then immediately set about remaking the neglected botanical garden into his own abridgment of the world entire. Banishing the grazing cows and clearing the overgrowth (which by now had killed off all but fifty-three original specimens), he divided the new garden into two areas, one for spring-blooming plants and one for autumn-blooming ones. These he subdivided into annuals and perennials, then divided those further still into the twenty-six categories of vegetable letters in his sexual system. The effect was a patchwork of greenery more irregular than ornamental, as neighboring plants grew at different rates and to different sizes. In the rear of the garden he added three ponds of different construction—one to simulate a stream, one a lake, and one a marsh—fringing them with plants native to each environment. Behind these he built two adonis houses, one an orangery and one for exotic plants that could not survive Swedish winters. There was a large stone building on the periphery, abandoned for over a decade and "more like an owl's nest or a robber's den than a professor's home," as he described it. This he rehabilitated into a proper residence. His family occupied the ground floor and a portion of the second, with the remainder of the rooms reserved for private lectures and his personal museum.

In the middle of the garden, just before the central pond, he installed a marble statue of Venus, rendered in a posture of unembarrassed nudity. It was a subtle nod to his prior success as a physician *in castris Veneris*, but the creator of the sexual system also intended it as a constant reminder: Nature and sexuality were permanently entwined.

The broad variety of mini-environments in his revamped garden reflected Linnaeus's ambitions, which were economic as well as academic. In many ways, the discipline of natural history in the eighteenth century was roughly analogous to technology today: a means of disrupting old markets, creating new ones, and generating fortunes in the process. The freshly minted professor did not rule out getting rich. "All that is useful to man originates from these natural objects," he'd written in the first edition of *Systema Naturae*. "In one word, it is the foundation of every industry." European mercantile empires were rising on the creation and feeding of new appetites, flourishing sectors of the economy that had naturalists' insights at their core.

One of the foremost examples of this was sugar, at the time produced solely from sugarcane. Introduced to western Europe by eleventh-century soldiers returning from the Crusades, the reed that "gives honey without bees" was for centuries an extremely expensive novelty item, imported from India and Arabia in such small quantities that it was considered a spice. Attempts to cultivate sugarcane in Europe failed (the crop required near-tropical conditions), but early test plantings in the Americas were promising: An intriguing percentage of the New World shared latitudes with India and Arabia. The first sugar crop in the New World was harvested in 1501. By 1550 there were more than three thousand sugar mills operating there, primarily on Caribbean islands and the South American coast.

In Europe, sugar went from being a costly rarity to a standard ingredient. This new abundance in turn fueled demand for complementary imports such as coffee, tea, and especially chocolate (which was a bitter substance, usually consumed as a beverage, until the Spanish began mixing it with sugar). In Linnaeus's day, sugar comprised one-fifth of all imports to Europe—an economic mainstay made possible by botanists, who were the first to observe the crop's dependence on environment, and to postulate where and how it might be transplanted.

Nor was the commercial bounty of natural history confined to plant life. At the time, Mexico's second-most valuable export (after silver) was cochineal, a tiny insect whose crushed shells were used to make carmine, the most vibrant and durable red dye then available. Cochineal was in such demand that it was listed on the commodity exchanges of Amsterdam and London, yet no living specimens had made it to Europe—it flourished exclusively on prickly pear cactus plants in the Mexican high desert. Linnaeus would long dream of obtaining such a cactus complete with its cochineal colony, believing that a fortune could be made either by cultivating prickly pears in Europe or by finding a new host plant for the insects. What other dazzling colors might be extracted from nature? What new commodities might flourish under transplantation, either to or from the New World? With the garden as his dedicated laboratory, Linnaeus was determined to find out.

Monsieur Buffon, intendant du Jardin du Roi

ELEVEN

The Quarrel of the Universals

"HE CARRIES HIMSELF MARVELOUSLY," VOLTAIRE OBSERVED OF the Jardin's new director. "The body of an athlete and the soul of a sage, that is what is needed in order to be happy." He was not alone in admiring Buffon, who at twenty-seven seemed to somehow project a formidable presence while simultaneously putting people at their ease. Well aware of the resentments swirling around his appointment, Buffon was swift to cultivate a disarming air of deference that counterbalanced his obvious gifts. "His manner to public men . . . was conciliatory and tactful," observed another Parisian savant, "and to his subordinates he was modest and unpretending." It was a charm offensive by a young man in a hurry. Soon after his appointment he was omnipresent in the Jardin, learning its ways and judiciously enacting changes. He ordered an expansion of the Cabinet's cramped spaces, even though it meant surrendering a portion of the rooms reserved for his private apartment. In a remarkable act for the era, he hired a woman for a senior position, naming the artist Madeleine Françoise Basseporte as the Jardin's official painter and chief botanical illustrator, an important role in the era before photographic documentation.

Originally a portraitist working in pastels, the thirty-eight-year-old Mademoiselle Basseporte switched to the more precise lines of watercolors for her role in the Jardin. With Buffon's approval she also began studying botany under Jussieu's tutelage, the better to understand her subject. Over the next four decades she would become a respected and beloved fixture of the Jardin, serving as painting instructor to the royal family and as a decorating consultant to Madame Pompadour. She would also bring her craft to new heights, achieving an accuracy that would remain unsurpassed until the invention of photography. "Nature gives plants their existence," Jean-Jacques Rousseau wrote in tribute, "but Mademoiselle Basseporte gives them their preservation."

As to the gardens themselves, Buffon left them alone. He knew that Jussieu and several of the other demonstrators had supported another candidate for his job and viewed him warily as an outsider. He began to win them over by using his influence at court to establish the formal honorific *Correspondent de la Jardin du Roi*, a mark of distinction that the intendant could bestow upon anyone he deemed to have made a useful contribution to the Jardin. It was an honor in name alone, but it nonetheless encouraged savants and amateur naturalists around the world to send seeds, samples, and curiosities to the Jardin, increasing the number of planting beds and further swelling the inventory of the Cabinet.

Syringa vulgaris,
by Madeleine Françoise Basseporte

The existing inventory posed another challenge. Buffon further endeared himself to Jussieu by relieving him of the ancient post of Defender of the Cabinet of Drugs, the titular overseer of the collection. Far more interested in living plants than preserved ones, Jussieu had considered the post a dreary clerical task, particularly since it involved the obligation of compiling and publishing an inventory of the Cabinet, a task he had successfully put off for decades. With the help of Louis Daubenton, an assistant he'd imported from Montbard, the new intendant started assessing its contents. Buffon, a man who looked at scattered needles and saw probability theory, once again set to the task of seeking order out of chaos.

The Cabinet was not a carefully curated collection. It was a hodge-podge storehouse of as many items as his predecessors could cram into a series of ground-floor rooms, from precious gemstones to desiccated plants and assorted creatures, stuffed or floating in spirits. In addition to the original Cabinet of Drugs, there was now a companion compendium of poisons. There were drawers upon drawers of pebbles, gathered from various locations. There were displays of ancient furniture, antique weapons, and what was described as a "Giant's Bone," contributed by the Duke D'Orleans upon his death in 1660 (it was later determined to be part of a giraffe).

The Cabinet was open to the public two days a week, but no special efforts had been made to arrange informative exhibits. Many of the rarer items were hidden from view, to guard against pilfering visitors. Four generations of ledger-keepers had documented items as they entered the collection, but the items' present whereabouts, or even existence, was another question entirely. "The king's cabinet is not rich in insects, in minerals, or in birds," observed René de Réaumur, one of the most prominent naturalists of the day. "The collection of birds consists of sixty or eighty [specimens] that they had prepared in Strasbourg and which were most eaten by worms last year, because they did not know how to preserve them."

Buffon would eventually reorganize the physical Cabinet, taking advantage of the extra space he was allocating for the collection. But cataloguing came first. This required a conceptual structure, a system of organizing principles to adhere to during inventory. Fortunately Jussieu was familiar with a Swedish fellow, a naturalist who'd recently devised a system designed to sort specimens—animal, mineral, and vegetable—into logical, comprehensive categories. If ever there was a customer for the grand schematics of Carl Linnaeus, it was Buffon. He obtained a copy of *Systema Naturae*, read it intently, and began taking notes.

. . .

Within the first few pages of Linnaeus's work, Buffon found himself enmeshed in an ancient debate. Since antiquity, philosophers in the field of epistemology—the analysis of knowing and understanding—have been arguing a question known as the Quarrel of the Universals, which might be succinctly described thus: Do abstract concepts get us closer to making sense of reality, or do they distance us from genuine understanding? When we use terms like *bird*, *beetle*, or even *human*, we're trafficking in "universals," generalizations based on observed commonalities. But do universals truly exist? It may seem absurd to ponder whether *humanity* is real, but the fact is we have never been able to directly perceive it. We can only perceive what we believe to be a succession of examples—you, me, your first-grade teacher, Thelonious Monk, Abraham Lincoln, Lao-Tzu, Queen Hatshepsut. In nature, only individuals exist.

Plato believed that this represented an impassable gap between perception and reality. He held that no matter how many humans (or birds, or beetles) we study, our cumulative observations comprise a pile of opinions, not truths, for truth itself exists only on the plane of abstraction. His pupil Aristotle disagreed, concluding that it was possible to grasp the "essence" of something through continued examination of its properties, thereby proceeding from the specific to the universal. The essence of *cup*, for instance, includes the property of having a bottom, a property absent in the essence of *tube*.

The logic of such an approach is difficult to dispute, but in the face of complexity complications ensue. What is the essence of *chalice*, arguably a species of cup? What is the essence of *container*, arguably the genera to which cups belong? By the sixth century, the philosopher Boethius was giving up on deciding whether species and genera "subsist (in the nature of things) or in mere conceptions only . . . for such a treatise is most profound, and requires another more extensive investigation."

Reading *Systema Naturae*, Buffon perceived that Linnaeus had jumped wholeheartedly into the Quarrel of Universals, landing firmly in the essentialist camp. For Linnaeus, the matter was simple: If it looks like a dung beetle, it *is* a dung beetle, meaning that its essence begins and ends with physical appearance. Buffon, a constant practitioner of order and system in his daily life, could see the appeal of such an approach. But he could also see the potential for illusions and distortions.

As a mathematician, Buffon knew that numbers were abstract: Any instance of, say, the number three is the equivalent of any other. It makes no difference if it is rendered as three, 3, iii, or III—none is *more* three or *less* three than the others. But it struck Buffon as false to treat physical objects as if they were abstractions, which was what Linnaeus was doing when he looked at a single specimen and used it to define an entire species. Linnean systematics demanded the designation of a single physical instance (or in some cases, a drawing of a single instance) as the type specimen, the literally definitive example of a species. The majority of these were specimens that happened to make their way to Uppsala. They were "definitive" because Linnaeus used them to define, conveniently deriving the quintessence of a species from the specimen he had on hand. Furthermore, Linnaeus was choosing which physical characteristics constitute definitive features, expressions of their "essence," and which were mere inessential variations. The pattern of spots on a moth or butterfly mattered immensely, he'd decided. The pattern of spots on a cow did not.

Buffon found himself landing on the opposite side of the Quarrel. "The abstract does not exist," he wrote in his notes, which were taking on the shape of an essay. "There are no simple principles. All is compound." He believed that there were likely organizing patterns in life, but it was another thing entirely to assume humans were capable of readily discerning them. They would emerge only after a great deal of observation, if they emerged at all, and needed to be approached with patience and humility. To build neat boxes, nested or otherwise, was to "make suppositions which are always contrary to Nature, to strip the subject of most of its qualities, and to create an abstract being that has no resemblance to the real being."

Buffon was not arguing for chaos. He still had a Cabinet to organize. But he found himself agreeing with the Swiss naturalist Charles

Bonnet, who argued that systematics should be applied cautiously and conditionally. "If there are no cleavages in nature, it is evident that our classifications are not hers," Bonnet had warned. "Those which we form are purely nominal, and we should regard them as means relative to our need and to the limitations of our knowledge."

Systema Naturae would definitely not do as a template for his catalogue of the Cabinet. All systematic approaches to nature were flawed, he wrote to a friend, but "Linnaeus' method is of all the least sensible and the most monstrous."

. . .

In August of 1743, Buffon delivered to the Académie des Sciences the manuscript of an essay entitled *On the Manner of Studying and Considering Natural History*. It was a point-by-point dismantling of systemist thinking in general, and of Linnaeus's work in particular. "The study of nature supposes two qualities of mind which are apparently in opposition to each other," he wrote. "Intense intellectual power which takes in everything at a glance, and the detailed attention of an instinct which concentrates laboriously on a single minute detail." It was from this opposition that the impulse to invent systems arose, Buffon contended. It was entirely human to want to perceive large-scale patterns in nature (the everything-at-a-glance view) in order to place small-scale observations (the laborious, minute details) into a meaningful context. "Methods are very useful, when applied with appropriate restrictions," he conceded. "They shorten the work, assist the memory, and offer to the mind a series of ideas composed indeed of objects which differ among themselves but which nevertheless have certain common relations." They were dangerous nonetheless.

> We think that we know more because we have increased the number of symbolic expressions and learned phrases. We pay hardly any attention to the fact that these skills are only the scaffolding of science, and not science itself. We ought to use them only when we cannot do without them, and we ought always to be careful lest they happen to fail us when we wish to apply them to the edifice of science itself.

The danger lay in mistaking the scaffolding for the edifice. Methods and systems brought us closer to natural truths, but they did not themselves comprise natural truths—they were human constructs, and thus inherently flawed and incomplete. This was Linnaeus's crucial mistake, Buffon contested, "the primary error of confusing arbitrary genera, as entities of reason, with physical and real genera." While a species might at least be described, proceeding beyond the concrete to the conceptual—the hierarchical groupings of genera and beyond—was a journey to be approached with caution, keeping one's mind open to the dangers of treating such groupings as if they were real physical lines within nature, not just abstractions created for convenience. They were notional constructs, not natural ones.

What was the difference? Consider, as a random example, the state of Vermont. It clearly has a physical presence. That presence is firmly documented, and that documentation constitutes an agreed-upon definition of "Vermont" around the world. There are some fuzzy aspects to its existence: Its eastern border is defined by the Connecticut River, and that river fluctuates with time and the seasons. Also, the region has gone by other names: the Vermont Republic, the Dominion of New England, the territory of New France, Abenaki ancestral land.

Is Vermont real? Of course it is. But only a political map of the North American continent would include its borders. A geographical map would not. As an "entity of reason," it definitely exists. As a fact of nature, it does not.

Linnaeus did not seem to make this distinction. What's more, he built a tower of five such conceptual constructs, stacked upon one another, requiring the identification of life to be an act of fivefold certainty. Buffon considered this a monument to hubris, serving only to compound perceptual errors. He warned against

> the tendency to overextend or to unduly constrict the chain of connections, to wish to subject the laws of nature to arbitrary laws, to wish to divide this chain where it is not divisible, and to wish to measure its strength by means of our weak imagination . . . to judge of the whole by a single instance, to reduce nature to the status of petty systems which are foreign to her.

"There are two equally dangerous positions," Buffon concluded. "The first is to have no system at all, and the second is to try to relate everything to a restricted system." Life was more finely calibrated than labels. "Man will see with astonishment," he wrote, "that it is possible to descend by almost imperceptible gradations from the most perfect of creatures to the most formless matter."

Having warned about the folly of clinging to systems, Buffon moved on to a blistering critique of "that system of Monsieur Linnaeus, which is the newest of such methods. . . . If the general characters which Monsieur Linnaeus employs and the manner in which he makes his particular divisions are examined, even more essential shortcomings will appear."

Buffon devoted no time to analyzing *Systema Naturae*'s taxonomy of the mineral kingdom, which even ardent Linneans tended to ignore as incomprehensible. Addressing the vegetable kingdom, Buffon made short work of Linnaeus's sexual system, pointing out that plants, unlike animals, could propagate perfectly well through grafts and cuttings. "As the generation of plants can be accomplished by several methods which have no dependence on sexes, or the parts of fructification, this opinion has not been successful," he wrote. "It is only by the misapplication of an analogy, that the sexual system has been pretended to be sufficient to enable us to distinguish the different species of the vegetable kingdom."

Buffon reserved his most eloquent disdain for *Systema Naturae*'s treatment of the animal kingdom, purportedly a natural system. Linnaeus, he observed, divided it into six classes: quadrupeds, birds, amphibians, fishes, insects, and worms. Where in this schema, exactly, did lobsters fit in, or oysters, or snakes for that matter? "It appears at first glance that they might have been overlooked," Buffon stated, before peering closer to find that Linnaeus kept to his tidy boundaries by declaring that snakes were amphibians, oysters were worms, and lobsters were insects—"and not only insects, but insects of the same order as lice and fleas."

> Instead of making only six classes, if this author had made twelve or more of them . . . he would have spoken more clearly, and his divisions would have been more accurate and less arbitrary. For, in general, the more one augments the number of divisions of the

production of nature, the more one approaches the truth, since in nature only individuals exist, while, genera, orders, and classes only exist in our imagination.

To Buffon, Linnaeus's imagination seemed to manifest "a mania for classification," one that regularly produced absurdities. Take the order *Ferae*, which Buffon noted was Latin for "savage beast." Monsieur Linnaeus, he wrote,

> indeed begins with the lion and the tiger, but he then continues with the cat, the weasel, the otter, the seal, the dog, the bear, the badger, and ends with the hedgehog, the mole, and the bat. Would one ever have believed that Ferae was applicable to the bat, the mole, or the hedgehog? Or that domestic animals such as the dog and cat might be savage beasts? Isn't that just as careless a use of ideas as it is of the words that represent them?

As Buffon pointed out, physical distinctions were inconsistently applied. If everyone agrees that a striped cat and a spotted cat are equally cats, why were we so certain that tigers and jaguars are different species? Tigers were found chiefly in India and jaguars were native to the Americas, but did physical separation justify species separation? Linnaeus classified tigers as *Tigris* and jaguars as *Onca*, within the genus *Felis*. Since *felis* meant cat and both were placed alongside *catus*, the domestic cat, Linnaeus was simultaneously asserting that coloration was not an essential difference (in the varieties of domestic cats) and that it was (as the differentiation between tigers and jaguars).

Buffon continued to meander across Linnaeus's demarcations, discovering instance after instance of strained conformity and bizarre juxtaposition. The order *Glires*, meaning mice: "These mice of Monsieur Linnaeus are the porcupine, the hare, the squirrel, the beaver, and the rat." The order *Jumenta*, meaning beast of burden: "These beasts of burden are the elephant, the hippopotamus, the shrew-mouse, the horse, and the pig." In other corners of the hierarchy,

> one finds that the lynx is only a species of cat, the fox and the wolf species of dog, the civet but a species of badger, the guinea pig but a species of hare, the water rat a species of beaver, the rhinoceros a species of elephant, the ass a species of horse, et cetera. And all

that because there are some small resemblances between the number of mammary glands and teeth of these animals, or some slight resemblances in the form of their horns. . . . Is it necessary to go any further to make it apparent that all these divisions are arbitrary and this method is not justifiable?

For good measure, Buffon included a footnote quoting Linnaeus's old nemesis Siegesbeck: "I have found this system, Linnaeus's that is, to be not only most contemptible and inferior . . . but also too strained, uncertain and fallacious and I would say even insignificant."

The shot across Linnaeus's bow could not have been more thunderous. Several members of the Academy were shocked, not by Buffon's argument but by the effrontery of making it in the first place. "It is not in such terms that one criticizes a generally respected author," said Guillaume de Malesherbes, soon to become the chief censor of France. "Especially when the one who criticizes him (whatever talents he may have) is still a man new to science." But Buffon was just getting started.

. . .

News of an articulate and powerful critic soon reached Linnaeus, but he had yet to learn the full scale of Buffon's ambitions. That came the following year, when public notices announced Buffon's work in progress: *Histoire Naturelle, Générale et Particulière, Avec la Description du Cabinet du Roi*. As implied by the title, an inventory of the Cabinet would be a secondary goal. The primary one would be a grand tour of "all the objects presented to us by the Universe," in a work projected to fill fifteen volumes.

It was a brilliant ploy. The cost of printing a book of any substantial dimensions was prohibitively expensive in 1744, as the process involved individual plates inked by hand, then manually pressed on single sheets of paper subsequently hung up to dry. A fifteen-volume work was practically unheard of: The first French encyclopedia, under development at the same time, was expected to run to only five volumes. But Buffon, using his influence at court, had convinced the king to underwrite a massive discursion on natural history, functioning as the required catalogue but delivering so much more. Newspaper descriptions went into considerable detail.

The [first three] parts of this great and magnificent work we have on hand belonging to the description of the Cabinet, will be followed afterwards by a fourth and a fifth, in which all the Quadrupeds, and all the animals living [both] on land and in water, will be described. In the sixth will be described the Fishes; in the seventh the Mollusks; in the eighth the Insects; in the ninth the Birds; and in the tenth, eleventh and twelfth, the Plants. In the thirteenth, fourteenth and fifteenth, the Minerals will be described. In the order will be disclosed the whole extent of Nature, and the entire Kingdom of Creation that is subject to man and created for his contemplation.

While Buffon would employ his assistant Daubenton to compile many of the Cabinet descriptions (and credit him for it), the remainder of the work would be entirely his. Could a single person possibly write an account of the entire Kingdom of Creation? To the man in Uppsala already attempting to do so, this did not read as hyperbole. Up until now, Linnaeus had garnered critics, not competitors. Buffon had ample resources to command—his own fortune, his large staff of assistants and subordinates, and the prestige and backing of King Louis XIV. French naval officers now had standing orders to collect specimens for the Jardin during their voyages; all French physicians working abroad were strongly encouraged to submit specimens as well. Linnaeus had a significant head start, but Buffon could simply outwork him, fitting all the pieces together in a more consistent and logical manner. His own *Systema Naturae* was, after all, not one system but three, each with its own set of rules.

So Buffon would begin with quadrupeds, presumably introducing his own taxonomy in the process. "I look forward to the subsequent work of Monsieur Buffon, who from a Dog and a Horse begins a natural method," Linnaeus wrote to Jussieu. "This will certainly do as an experiment. I see his theory. What I'd like to see is a practical path."

PART II

This Prodigious Multitude

Little flower—but if I could understand
What you are, root and all, and all in all,
I should know what God and man is.

—ALFRED, LORD TENNYSON

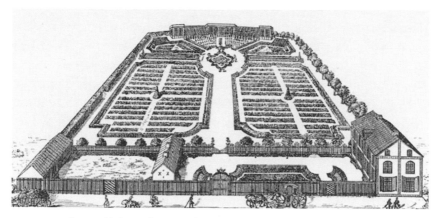

The remodeled Uppsala Botanical Garden, with Linnaeus's residence on the right

TWELVE

Goldfish for the Queen

ALMOST AS SOON AS HE ARRIVED AT UPPSALA UNIVERSITY IN February of 1743, Peter Lofling developed a habit of drawing as little attention to himself as possible. "He came to Uppsala quite young, of such simple manners as might be taken for stupidity," one of his classmates recalled, by way of explaining his desire to fade into the woodwork. Just fourteen when he began his university studies, Lofling was not a prodigy but the product of a hurried education. His father, a bookkeeper at an ironworks in the province of Gestrikland, on Sweden's eastern coast, had opted to skip the cost of boarding at a Trivial School by bundling him off to Uppsala as soon as possible. There, Lofling was supposed to study theology, a plan he dutifully followed for his first two years. But he could not resist being drawn into the bustle of activity that swirled around the celebrated Professor Linnaeus.

Foremost was the spectacle of the Uppsala botanical garden, technically belonging to the university but clearly the professor's personal domain. The once-neglected thicket was now the city's most popular tourist attraction, drawing visitors from afar not only to admire the plantings but to gawk at the exotic animals that Linnaeus, in true showman style, had acquired and placed on display. The menagerie began when Sweden's Crown Prince Adolf Frederick presented him with a living specimen of an "American bear," otherwise known as a raccoon. Linnaeus named him Sjupp, and his species *Ursus cauda elongata*, or "long-tailed bear" (he would change it to *Ursus lotor*, "washerman bear," in a later edition of *Systema Naturae*), and provided him with a specially designed lair that, unfortunately, did not prevent a neighborhood dog from killing him two years later. The ill-fated American bear was only the first of a number of zoological residents added to the garden: guinea pigs; a gray parrot named Diana, fond of calling out "Blåsa din näsa!" (Blow your nose!); an unnamed weasel fitted with a bell collar; and a capuchin monkey named Grinn. At least a dozen other monkeys would come to call the garden home, although unlike Grinn they were confined to shelters perched on high poles dispersed throughout the garden, shackled to prevent their escape.

Then there was the excitement of Professor Linnaeus's lectures, which packed the university's amphitheater or, in fair weather, filled the garden itself. Unlike other professors who zealously limited attendance to students enrolled in their department, Linnaeus permitted any student to take part, both in his lectures and in the extensive question-and-answer sessions he held afterward. More dramatically, he invited everyone to take part in organized field excursions, events he called "herbaciones." These proved so popular he'd begun charging for participation, providing himself with both additional income and an outlet for his well-honed showmanship.

Linnaeus had always relished adding military flourishes to his work. One early notebook was labeled *Spolia Bellum* ("The Spoils of War"), and *regiment, battalion, company, platoon, soldier* had become a favored metaphor for explaining his five nested hierarchies. He'd adopted *phalanx* as the standard collective term for all taxonomic units (the modern term is simply *taxon*). In writing the second of his autobiographies, he'd awarded ranks to thirty-three of his fellow botanists, naming them "officers of Flora." Naturally Linnaeus placed himself at the top of the list, ranking himself as both a general and *omnium seculi sui botanicorum princeps*—Prince of All Botanists of His Day (his old adversary Siegesbeck ranked at the very bottom as *fältväbel,* or sergeant major).

In his herbaciones, he continued to indulge his martial fantasies.

On warm weekend mornings the residents of Uppsala had grown used to the sight of upward of three hundred "troops" sweeping through the city, some brandishing guns, others armed with umbrellas and nets, fitted out in a uniform of their commander's design: a short jacket and broad sailors' pants, topped off with a wide-brimmed hat. They moved through the streets with military precision, organized into companies under the command of lieutenants and captains. Upon their leader's signal at a little after seven o'clock in the morning, they filed out of the city into the surrounding fields. Linnaeus himself marched at the head of the procession, flanked by flags and a corps of horns and kettledrums.

The gun bearers, officially designated Sharpshooters, readied their weapons. The umbrella soldiers crawled through the underbrush on their hands and knees, stuffing objects of value into sacks. The net bearers snagged their prey, then killed them with jabs of long, sharp pins. Hours later, the army streamed back into Uppsala, triumphant, slightly sunburned, laughing and singing. They laid their spoils before the feet of special officers named Annotators, who itemized it all, identifying certain items as fit for further study. For the Sharpshooters, the objects submitted for inspection were birds; for the Umbrellas, a flower, leaf, or lizard; and for the Nets, a moth or butterfly. Upon discovery of a true rarity a trumpet was sounded, and the specimen was rushed into the presence of Linnaeus himself.

It is unlikely that Lofling, a theology student with little money to spare, purchased many tickets to these increasingly ritualized herbaciones. But he did begin slipping quietly into the perimeter of the professor's lectures. Linnaeus usually ended these sessions by prompting his audience to ask him questions, either verbally or, if the questioner needed time to compose their thoughts, in written form. Over the summer of 1745, Linnaeus found himself regularly fielding letters from Lofling, respectful queries that betrayed a deep engagement with his lectures. When he learned that his correspondent was only sixteen, it occurred to him that such a youth could serve as both tutor and friend to Carl Junior. "I decided to take him into my own house as a companion for my son," Linnaeus later wrote. "I could foresee that Lofling's enthusiasm for his beloved natural history would fan rather than extinguish the flame of interest." That flame was, in the father's eyes, disturbingly weak at present; the son evinced little interest in natural history, yet Linnaeus, determined to raise him as his successor, was planning to enroll him in university that fall. At the time, Carl Junior was only nine years old.

Lofling accepted the role of instructor/companion, changing his studies from theology to medicine without telling his parents. "He lived with me in the greatest trust," Linnaeus would write of his youngest acolyte. "He possessed a soul as pure as gold." He nicknamed him the Vulture, not for his appearance but for his keen eye, capable of discerning specimens at a distance.

It is understandable that Lofling concealed the move from his family. He was joining the household of a charismatic but controversial figure, who straddled the line between notoriety and celebrity. Linnaeus's work had become well enough known to spark public parodies. *L'Homme Plante*, written by the French satirist Julien Offray de la Mettrie and whimsically dedicated to Linnaeus, poked fun at his sexual system by taking it to near-pornographic extremes:

> I have botanically described the most beautiful plant of our species, I mean the woman; if she is wise, though transformed into a flower, she will not be easier to pick. . . . Women make Monagynia, because they have only one vagina. Finally, the human race, from which the male is separated from the female, will increase the class of Dieciea: I use words derived from Greek, and imagined by Linnaeus.

Linnaeus was unruffled by efforts to paint his work as licentious—he had, after all, lived down the "loathsome harlotry" accusations of Johann Georg Siegesbeck. But the revengeful act of naming a foul-smelling plant *siegesbeckia* after his detractor was beginning to take on serious repercussions. Siegesbeck had counterattacked with a fresh assault, this time basing his arguments not on morality but on observable reality.

The "vegetable letters" of Linnaeus's sexual system were determined by the number and arrangement of stamens and pistils. Yet Siegesbeck had identified (and independently named) *Anandria siges-beckioides*, a plant that seemed to bear no stamens or pistils at all—hence the forename *Anandria*, Latin for "lack of masculinity." This was, he asserted in a pamphlet, definite proof that Linnaeus's work was objectively wrong. If some plants lacked sexual characteristics, how could plant life be characterized by sex?

A worried Linnaeus called in favors to acquire seeds of the plant, which he germinated and carefully tried to bring to blossom. When weeks passed with no visible changes, *Anandria sigesbeckioides* began to look like a far more formidable threat to Linnean systematics than the Hydra of Hamburg had ever been; this one could not be consigned to the ash heap of *Paradoxa*.

But he did not concede defeat. After a few more anxious weeks he began to discern buds—minuscule ones, but buds nonetheless. In time they began to bloom, revealing miniature stamens. Triumphantly he published a thesis entitled *De Anandria*, refuting his adversary and reclassifying the plant into the genus *Tussilago*. It was a pointed designation: The word meant "cough suppressant." It was time for Siegesbeck to stop making noise.

Still, the attack deeply disturbed Linnaeus. It was "a public malice without cause," he wrote to a friend. "It cannot be wiped out in the mortal heart." While he'd proven his adversary wrong about this particular plant, it was impossible to prove that *all* plants had sexual characteristics. New botanical discoveries might produce irrefutable specimens in abundance. Besides, squinting at minuscule pistils and stamens undermined one of the perceived advantages of his system, namely that it was handy and practical for field identification.

Other critics were harder to quell. Linnaeus was also busy appeasing natural historians throughout Europe, who'd grown upset over his practice of dismantling their species identifications and substituting his own. Albrecht von Haller, the prominent Swiss botanist, had long been one of Linnaeus's most vocal supporters, but after seeing his contributions to nomenclature systematically and causally erased, he'd had enough. "You gratify your enemies, who are neither few nor impotent, when you thus attack your friends," Haller wrote, in a heated letter to Linnaeus. "Yet you assault me, not with some signs of contempt, and with an evident intention to harm me. . . . If I were so inclined, it would not be difficult for me to be troublesome to you."

Linnaeus's rush to smooth things over only made matters worse. "I have ever loved you, commended you," he wrote in reply, before displaying the same self-centeredness that had infuriated his colleague in the first place. "You do not seem to be aware, my distinguished friend, what mischief your censure may do to my fame; nor how gratifying it will be to my rivals and enemies. . . . Few people examine into the merits of a quarrel."

The imperious tone further alienated Haller. "I had rather been without his apology," he complained to Nils Rosén, Linnaeus's former nemesis. "The man is active I cannot deny . . . but his character has for me a something—I know not what to call it, of asperity, fickleness, and unevenness."

Haller was not alone in his disquiet, although other colleagues couched their reproaches in politesse. "We that admire you are much concerned that you should perplex the delightful science of Botany with changing names that have been well received, and adding new names quite unknown to us," cautioned the naturalist Peter Collinson, addressing Linnaeus on behalf of an unspecified British delegation. "Thus Botany, which was a pleasant study and attainable by most men, has now become, by alterations and new names, the study of a man's life, and none now but real professors can pretend to attain it. As I love you, I tell you our sentiments."

Linnaeus was not about to change his methods. But he did open a new phase in his work, one aimed at expanding *Systema Naturae* while bypassing the qualms of colleagues. Up until now, he'd chiefly acquired specimens in the same manner they did: through one another. Most natural historians of standing, academics and amateurs alike, shared seeds and cuttings as a matter of professional courtesy. Jussieu in Paris had always responded generously to Linnaeus's requests (a practice Buffon prudently chose not to discontinue), and even Haller's angry letter closed with "to the seeds for which you enquire, I now send all I could procure." But such specimens arrived pre-labeled, pre-interpreted, as threads of a complex web of obligations among a status quo—a status quo that Linnaeus was increasingly upsetting. How much better might it have been to encounter *Anandria* prior to Siegesbeck? How much better still to be the first to classify specimens entirely new to European savantry? It was time to expand into the unknown.

• • •

Whereas Peter Lofling was the youngest of the group of students beginning to coalesce around Linnaeus, Christopher Tarnstrom was the
eldest. He was older than Linnaeus himself, a married thirty-four-
year-old teacher with a degree in divinity. It was this degree that allowed him to become the first in a series of amateur expeditionaries
that Linnaeus, with characteristic immodesty, would come to call his
apostles. Over the next three decades he would anoint fourteen of his
students as apostles, sending them out on quests to gather specimens
across the globe.

Here Linnaeus can be seen as naïve, overly optimistic, dangerously
foolhardy, or some combination of all three. While the dispatched students departed willingly, they were scholars, not explorers, ill-suited
and ill-equipped for the task at a time when four-fifths of the planet
remained uncharted. Since the Swedish government had no interest
in funding properly equipped foreign expeditions, most of the apostles' journeys were improvised, undertaken with minimal funds and
more zeal than resources. The results would all too often prove lethal.

The apostle program began in 1746, when the Swedish East India
Company granted Linnaeus the privilege of one free return passage
per year, to be awarded to the pupil of his choice. That cut down on
travel costs, but there was still a need for ingenuity—it only covered
the return passage, not the outward voyage. Since Tarnstrom had a
divinity degree, Linnaeus immediately secured for him a position as
ship's chaplain onboard the *Calmar*, a Swedish East India Company
vessel bound for China. Now that Tarnstrom was able to earn his outward passage, Linnaeus packed him off with a mission brief that reads
more like an elaborate shopping list:

1. To acquire a tea bush in a pot or at least seeds thereof to
   be kept according to the oral instructions he has received
   from me.
2. Seeds for the Chinese mulberry tree with palmated leaves.
3. The many undescribed fish that are to be found everywhere in
   the East Indies are preserved in brandy, to be described in their
   natural state.
4. As many plants as possible shall be collected and preserved. If
   possible with flowers and fruit.

5. Seeds of as many plants as can be found shall be collected.
6. Bulbs and tubers of the roots of lilies are to be kept in sand or moss; also applies to succulents.
7. Insects are to be preserved on pins but zoophytes in brandy. [A zoophyte was an animal resembling a plant, such as the sea anemone.]
8. All types of snakes are wanted, especially cobras.
9. A sample of unused soil for the making of genuine porcelain.
10. The following unknown drugs: Anisum stellatum, Gummi Ammoniacum, Catechu, Lignum, Aloes and Myrobalani should especially be sought which tree they are taken and method of fructification.
11. A good botanical description of nutmeg is required.
12. Ripe fruits from as many types of palms as can be found.
13. Live goldfish for Her Majesty [Sweden's Queen Ulrika Elenora].
14. Measurements night and day with a thermometer south of the Equator and in Canton.

That Linnaeus believed a ship's chaplain could return from China with a stock of live goldfish was a tour de force of misplaced confidence. Capturing a venomous cobra was a potentially deadly undertaking. The soil used in genuine porcelain was a closely guarded secret. No one had ever brought a living "tea bush" capable of propagation to Europe; the international tea trade revolved around the fact that such plants failed to thrive outside their indigenous Asian climates.

Tarnstrom only fulfilled a single item on Linnaeus's shopping list, and at the greatest possible cost. Sailing in February of 1743, he adopted a hybrid tactic of acquiring specimens: At ports of call on the outbound voyage he'd purchase living creatures, then strive to keep them alive as long as possible onboard the Calmar. Most could not be expected to survive the entire journey, but at least he could forestall the occasion of having to preserve them in brandy. By December he'd populated the decks and his cramped quarters with lizards, turtles, and peacocks. Then, on a stopover at the island of Pulo-Condore, in what is now Southern Vietnam, he contracted a fever and died.

The menagerie disappeared, as did Tarnstrom's belongings. The only items that made their way back to Uppsala were a handful of seed packets and a few dried plants, most of which he'd collected not in Asia but in Spain.

Tarnstrom was not a footloose youngster with nothing to lose; he left behind a wife and children. His grieving widow Brigitta accused Linnaeus of ruining her life. Although Tarnstrom had in fact begged to go, Linnaeus accepted the blame and tried to make amends by offering to find a new husband for her. He commemorated his first apostle by naming an entire genus after him: *Ternstroemia*, which (ironically or poetically) are notoriously hardy, sporting leaves of evergreen. In selecting future apostles, Linnaeus resolved to opt for younger, single men.

Pragmatically acknowledging that specimens might need to outlive their collectors, he experimented with improved methods of preservation. Apostles could dry and press plant specimens between sheets of paper, but the process distorted their fruits and root systems; these were more intelligibly documented via detailed sketches. Keeping seeds alive for later cultivation, without germinating them in transit, was trickier. Linnaeus found that the most successful method was to dip calico cloth into melted wax, then press the seeds into the wax as it cooled, then roll up the cloth and enclose it in a cylinder of beeswax.

Animals were more problematic. Brandy not being strong enough, Linnaeus now recommended storing smaller creatures whole in bottles filled with aquavit, a high-proof Swedish spirit. This had its limitations: The bottles were bulky and fragile to carry, the alcohol evaporated and needed refilling, and the result was highly flammable. Fish, he concluded, should be skinned, brushed with turpentine, and pressed between papers just like plants. Quadrupeds were to be skinned and stuffed with hay or cotton, mixed with snuff, saltpeter, or aromatic balm, then "wrapped up in leaves or paper, dried in the sun or by a fire and laid in boxes which are to be filled with the same snuff, lime or other aromatic things so that moths may not gain entry." This was not true taxidermy: Even when a specimen survived long enough for transportation back to Sweden, the resulting item was often distorted, discolored, and given to crumble under examination. (The majority of quadruped specimens collected according to Linnaeus's instructions would not survive the eighteenth century.)

To launch his second apostle, Linnaeus resorted to a deft maneuver. In 1747 the Swedish Academy commissioned him to plan another expedition to Lapland, a follow-up to his own travels there in 1732. But he was not interested in another naturalist retracing his steps, particularly since some of those steps had been exaggerated in his official account. Instead he cast his eye toward North America, which he found intriguing not for its diversity but for its similarity to Europe. "You would scarcely believe how many of the vegetable productions of Virginia are the same as our European ones," he wrote. "There are Alps in the country of New York, for the snow remains all summer long on the mountains there." He suggested that the Lapland grant be used instead to send an apostle to the wilderness of Canada. It was, he explained, along the same latitude: The ecological landscape could therefore be expected to approximate that of Lapland, and therefore to yield transplantable resources.

For this mission he chose a student named Pehr Kalm, a large, implacable though somewhat plodding young man, who as a native of Finland was presumably inured to the cold. "Now is the time," Linnaeus wrote to the Academy, urging them to underwrite the expedition. "Another time he will be heavy footed, lazy, and comfortable, and too fat to run like a hunting dog in the forests."

. . .

Meanwhile Buffon, two years into the writing of his multi-volume work on natural history, was not yet devoting all available hours to the project. He was still exploring other avenues of inquiry, notably conducting experiments in self-propelled projectiles (a discipline that would later be called rocketry) and the practical applications of optics. Already well known in the French community of savantry, he vaulted into international fame by solving a centuries-old mystery.

During the Second Punic War in 212 B.C., the Roman fleet under Consul Claudius Marcellus besieged the Sicilian city of Syracuse, punishing the city for its alliance with the Carthaginians. The siege was ultimately successful, but not before Archimedes, Syracuse's most famous citizen, supposedly launched a startling counterattack. He constructed "a sort of hexagonal mirror," as one of the earliest surviving accounts reads, and "placed at proper distances from the mirror other smaller mirrors of the same kind, which were moved by means of their hinges and certain plates of metal.

He placed it amid the rays of the sun at noon, both in summer and winter. The rays being reflected by this, a frightful fiery kindling was excited on the ships, and it reduced them to ashes, from the distance of a bow shot. Thus the old man baffled Marcellus, by means of his inventions.

Was this reality, or legend? The "frightful fiery kindling" of Archimedes's weaponized mirrors became a much-debated topic over the centuries. Descartes, for instance, contended that such a device would have to be no less than eight hundred feet long on a side, making it "extremely large, or more likely mythical."

Buffon proved otherwise. He built several prototypes, ultimately developing a device only six feet in diameter. On a wooden frame he mounted 168 mirrors, each fitted with movable screws, then carefully adjusted each to converge on a single focal point. Fully aligned and in full sunlight, the array produced a surprisingly powerful beam. In a series of experiments, Buffon was able to set ablaze a creosoted plank at 66 feet by aligning just 40 mirrors. When he aligned 128 mirrors, a pine plank 150 feet away burst instantly into flames. In another experiment, an array of 45 mirrors melted six pounds of tin at 20 feet. Buffon hauled his device to the Chateau de la Muette, where Louis XV and his entourage watched with rapt fascination as he generated powerful beams of fire.

Newspapers hailed Buffon as a "new Archimedes." Frederick the Great of Prussia sent him a personal letter of congratulation. One French periodical published a poem:

Buffon! There is nothing that does not cede
To your ingenious efforts.
What! The miracles of Archimedes
Are merely the games of a studious leisure for you.

As Edward Gibbon wrote in his *The Decline and Fall of the Roman Empire*: "Without any previous knowledge . . . the Immortal Buffon imagined and executed a set of burning-glasses. . . . What miracles would not his genius have performed for the public service, with royal expense, and in the strong sun of Constantinople or Syracuse?"

Resurrecting Archimedes's mirror was a brilliant show of public savantry, but like Buffon's Needle it was a one-off. Having achieved his goal of fostering public interest in his work, Buffon disposed of the mirror by presenting it as a gift to the king. Then he returned to his manuscript in progress.

Linnaeus could not ignore the growing celebrity of his most out-spoken critic, nor the fact that Buffon was making steady progress toward publication. In 1748, Linnaeus released a sixth edition of *Systema Naturae*, this time adding a prologue dismissing his detractors. If his systematics "do not displease the Divine teacher of true method," he wrote, referring to God Himself, "I will welcome the fables of ac-tors and the barkings of dogs, with tranquility of soul." But that same year, he dubbed an entire genus of plants *Buffonia*. It was a group of grasses, notable for their slender leaves. According to a later disciple, Linnaeus was not above commemorating Buffon's "very slender pre-tensions to botanical honor."

Opening illustration of Histoire Naturelle, *volume 1 (1749)*

THIRTEEN

Covering Myself in Dust and Ashes

IN SEPTEMBER OF 1749, THE FIRST THREE VOLUMES OF BUF-
fon's *Histoire Naturelle, Générale et Particulière, Avec la Description du Cabinet
du Roi* arrived at booksellers throughout France. Even before reading a
word, customers could not help being impressed. The volumes were
hefty (417,600 words long, spread across 1,600 cumulative pages),
handsomely bound and printed, and bearing the pedigree of the king's
own Imprimerie Royale. Yet despite their length and detail, they examined only a single animal. After an opening salvo on systemists in
general and Linnaeus in particular, Buffon spent most of the first volume and all of the second putting forth his theories of the Earth's
formation and development. It was not until the third volume that he
took up the subject of what Linnaeus had labeled class *Quadrupedia*,
order *Anthropomorphia*, species *Homo*.

Where Linnaeus had declared four variations, Buffon provided a
spectrum. Drawing from what he considered credible sources "established on undoubted testimony," he led the reader on a tour of ethnic
groups across the globe, starting in the north and meandering across

temperate zones. Surveying more than two hundred distinct cultures, he compiled travelers' reports on dominant physical characteristics— passing through the arctic, he cites Linnaeus's description of the Lapps as having a "snub & stubby nose, iris yellow-brown and tending towards black." Through his choice of sources and citations Buffon reveals his prejudices, but they are primarily aesthetic in nature. The people of Kashmir are "celebrated for their beauty," he reports, but so are the denizens of the Dara province of Morocco. As regards the women of Java, "their complexions beautiful . . . their hands delicate, their air soft, their eyes brilliant, their smile agreeable."

These are not his words: He is quoting the explorer François Leguat, who visited Java in 1708. He quotes the privateer William Dampier in describing the natives of the coast of New Holland (present-day Australia) as "of all mankind, perhaps the most miserable. . . . They have no beard; their visage is long, nor does it contain one pleasing feature." Buffon would later disavow many of these travelers' accounts, noting their tendency to exaggerate in one direction or another. But his overall eye is clinical, assessing even his fellow Europeans in terms of regional variations.

> Black or brown hair begins to be unfrequent in Britain, in Flanders, in Holland, and in the northern provinces of Germany; and in Denmark, Sweden, or Poland, it is seldom to be met with. Linnaeus informs us that the Goths are tall, their hair smooth and white as silver, and the iris of their eye is bluish. The Finlanders are muscular and fleshy, the hair long, and of a yellowish white, and the iris of the eye is of a deep yellow.

Far from falling into a handful of categories, humanity offered a panoply of features and colors. Why was this? Buffon suspected adaptation was at least a large part of it. People in hotter climates tended to have darker skin than those in more temperate zones, but this did not seem to apply consistently across the globe. "If blackness was the effect of heat, the natives of the Antilles, Mexico, Santa Fe, Guiana, the country of the Amazons, and Peru, would necessarily be so," he wrote, "since those countries of America are situated in the same latitude with Senegal, Guinea, and Angola." One possible factor was that the Americas had been settled more recently than other continents, by humans migrating across the Arctic Ocean.

> I am inclined to believe, therefore, that the first men who set foot on America landed on some spot northwest of California; that the excessive cold of this climate obliged them to remove to the more southern parts of their new abode; that at first they settled in Mexico and Peru, from whence they afterwards diffused them-selves over all the different parts of North and South America. . . . Many ages might elapse before a pale race would become alto-gether dark, but there is a probability that in time a pale people, transported from the north to the equator, would experience that change, especially if they were to change their manners, and to feed solely on the productions of the warm climate.

Adaptation manifested slowly, and perhaps was unfolding still. Humanity was a work in progress. As Buffon acknowledged, none of this accounted for individual variations in the human organism: height, hair color, facial features. These "ought to be considered as accidental, since we find in the same country, and in the same town, men whose hair is entirely different from one another." But "acciden-tal" was a concept he would come to explore further. In keeping with the fact that it was ostensibly a catalogue, the volume closed out with "Description of the part of the Cabinet which relates to the Natural History of Man," inventoried thusly:

1. Bones
2. Bones cut to expose interior parts
3. Deformed bones
4. Bones deformed by defect of conformation
5. Hunchback & Stunted skeletons
6. Bone spurs and tooth decay

Despite its dry, clinical ending, the first printing of *Histoire Naturelle* sold out in six weeks. It was the beginning of a publishing phenome-non that would make Buffon the most popular French author of his lifetime, outselling Voltaire, Rousseau, and Montesquieu. The first three volumes, and volumes to follow, would become a ubiquitous presence on European bookshelves, remaining in print in various editions and translations for more than 150 years. Demand was such that at one point Buffon was overseeing six editions that differed not in content but in page size, the better to fit on a variety of bookshelves. Buffon was already a respected public figure prior to publication, but *Histoire Naturelle* turned him into the embodiment of a new category of fame: the celebrity savant. It was a role he would play for the rest of his life.

Buffon's indictment of systemist thinking, reaching a general audience for the first time, struck critics as particularly effective. "This attack is directed straight at the celebrated Linnaeus, an author, it is known, of a new system which destroys all previous ideas," commented the *Journal of Trévoux*, which considered this section of *Histoire Naturelle* so persuasive it formally retracted its earlier praise of Linnaeus. "He [Buffon] shows the defects, even the ridiculousness of such a method."

> Those who read this section of the discourse will see that energy and freedom still remain among the writers of natural history. We add that despite our esteem for the learned Swedish botanist, we are not very sorry to see his system disturbed. . . . By virtue of the discourse of M. De Buffon we are supported, so to speak, in our manner of thinking, and we acquire strong arms to defend us against the partisans of novelty.

. . .

Acclaim, however, was far from universal. The same *Journal of Trévoux* that praised the *Histoire* as a masterstroke against Linnaeus later published a critique angrily taking issue with Buffon's assertion "that it is possible to descend by almost imperceptible gradations from the most perfect of creatures to the most formless matter." The theological implications, it argued, were disturbing: If life was a continuum, there could be no clear leap between ensouled beings and those without souls.

"Everyone knows that spirit, which is immortal, and matter, which contains the seeds of its own dissolution, are two incommensurable orders of beings," opined the article. "Thus there can be no gradation so happily placed as to serve as the imperceptible passage from one to the other." (The *Journal*, a Jesuit publication, thereafter curtailed its praise of Buffon.) More religious objections arose in the pages of *Ecclesiastical News*, which attacked Buffon's postulation that the planets had been formed of molten matter ejected from the sun, "which heat necessarily underwent a gradual decay: it was in this state of fluidity that they took their circular forms."

The offending words here were "gradual decay." Buffon was stating that the Earth was created not as the singular act described in Genesis but as a process—a process that, he impiously concluded, continued for millennia thereafter. It was a blatant attempt to break the lens of fixity, to impart a non-biblical timescale to creation. This was enough for the *Ecclesiastical News* to condemn *Histoire Naturelle* as "a book whose venom we believe ourself obliged to expose," but Buffon went further still:

> So the sun will die out probably for the same reason, but in some future age, and proportionately as far from the times that the earth and the other planets become extinct.

The death of planets, and of the sun itself? "Can one without blame leave uncriticized a work as pernicious as this?" the *News* thundered. "How, then, will Monsieur Buffon begin to lead these unbelievers back to the salutory yoke of faith? Beyond the injury which this book does to God, it dishonors the name of the king, to whom it is dedicated." The anonymous reviewer quoted the book of Job (*therefore I reprehend myself, and do penance in dust and ashes*), urging Buffon to renounce his heresy in a similar fashion.

Buffon intended to do nothing of the sort. "I think of acting differently, and shall not reply by a single word," he confided to a friend. "Everyone has his delicate shade of self-esteem. Mine goes so far as to think that certain people cannot even offend me." But the controversy amplified, to the point that the Sorbonne—the Jardin's ancient nemesis—decided to weigh in. Its theology faculty notified Buffon that *Histoire Naturelle* was officially under censure, "because it contained principles and maxims which are not in accordance with those of Religion."

Silence was no longer an option. In March of 1751, Buffon replied with an explicit show of capitulation. "I disapprove of my behavior and I repent, by covering myself with dust and ashes," he wrote, intentionally echoing the *Ecclesiastical News*' invocation of Job. "I abandon whatever in my book concerns the formation of the earth." To further appease, he declared his theories about the Earth merely "pure philosophical speculation," assuring the theology faculty he believed wholeheartedly in the biblical account of Creation, "both as to the order of time and the circumstances of the facts."

He included the Sorbonne letter and his response, verbatim, in the next volume of *Histoire Naturelle*. He did not, however, change or remove in subsequent printings a single word of what had already been published. The work rolled on.

To Buffon's surprise, the gesture worked. "I have extricated myself to my very great satisfaction," he wrote to a friend. "Of one hundred and twenty assembled doctors, I had one hundred and fifteen, and their decision contains words of praise for me, which I didn't expect." It was an act of what he confessed as "persiflage," a flattering but patently insincere flutter of words. The incident inspired one of Buffon's more famous quips: "It is better to be humble than be hung."

It also posed the challenge of how to steer clear of religious sensibilities in the future. Buffon realized his mistake: He had employed no rhetorical safeguards. Commonly used by other authors whose work touched upon theologically sensitive matters, these were prominent passages carefully crafted to deflect religious censure, inserted in the text to buffer subsequent pages from controversy. Even a work as innocuous as *De Uitlandsche Kapellen*, Caspar Stoll's four-volume study of butterflies, took pains to reassure the reader that it had been written "without losing sight of the all-powerful hand of the Creator." Linnaeus, in *Systema Naturae*, had also used rhetorical safeguards, going out of his way "to attribute this progenitorial unity to an Omnipotent and Omniscient Being, namely God."

It had bordered on naïve for Buffon to toss out a theory of planetary creation in the absence of similar caveats, or at least language framing it as an idle thought exercise. Mindful of the delicate roles he played, both as a member of King Louis XV's court and as director of the already-controversial Jardin, Buffon took the lesson to heart. Future volumes of *Histoire Naturelle* would emerge from the presses generously laced with strategically placed rhetorical safeguards, some of which would puzzle later generations of readers, unaware of their original context.

In a move that further served to reestablish his respectability, Buffon got around to marrying. On September 22, 1752, he wed Marie-Françoise de Saint-Belin-Malain, a woman he had been courting for the past two years. She came from a noble family that had descended into poverty; the two had met when she was a charity student at Montbard's convent school, an institution run by Buffon's sister, who had become a nun. Marie-Françoise, described as "charming, gentle, pretty rather than beautiful," had no dowry or useful connections. She would make no splash in social circles, much less in the royal court, preferring like her husband to remain in Montbard as much as possible. Their union would prove harmonious, and was, by all accounts, rooted in genuine affection. While eyebrows raised at the unlikely pairing, Buffon was undisturbed. "I will worry even less about criticisms of my marriage than those of my book," Buffon quipped, before getting back to work.

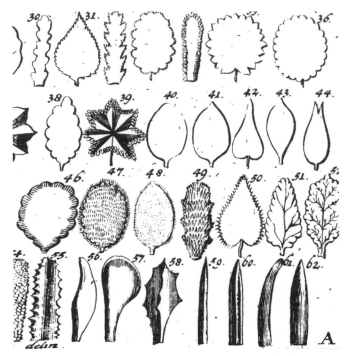

Leaf types, from Linnaeus's Philosophia Botanica *(1750)*

FOURTEEN

The Only Prize Available

WHILE OTHER STUDENTS WERE OUT FORAGING FOR WILD strawberries, Peter "The Vulture" Lofling sat next to a bed on the second floor of the Botanical Garden residence, transcribing the words of his ailing professor. "The Ariadne's thread of botany is system, without which botany is chaos," Linnaeus dictated. "Honor lasting for all time will be given to the systematists who apply this thread."

The last two years had not been banner ones for Linnaeus. His militaristic herbaciones had proven too popular: Uppsala University's rector, concerned with the "instruction of a uniform and a new way of life that turns the youths' minds from all other duties and tasks," banned them in 1748. "We Swedes are a serious and slow-witted people," the rector apologetically added. "We cannot, like others, unite the pleasurable and fun with the serious and useful." Then came the pyrotechnic debut of Buffon's *Histoire Naturelle*, after which Linnaeus took to his bed, complaining of an attack of gout that "incapacitated my mind and spirit, as well as my bodily strength." The only thing that seemed to alleviate his symptoms, even temporarily, were wild strawberries, as freshly picked as possible. While other students scoured the countryside, he drew Peter Lofling aside and began dictating what he called "the digest of the Science of Botany," to be entitled *Philosophia Botanica*.

Despite its title, *Philosophia Botanica* is not a philosophical work. It is a field guide, a codification of Linnaeus's core tenets expressed in a series of 365 conveniently numbered rules and definitions for aspiring and practicing botanists. The basics of the sexual system are memorably condensed in number 145: "The calyx is the bedroom, the corolla is the curtain, the filaments are the spermatic vessels, the anthers are the testicles, the pollen is the sperm, the stigma is the vulva, the style is the vagina.

"We reckon the number of species as the number of different forms that were created in the beginning," number 157 asserts. "That new species can come to exist in vegetables is disproved by continued generations, propagation, daily observations, and the cotyledons [embryonic leaves]." He underscored this with number 162, stating flatly that "the species are very constant." He was equally clear-cut in classifying his fellow systematists, breaking them into "heterodox" and "orthodox" categories (24), the latter being those who "have established all the classes of vegetables according to a genuine system." He listed his name among the orthodox, as part of a roll call of naturalists past and present. But when it came to defining *taxonomists* (40) as those who "have determined the truly proper names for genera and species," the only example he gave was himself.

"If you do not know the names of things, the knowledge of them is lost too," Linnaeus proclaimed in rule 210, which explains why nearly a third of the entries in *Philosophia Botanica* addressed the giving of names. It is here that Linnaeus moved confidently to a seat of authority, a powerful central position within natural history. Up until then, botanists had frequently listed commonplace names in their catalogues of plants, but they also provided a *differentia specifica*—a long, unwieldy label that served to describe the entity as completely as possible. One example was *Mimetis cucullatis leucosopermun hypophyllocarpodendron*, a South African shrub, whose label translated roughly to "imitating-monk's-cowl-white-seeded-water-liking-wrist-leaf." In *Philosophia Botanica*, Linnaeus rejected *differentia specifica*, saying that "words one and a half feet long are actually painful to pronounce, and liable to damage the throat of the speaker." He disdained as "disgusting" the use of words containing more than twelve letters, a guideline that would cause him to scrap *hypophyllocarpodendron* and substitute *protea* in its place.

As to the names themselves, Linnaeus established a new structure. Each would henceforth contain two parts: a *generic* name, identifying the genus, and a *specific* name unique to the species. This had been implicit in *Systema Naturae*: One need only read the genus and species simultaneously to arrive at *Homo diurnus*. But to strictly enforce hierarchy as identity, to sweep away all linguistic flourishes in favor of a simple two-part label—that was an innovation that set natural history on the steel rails of standardization. "Without the concept of a genus, there is no certainty of the species," he wrote. "A specific name without a generic one is like a clapper without a bell."

Linnaeus was not the first to employ binomial nomenclature. In 1576 the French botanist Carolus Clusius coined two-name words such as *Genista tinctoria* and *Dorycnium Hispanicum* for his study on the flora of Spain. However, Clusius was not putting forth an organized system; he also experimented with simply numbering plants. And in the vernacular it is common practice to identify lifeforms with a two-part name: the first part separating a group from the mass, the second part making a distinction within the group itself. The label *owl*, for instance, cordons off those birds that normally appear short-beaked, flat-faced, and nocturnal, while *snowy* or *spotted* differentiates two kinds from each other.

But problems quickly arose with such folk usages, which were meant to have meaning in local contexts rather than in the universal abstract. While *snowy owl* might seem an appropriate name for a species with mostly white plumage, not all snowy owls are the snowiest owls around; on the whole, the more common barn owl can be paler. Other qualifiers are limited in their descriptiveness as well. Calling a species a "barn owl" expresses no quality of the species itself. Additionally, folk terminology tended to overlap. The name "screech owl" referred to one species in England and another one entirely in North America. By applying his rules of binomial nomenclature Linnaeus could bring the clarity of a clean slate, naming the North American screech owl *Megascops asio* and the English screech owl *Tyto alba*.

The imposition of binomial nomenclature began with eliminating preexisting compound names, which "are to be banished from the Commonwealth of Botany." Thus Tournefort's *Bella donna* and *Corona solis* became *Atropa* and *Helianthus*, respectively. For further concision, he limited the formal description of each binomial species to no more than twelve words of Latin, which he considered sufficient length to document what he called the "essential character": not necessarily the most obvious or memorable aspects of their physical construction, but those that differentiate them from other lifeforms. As rule 259 spelled out, "the specific name ought to be derived *from parts of the plant that do not vary*" (italics his). If something is capable of variation, then it logically constituted a "variety," a distinction below the level of classification.

In the choosing of names, he rejected subjective descriptive criteria, such as smell and taste, as "mostly variable and rarely constant," explaining that "scents do not allow for fixed boundaries and cannot be defined. . . . Taste is often variable, according to the person doing the tasting." He retired the names of several species identified by such measures, calling them the result of a "zeal for subtleties" and "contagious madness among lovers of flowers." He further declared that "generic names should not be misused to gain the favour, or preserve the memory of saints, or of men famous in some other art. It is the only prize available to botanists; therefore it should not be misused." However, "I retain generic names derived from poetry, imagined names of gods, names dedicated to kings, and names earned by those who have promoted botany." Even more important, "names that have been formed to perpetuate the memory of a botanist who has done excellent service should be religiously preserved."

More rules: Only "genuine" botanists had the ability to apply names to plants, although Linnaeus did not specify who qualified as such. This was necessary because "private individuals have applied absurd names" to plants such as the flowering *Noli me tangere* (touch me not), which he renamed *Impatiens* in reference to its seed pods, which when ripe can explode upon being touched. He reiterated a dictum from an earlier pamphlet: Under no circumstances could a plant be allowed to retain a native name—in other words, a name already given to it by an indigenous culture. "No sane person introduces primitive generic names," he wrote, adding that "all barbarous names are regarded by us as primitive, since they are from languages not understood by the learned." He banned all names that were not Latin and/or Greek, and decreed that even Greek names must thenceforth be rendered in the Latin alphabet. Hence Tournefort's *Ketmia* must be renamed *Hibiscus*, since the original name was Turkish. As a final touch, he banned all words conveying usage, or even warning. Not only could *Anoctium* no longer be *salutiferium* (health-giving), but the deadly nightshade could no longer be called *Solanum lethale*, but only *Solanum*.

There was also a dictate against vanity: The discoverers of new species should not be allowed to name them after themselves. This, of course, would have the effect of naturalists sending specimens directly to him, in the hopes that *he* would bequeath their names to the species, a dynamic he reinforced with a request: "As I am now occupied in collecting species of plants, I vehemently request and implore the most eminent botanists throughout Europe to send me plants complete with flowers, if they have in duplicate any of the comparatively rare plants I have not described."

As the diet of wild strawberries restored his health, Linnaeus capped off *Philosophia Botanica* by establishing standards of measurement, defining a thumb, a palm, an arm, even a hair as legitimate descriptors of size—inexact measures, but serviceable enough for someone accidentally encountering a species to describe it with sufficient accuracy. A final section lays out precise, step-by-step instructions on a number of tasks for beginners: how to study botany, how to build an herbarium or lay out a botanical garden, and even how to travel ("The starting-point must be to marvel at all things, even the most commonplace").

While *Systema Naturae* had charted out the worldview Linnaeus was constructing, *Philosophia Botanica* was an invitation to join in that construction—a re-creation of his now-banned herbaciones on a global scale. Reprinted at least a half-dozen times during his lifetime, and translated from the original Latin into English, Dutch, Spanish, German, French, and Russian, *Philosophia Botanica* became Linnaeus's best-selling book, and the foundation of his international reputation. As a follow-up, in May of 1753 he published the first edition of *Species Plantarum*, a catalogue of the vegetable kingdom that thereafter served as the botanical subvolume of *Systema Naturae*. In a flurry of naming, he applied 5,940 binomial coinages to plants, from *Acalypha australis* to *Zygophyllum spinosum*. Then he arranged the species in nearly a thousand genera, themselves described in another spinoff volume, *Genera Plantarum*, which conformed the whole to the original classes of his sexual system. The botanical hierarchy was still glaringly artificial—if anything, the increased population of the categories produced more awkward species juxtapositions than before. But binomial names, as conceived by Linnaeus, would prove a lasting innovation. Within a few years of the publication of *Species Plantarum*, most botanists had at least grudgingly accepted binary nomenclature and the author's broad reforms. While criticisms of Linnaeus's taxonomy as delineated in *Systema Naturae* would continue, the rules established in *Species Plantarum* quickly comprised a lingua franca, a grammar for consistent communication of ideas and discoveries. Linnaeus's monumental self-assurance had, for once, paid off.

Buffon's physiognomic inventory of a horse

FIFTEEN

Durable and Even Eternal

WITH THE FOURTH VOLUME OF *HISTOIRE NATURELLE*, PUBLISHED in 1753, Buffon at last turned from humanity to the rest of his advertised prodigious multitude. He was determined not to impose a system upon nature, but the massive scope of the work still required some sort of organizing principle. Buffon decided on an approach calculated to strike readers as comfortable, even familiar. He began by posing a question: How did an ordinary person, a non-savant, come to learn about the natural world?

As Buffon pointed out, they began with their immediate surroundings, gaining familiarity with domesticated animals—dogs, horses, oxen—then proceeding outward from there into the wider world. They learned about deer, hares, and other wild inhabitants of nearby forests and fields, before expanding their scope to more distant climates, more exotic creatures: elephants, dromedaries, et cetera. "The case will be the same with fishes, birds, insects, shellfish, plants, minerals, and all other productions of nature," Buffon wrote. The common man "will study them in proportion to their usefulness; he will consider them to the extent that they are familiar to him, and he will rank them in his mind relative to the order of his acquaintance with them. . . . This order, the most natural of all, is what we believe ought to be followed."

Histoire Naturelle therefore began its grand tour of nature in the barnyard, describing only three animals—the horse, the donkey, and the bull—and taking 544 pages and 142,000 words to do so. This was a distinct departure from the usual approach of general-interest books on natural history, which sought to boost sales by trumpeting the exotic as quickly as possible, but it deftly performed three feats. First, it placed humans within the natural world—not on a separate tier, as in the schema of the Great Chain of Being, but at its gravitational center. By surveying other life in terms of its relationship to humanity, Buffon was able to transition seamlessly to discussing humans as part of the natural equation, neither aloof nor apart. Second, it accessed a large body of preexisting knowledge, since domesticated creatures were the most studied of all. Third and most pointedly, it embodied Buffon's core objection to the Linnean hierarchy and other systems—namely that they were clinical constructs, seemingly designed to isolate life and remove it from context. "We believe we have had sufficient reason for giving the preference to our method," Buffon wrote. "Our divisions are based solely upon the relations which things seem to have with us."

The nature of each entry was also markedly different. Whereas Linnaeus simply labeled a species and attached a terse description before slotting it into his hierarchy, Buffon commissioned dry-point engravings of each animal in its natural habitat, painstakingly reviewed for anatomical accuracy. These he matched with deeply researched verbal portraits, attempting to capture the essence of each species by providing two views of the subject, first "in a state of rest" and then "in a state of movement." Many of these descriptions ran to several thousand words. At equal turns poetic and precise, Buffon's writing style vividly invoked not just physiognomy but presence and even personality as well. His entry on the horse ("whose natural qualities have been matured by art, and turned with care to the service of man") incorporates a detailed map of equine physiognomy, but also this description:

> He exults in the chase and the tournament; his eyes sparkle with emulation in the course. But, though bold and intrepid, he suffers not himself to be carried off by a furious ardor; he represses his movements, and knows how to govern and check the natural vivacity and fire of his temper. He not only yields to the hand, but seems to consult the inclination of his rider.

Buffon's prose can strike contemporary readers as ornate and long-winded, achieving clarity less through concision than profusion, holding forth on so many facets of his subject that the reader is pummeled into understanding. But the style was popular in the day, and most important, it *was* a style. Most works of natural history were dry compilations, conveying no sense of audience beyond academia. "Buffon had first dazzled as a poet; the charm and power of his words were both incendiary and seductive," one critic wrote. Another called his prose "noble, dignified, with a magnificent appositeness and a perfect clearness." For all its exhaustive scope and intellectual rigor, *Histoire Naturelle* registered as a work of literature, and an engrossing one at that.

· · ·

Histoire Naturelle's approach to the animal kingdom, however, did not escape criticism. "I will say nothing against a method that has revolted everyone who studies natural history," wrote the natural historian Réaumur. "If this order, as this method has it, were arranged according to the relations that animals have with us, everything would be arranged in a manner far too bizarre. . . . The Canadian naturalist will put the beaver in the first place. In China the pig will be placed before the cow." This was a valid point, but its very validity demonstrated Buffon's primary objection to systemist thought, his landing point in the Quarrel of Universals. There *was* no abstract, Olympian perspective, no rarified approaches to nature removed from cultural context.

The public did not object. In the four years since its debut, *Histoire Naturelle* was continuing to sell briskly. It was also increasingly accepted as a work of not only savantry but art. The year 1753 also brought news of Buffon's election to the Académie Française, both the nation's highest literary honor and a position of authority: The Académie was (and remains) charged with safeguarding the beauty and majesty of the French language itself. This immensely pleased Buffon, who agonized over verbal elegance as much as accuracy. "Well-written works are the only ones which will go down to posterity," he said in a speech composed for the occasion. "Quantity of knowledge, singularity of facts, even novelty in discoveries, are not certain guaranties of immortality."

> Knowledge, facts, discoveries, are easily abstracted and transferred. Those things are outside the man; the style is the man himself; the style, then, cannot be abstracted, or transferred, or tampered with; if it be elevated, noble, sublime, the author will be equally admired at all times, for it is only truth that is durable and even eternal.

These were lofty ideals, and despite the accolades Buffon did not deceive himself that he had fully achieved them. "I am every day learning to write," he would say, a quarter-century later. After pacing through the Jardin or his private park, composing in his head, Buffon would dictate the first version of a text to a secretary, then read the draft and dictate further revisions. He preferred to orate his sentences, speaking aloud as if to an audience—another reminder that he was addressing posterity. When the work had progressed through several revisions, he usually sought out an actual audience of visitors and friends and took note of their reactions. Did the listeners seem confused by one passage, bored by another?

He did not, however, ask them for their opinion. Instead, he asked them to paraphrase the work back to him. If their responses indicated he had not got his point across sufficiently, the rewriting cycle began again. The text might progress enough for Buffon to send it to the Imprimerie Royale for typesetting, but the return of page proofs was only a signal for further rounds of editing and emendation. At least

Buffon's portrait of the bull (above) and its skeleton (below)

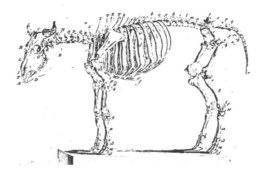

one section of *Histoire Naturelle*, a discourse on geologic time, went through no fewer than seventeen revisions.

Buffon was no less stringent in the dogged accumulation of data. For instance, volume four's entry on the cow included accurate dry-point illustrations not only of the living animal and its assembled skeleton but eight equally meticulous illustrations of its kidneys, spleen, lungs, and other internal organs. If the reader wished to know the exact length and width of the foreleg bones of the average mature bull, they would find such dimensions documented on page 526. If one's curiosity remained unsated, the section ended with a description of bovine bones on display in the Cabinet, closing with the "skull of a monstrous calf." It was an unprecedented progression from high-flown prose ("all the labour of our country depends upon him") to mundane statistics, and Buffon allowed himself no shortcuts. Genius was, after all, the greater gift of patience.

Adansonia digitata

SIXTEEN

Baobab-zu-zu

BUFFON DID NOT RECRUIT APOSTLES, BUT THE PRODIGY MI-
chel Adanson became one nonetheless. Born in 1727, Adanson had
shown early signs of brilliance, mastering Latin by the age of seven,
Greek by nine, and artfully translating Horace's lyric poetry at eleven.
But by then he was also facing a hard truth: He was the eldest of ten
children, and his father worked as a minor attendant in the household
of the Archbishop of Paris. The Archbishop was charitably paying his
tuition at a conventional school, but he expected the boy to end his
education at the age of thirteen and enter training for the priesthood.
For the next two years, knowing his family could afford no other path,
young Adanson made the most of his remaining time in the classroom, resigning himself to a future far more limited than his abilities.
Then he discovered that the gates of the Jardin swung open for everyone, welcoming even a precocious thirteen-year-old boy to its formal
grounds and informal courses of instruction.

Adanson began spending as much time there as possible. The staff grew accustomed to the sight of a small redheaded boy, looking even younger than his years, occupying a seat in the auditorium, roaming the Jardin's green paths, and asking increasingly sophisticated questions. By fourteen he was a valued volunteer assistant to Bernard de Jussieu, accompanying him on specimen-hunting forays around the Paris countryside. By fifteen, he'd been spending so much time in the Jardin library that a visiting British botanist gave him the present of a small microscope, an expensive rarity at the time. "Young man, you have studied enough of the books," the visitor said. "Your future path will be among the works of nature, not of man."

Adanson took the hint. His ambition turned beyond the Jardin walls to the possible glories of the unknown. As far as he was concerned, Europe had few remaining surprises to offer the field naturalist. The true lodes of discovery were at least a continent away.

France had thus far launched only a single scientific expedition, the Geodesic Mission of 1735 to measure the circumference of the Earth, and it had ended in failure—nearly a decade later, several members of the mission (including Jussieu's younger brother, Joseph) had yet to return. In 1748, with no other way to finance his travels, Adanson signed on as a bookkeeping clerk in the Compagnie des Indes, a French mercantile group that maintained a trading post in Senegal, on the West African coast. The job paid almost nothing, but the distant posting exuded adventure. "Senegal is of all white settlements the most difficult to penetrate, the hottest and most unhealthy to live in, the most dangerous in all respects," Adanson announced, "and so known the least to naturalists."

The Adanson who left was a bookish, unworldly twenty-year-old. The one who returned to the Jardin in 1754 was not only six years older but a man transformed. As one description noted,

> He was . . . robust and healthy. He enjoyed dancing and was a good swordsman, a fine pistol shot, and a smart rider, having had a long training while in Africa in all those sports. He wrote that his blood was impetuous and swift, causing him to expend his energies in physical activities. He appreciated Gluck, ballet, and the opera. He was far from disdaining the attractions and charms of the ladies.

This swashbuckling figure also brought with him an astonishing trove of specimens: 350 birds, 40 quadrupeds, 600 insects, 700 shells, 400 fossil snails, 600 dried plants, and the seeds of 1,000 species. Much of it was utterly new to European science. He'd also collected hundreds of samples of minerals, resins, and exotic woods, logged astronomical observations, charted maps, and compiled a dictionary of the indigenous Wolof language.

Adanson moved in with the Jussieus, who had room to spare in their house on Rue des Bernardins, a fifteen-minute walk from the Jardin. To thank them, he prepared a special luncheon. As the Jussieus and five other invited guests looked on, he opened a box and removed several large, delicate objects packed in thick vegetable grease. He wiped off the grease to reveal African ostrich eggs, which, after several minutes of admiration, were whisked off into the kitchen to provide a memorable meal.

Adanson had enthralling stories to tell. He spoke of weeks of walking across desert sands so arid and hot that "the blood oftentimes opened itself a passage through my pores, which the sweat could not pervade," and of witnessing, but unfortunately not preserving, a serpent whose head was "the same size as that of a crocodile . . . more than wide enough to swallow a hare, or even a pretty large dog, without having any occasion to chew it."

Other stories reflected Adanson's marked changes in perspective. He'd arrived in Senegal sporting white stockings, knickerbockers, and a standard version of European condescension toward the natives. In time, both his appearance and attitude vastly changed. He let his hair grow long, nearly to his waist, tucking it up in his hat when necessary. The stockings grew dirty, tattered away, and were not replaced.

His walls of prejudice eroded as well. A turning point came on an evening in October of 1751, when his Senegalese guides joined him in admiring the night sky. He was startled when they pointed out not only constellations but several planets as well. "Nay, they went so far as to distinguish the scintillation of the stars, which at that time began to be visible to the eye," he recalled. "With proper instruments and a goodwill, they would become excellent astronomers." By the end of his stay he was a committed advocate for native rights, fiercely opposing the slave trade.

He'd emerged with a second epiphany. While always aware of Buffon's quarrel with Linnaeus, he'd kept his distance from the matter, knowing Jussieu's friendly relations with Uppsala and mindful of Linnaeus's self-appointed power to bestow species names. From Senegal he'd shipped Jussieu cuttings from a curious tree the locals called *Baobab*, adding in a cover letter, "I leave you the freedom to communicate the character of the Baobab to Mr. Linnaeus. . . . I further ask you to kindly ensure that this learned man learns of the infinite esteem I have for him and his works." Linnaeus did name the species *Adansonia digitata*, but by the time Adanson learned of this he no longer cared. The overwhelming diversity of life he'd witnessed in Senegal had convinced him: Buffon was right. "As soon as we leave our temperate countries to enter the torrid zone," he concluded, "there are always plants, but they have attributes so new that they elude most of our systems, whose limits hardly extend beyond the plants of our climate." In short, Linnaeus's neatly nested boxes of hierarchies were incapable of grasping life's true profusion.

The returned Adanson had a new purpose in life. He wished to devise a means of organizing life that maintained the complexist mindset, one that avoided the systemists' errors of over-simplicity and sweeping generalization. After all, it was theoretically possible to construct "an instructive and natural method," as Buffon had written. All that was necessary was "to place together those things which resemble one another and separate those which differ."

> If the individuals resemble one another perfectly, or if the differences are so slight that we have difficulty perceiving them, they will be of the same species. If the differences begin to be perceptible but there are still greater similarities than differences, the individuals belong to a different species but the same genus as the first. And if the differences are yet more marked, without, however, exceeding the resemblances, then the individuals will not only belong to another species but even to a different genus than the group to which the first and these second group belong. They will, nonetheless, belong to the same class, because they have more resemblances than differences.

To Buffon, a complexist taxonomy was possible but impractical, at least given the present state of technology. Differentiation would need to be made not in the manner of Linnaeus, via an arbitrarily selected handful of physical characteristics—it would have to be "all-inclusive," relying on as many characteristics as possible. In other words, a detailed knowledge of each organism, inside and out, was required: Otherwise, how could we find differences we didn't even know to look for? Yet in-depth knowledge of many organisms was an unachievable goal in the eighteenth century. In an era when only the shells of mollusks tended to survive transportation from distant countries, when microscopes were insufficiently powerful to discern minuscule details, "one sees clearly that it is impossible to establish one general system, one perfect method, not only for the whole of Natural History, but for even one of its branches."

But perfection was not necessary. Adanson had begun to see an approximative path toward taxonomy. "I have found a manner of describing, very different," he wrote, "not only because it includes absolutely all different parts of the natural body, but also in that it describes these parts in all the qualities that are their own." Adanson's vision was a system that took meticulous account of multiple characteristics, internal as well as external, and weighted them accordingly on a spectrum of difference. It would be complicated, and arduous to compile: Determining a species would seem more like a calculus equation and less like looking up a name and address in Linnaeus's directory. Yet Adanson thought it could be done, and that the result would be closer to the true natural system that even Linnaeus admitted was lacking in the botanical component of his work. Retiring the Linnean hierarchy by replacing it with something better might be a lifetime's work, but Adanson felt confident that he was the person to do it.

No one in the Jardin was about to discourage him. After all, he was their fondly regarded prodigy all grown up, returning with the largest bounty of specimens ever gathered by a naturalist on a solitary expedition. He had twelve years of experience in natural history under his belt, yet was teeming with youth and vitality, even charisma. He had the support of Buffon, who praised Adanson's collection as a "precious assemblage . . . worthy to be acquired by the King" and would ultimately quote him more than a hundred times in the pages of *Histoire Naturelle*. If anyone was going to dismantle the nested boxes of *Systema Naturae*, it was hard to imagine someone better suited to the task. Meeting in the Louvre in 1756, the Académie des Sciences approved the opening phase of his planned construction of a complexist taxonomy. "Let us multiply observations," he wrote in the introduction to his manuscript, "and not systems or books, which do more to increase the confusion in natural history than to instruct."

. . .

The following year, Adanson published the first volume of his *Histoire Naturelle du Senegal*, a preliminary work analyzing some of the specimens in his collection (which Buffon was still negotiating to purchase on behalf of the king). After six more years of work he published *Familles des Plantes*, a two-volume debut of what remained a work in progress. His use of weighted characteristics in *Familles* was still too tentative to evince much reaction, but the same could not be said for his unveiled system of reforming taxonomical names. It operated on three principles, which he dubbed the "rules of appointment."

The first one: *Use the historically oldest name for a species you can find.* That meant reverting Linnaeus's *Actea* back to the original ancient Greek term, *Akokorion*. But it also meant using indigenous names from the region of discovery. "Some modern botanists call barbarians . . . all the names Foreigners, Indians, Africans, Americans, and even those of some European nations," Adanson noted. "But if these Dogmatic Authors had traveled, they would have recognized that in these various countries our European names are treated alike as barbarians; they are such in relation to their way of pronouncing, as theirs are ours. Let us therefore judge otherwise the acceptance of such an improper term, and let us agree that all these names put in the balance are equivalent to each other, and that they should be adopted whenever they are neither too long nor too harsh or too difficult to pronounce."

The tree Linnaeus had named *Adansonia* in his honor? *Baobab*, the term he learned from the Senegalese, should be its formal name. "This is what made Monsieur de Buffon rightly say that the study of the appointment or modern nomenclature in Botany, is longer than the knowledge of Plants in itself," Adanson wrote. "So we consider it superfluous to quote anything other than the oldest or best primitive name."

He proposed taking the name of the "family" of plants (a name he preferred to genus), then adding an ingenious alphabetical code. The first species of baobab described would have the vowel *a* suffixed to it: *Baobab-a*. The next four would have the remaining vowels attached in succession: *Baobab-e, Baobab-i, Baobab-o, Baobab-u*. After that, the suffix would incorporate a cycle of consonants, preceding the vowel: *Baobab-be, Baobab-ke* (the *c* was changed to *k* for universal pronunciation), *Baobab-de*, et cetera. As he pointed out, this was a means of listing up to 110 species within a genus while keeping their names short, adding only two letters and one syllable. (He did the math wrong, counting 80.) He believed that would suffice for most botanical families, but the method stayed concise even if extended. Adding just two more letters of his twenty-one-letter phonetic alphabet produced 11,025 distinct variations, ending in the species name *Baobab-zu-zu*. That was not only pronounceable; one could look at it and know it had been discovered more recently than *Baobab-zu-ka*. In the tradition of Buffon, each species would also have a table of all historical names attached to it, sorted by rough chronology.

The second rule: *Name a genus after its most common or best-known species.* That was what earlier naturalists had done, after all, creating groupings like *Pinus* (pines), *Acer* (oaks), and the like. That would help the average person to begin conceptually zooming in on a species in question, as opposed to Linnaeus's practice of choosing opaque names.

The third and final rule: *Standardize spellings and pronunciation.* Since a name would be spoken and written by people around the world—people who speak very different languages—the names should be made easy to speak and write for everyone. This meant "to write as we pronounce, to delete letters which do not ring, to join together those which have the same sound." If a name contained silent letters, such as *Herba*, it should be streamlined to *Erba*. If it contained a letter that should be pronounced like another letter—the soft *g* of *gentle*, for instance, which sounds like a *j*—the letter most associated with that sound should be substituted. If a name uses the same letter twice in succession, as in *Buffonia*, simply eliminate one.

Linnaeus was swift in his dismissal. "I saw the natural method of Adanson," he wrote to a colleague, soon after the second volume of *Familles des Plantes* was published in 1764. "There is nothing more silly. . . . He gives copious notes, but nevertheless distinguishes nothing. He changes all names into worse ones. I wonder whether he be

sane or sober." In other letters that year he continued to rail against
the upstart ("He spoils and destroys, contrary to nature"), ultimately
concluding,

> Adanson's method is the most unnatural possible. . . . All my ge-
> neric Latin names have been deleted and instead come Malabar,
> Mexican, Brazilian, etc., names which can scarcely be pronounced
> by our tongues. . . . No one of his classes is natural, but a mixture
> of everything.

Adanson remained undeterred, both by Linnaeus's dismissal and
by the lackadaisical public reception of the work. As with any system,
the proof would be in its population. With any luck, an instructive
and natural method might be readied by the tenth volume, when Buffon planned to tackle the vegetable kingdom—the perfect occasion to
retire Linnaeus's admittedly artificial sexual system.

The master plan for *Histoire Naturelle*, however, was already undergoing modifications. Originally, the fourth and fifth volumes were to
describe "all the Quadrupeds, and all the animals living [both] on

Jocko, Buffon's chimpanzee

land and in water." The sixth one would turn to fishes. But over the course of 544 pages, volume four examined only the horse, the donkey, and the bull. Volume five, published in 1755, devoted 311 pages to the sheep, the goat (and Angora goat), the pig (including the boar), and dozens of varieties of dogs. Volume six, published the following year, took 343 pages to cover cats, deer, hares, and rabbits.

There were a lot of quadrupeds yet to go. Compounding the process was Buffon's insistence on examining an actual, living creature whenever possible, converting portions of the Parc Buffon into a private menagerie and spending freely to populate it. He paid 1,200 livres—roughly the equivalent of Linnaeus's annual salary—for a single Angola colobus monkey, imported from the African interior. He likely paid still more for his chimpanzee, which he named Jocko, and an unnamed lion that occupied a specially designed deep pit. Cages were detrimental to his purpose of observing animals in as natural a state as possible, and he was not averse to letting the more tranquil species roam freely across the grounds. Visitors report watching badgers pad indoors to warm themselves by the fireplace, a hedgehog helping himself to food in the kitchen, and the Angola monkey wreaking havoc in a reception room. Servants shooed them away and cleaned up in their wake, but only when Buffon's examining eye turned elsewhere.

Pehr Kalm's depiction of a Native American

SEVENTEEN

So Many New and Unknown Parts

LINNAEUS, MEANWHILE, CONTINUED TO ANOINT APOSTLES from among his students, scraping together a patchwork of free passages and scant funds, and dispatching them to their fate. There is a great deal to admire in the apostles themselves. They were, after all, risking their lives in the service of knowledge, a risk nobly undertaken and executed with sincerity. Linnaeus certainly did not force them into the field; they were eager to go. As demonstrated by the independent Adanson, the former prodigy of the Jardin, they were not the only naturalists straying far from home on improvised and slender means. But the first wave of Linnaeus's apostles produced an extraordinary number of harsh lessons, and cautionary tales.

In Pehr (Peter) Kalm, the second apostle, Linnaeus ran up against the protracted nature of the journeys, and the fact that even an acolyte ultimately wished to lead a life of his own. "I found that I was now come into a new world," Kalm reported, landing in Philadelphia in September of 1748. "Wherever I looked to the ground, I everywhere found such plants as I had not seen before.... I was seized with terror at the thought of ranging so many new and unknown parts of natural history."

His first stop was the home of a forty-two-year-old printer with a reputation as an amateur savant. "I invited him to lodge at my house," Benjamin Franklin would fondly recall, "and offered him any service in my power." Franklin and Kalm were soon debating North American geology, pondering evidence that seemed to suggest the continent had once been covered with water. Franklin showed him samples of a mineral called asbestos ("It came from New England, stones there are utilized for fireplaces, because it does not change or crumble in the least from the action of fire"), and together they experimented with feeding ants and designing more efficient candles. When Kalm moved into his own quarters for the winter, Franklin gave him one of his latest inventions, the improved Franklin stove.

Kalm enjoyed himself so thoroughly in the Colonies that the following spring he was reluctant to head north to Canada (with some reason, as the region was embroiled in the French-Canadian War), but leave he did in May of 1749. Accompanied by a manservant, he borrowed a birchbark canoe and paddled up the Hudson River, encountering by his account hostile natives brandishing scalps, and then even more hostile French soldiers, who suspected him of being a spy for the English. In a letter to Franklin, Kalm reported (in somewhat creaky English) on a sight only a few travelers had described before, the astonishing falls called Niagara.

> It is the most rapid water in the world.... When all this water comes to the very fall, then it throws it self there down perpendicular; the hairs will rise and stand upright upon your head, when you sees this; I can not with words express how amazing this is.

Kalm ventured farther north reluctantly, and never traveled as far as the expected Canadian latitudes that paralleled Sweden. His instructions from Linnaeus were to range as far north as Hudson Bay, but Kalm came to a halt only a little past Quebec City. The land was "five times worse than the Lapland region," he reported. "Nothing of value grows in the northern part of Canada; my time is too precious to waste it there." At any rate, the arctic beyond was dominated by the "Esquimaux" people, who, he reported, "are false and treacherous, and cannot suffer strangers among them. . . . They kill all that come in their way, without leaving a single one alive."

He did not mention the real reason it had taken him a year and a half to travel from Philadelphia to Canada: He had fallen in love with a widow in New Jersey, whom he was anxious to marry. When the stoic Finn and his American bride finally departed for Europe in 1751, Franklin was sad to see him go. "Our friend Mr. Kalm, goes home in this Ship, with a great cargo of curious things," he wrote. "I love the man, and admire his indefatigable Industry."

Realizing that he had likely disappointed the master, Kalm by-passed Uppsala and headed instead for his homeland of Finland. He delayed meeting with Linnaeus for months, until Linnaeus sent a pointed letter. "Take burning firebrands and throw them at Professor Kalm," he wrote, "so that he might come without delay to Uppsala, for I long for him as a bride for the hour of one o'clock at night." The bounty Kalm laid before Linnaeus's feet was, in fact, impressive. It included cuttings of sugar maple (the source of maple syrup), gourds of native squash, a North American variety of ginseng, a winter-hearty mulberry tree, and a quantity of blue lobelia, a plant touted by the Cherokee as a cure for syphilis.

Sadly, none of these plants delivered on its promise. Before its supposed cure could be tested, the lobelia withered and died. (Linnaeus labeled it *Lobelia siphilitica* anyway.) The mulberry and sugar maple refused transplantation. And while the squash yielded what Linnaeus called "beautiful and ripe fruit," he couldn't interest anyone in eating it, much less converting it into a commercial crop. He rewarded his apostle to North America with literal laurels, naming the entire genus of New World laurel trees *Kalmia* in his honor, but ultimately their relationship grew strained. Linnaeus's lofty expectations had not been met.

For the next several years Kalm begged Linnaeus to send him back to North America on a second expedition, to give him another chance and to assuage the homesickness of his American bride. His appeals were ignored. He accepted the snub, thanked Linnaeus for his "peculiar friendship and kindness," and settled down to teach botany in Finland for the rest of his days.

. . .

The third apostle provided a painful reminder that ranging great distances took a physical toll. The pupil Fredrik Hasselqvist ("modest, polite, cheerful and intelligent," according to Linnaeus) departed for the Middle East in 1749, having picked that destination after hearing Linnaeus complain about a lack of specimens from the region. He had a tenuous connection, a relative who was Consul-General in Smyrna, but he departed with a thin purse, and less than robust health. After five months in Egypt, Hasselqvist was destitute and suffering from a wasting disease that was likely tuberculosis. Linnaeus raised and sent sufficient funds to rescue him (a sum one-third greater than Linnaeus's own annual salary), but instead of heading straight home, Hasselqvist ignored his health and wended through Egypt, Palestine, and Syria, attempting to collect as many specimens as possible. He died in the field in 1752, two and a half years into his mission.

Linnaeus mourned the death of Hasselqvist, eulogizing him as "like a lamp whose oil is consumed." But he was further aggrieved to find that creditors had seized "all his collections of natural curiosities, observations and manuscripts, which they would not part with, until their demands were satisfied. . . . We knew no means of collecting on a sudden such a sum of money."

This was "a double death, since not only he has disappeared, but also all his work." Linnaeus appealed to the queen of Sweden, who paid the debt out of her own purse. Unfortunately, it was not an act of largesse but a purchase. She ordered Hasselqvist's collection delivered directly to her at Drottningholm Palace, where she exhibited them as curiosities; Linnaeus, who had counted on adding them to his collection, was instead compelled to examine them on visits. After the expense of over six hundred pounds, as well as a life, he swore off future expeditions, vowing never again to send an apostle abroad.

. . .

That vow was almost immediately broken. After seven years of faithful service in Linnaeus's household, Peter Lofling, affectionately nicknamed the Vulture, had grown into a physically impressive young man, and he was anxious to embark on his own adventure.

Lofling had completed his graduate thesis (on the taproots of trees) in 1749, and was voted into the Swedish Academy in 1751. As his mentor himself admitted, he seemed well suited for the rigors of exploration. "It was a matter of complete indifference to him whether he slept on the hardest boards or in the softest bed," Linnaeus wrote, "but to find a little plant or scrap of moss the longest road was not too long for him." It was time for him to go, but with no funds at hand Linnaeus kept him busy by sending him to Madrid, where King Fernando VI requested assistance in building a royal botanical garden.

In October of 1753, the Spaniards invited Lofling to join a surveying party, seeking to fix the borders between several South American possessions of Spain and Portugal. Linnaeus gladly gave him permission to go, cheered that at last an apostle of his would not be a solo traveler but part of a proper expedition. The master bade his student to preserve butterflies between the pages of books, and worms in brandy; Swedish aquavit would be in short supply.

After fifty-five days at sea, Lofling landed in Venezuela in April of 1754. On the long overland journey to Nueva Barcelona, he reported "colic and pains throughout my body . . . pains in the whole body and spine." He pressed on, but by the time the expedition had reached Guyana he was too weak to go farther. Racked with intermittent fevers, he died on February 22, 1756.

The expedition buried him under an orange tree, and marched on.

"The great Vulture is dead," Linnaeus lamented. "Lofling gave his life for Flora and her lovers, and they mourn his loss. . . . Such a profound and attentive botanist has never set foot on foreign soil, nor has there been a traveler who has had the opportunity to make the great discoveries that Lofling was able to make."

But those discoveries were irreparably lost. After spending decades preparing to make his mark in the world of natural history, Lofling had died before sending home more than a few field notes. Unknown hands shook the butterflies from his books and tapped into his brandy supply. Not a single one of Lofling's specimens survived.

. . .

Linnaeus was still mourning the loss of the Vulture a few weeks later, when a fourth apostle returned to Uppsala. Marten Kahler's account of his tribulations bordered on the fantastic, and demonstrated the dangers of not planning for the unexpected.

Despite being dispatched to South Africa all of five years earlier, Kahler had never gotten any farther than southern Europe. While sailing from Denmark, his ship had been caught in a storm so violent it flooded his cabin and swept away all of his possessions. Landing destitute on the coast of Bordeaux, he was forced to confront the fact that his charter from Linnaeus had come without contingency plans—if something unforeseen arose, there were no fallback instructions, no names of colleagues who might be enlisted to help him out.

Kahler begged and borrowed funds for another passage, only to yet again board a ship that became endangered as soon as it left port. In this case it was a pursuit by pirates, only narrowly escaped by a race to Marseille and emergency refuge there. His passage canceled, the apostle was once again stranded and broke. Not knowing what else to do, he began walking south, crossing into Italy and collecting specimens along the way. These he sent to Linnaeus via a third ship—which *was* successfully captured by pirates. The specimens never reached their destination. Kahler walked as far as Rome, where he abandoned any further hope of getting to South Africa and turned his attention to returning to Sweden.

But there remained no contingency plan. Why did Kahler not, like Hasselqvist, write to Linnaeus and inform him of his plight? Probably because Hasselqvist had been able to borrow money to live on in Syria while awaiting Linnaeus's reply. Finding no such beneficence in Rome, Kahler had begun walking home, all the way to Uppsala. It had taken him three years.

If the weary Kahler was expecting a hero's welcome, he did not receive it. Linnaeus hosted him for dinner and politely listened to his story, but otherwise seems to have mustered no sympathy or support. The former apostle had nothing to offer but a woeful tale; his very presence served only to remind the rest of Linnaeus's student circle that field assignments did not always yield glory and adventure. Kahler seems to have gotten the hint. He departed Uppsala, abandoned botany altogether, and signed up as a physician in the Swedish Royal Navy.

. . .

One month later, yet another wayward apostle unexpectedly arrived at Linnaeus's doorstep.

In 1754, a sugar plantation owner had approached Linnaeus, asking him to recommend a tutor for his children in the Dutch colony of Surinam. Jumping at the opportunity to land an apostle in South America, Linnaeus offered the services of his student Daniel Rolander, a twenty-nine-year-old from his own home province of Småland. Rolander had misgivings about taking the job—he was interested in botany and entomology, not teaching colonial children—but he'd been studying with Linnaeus for almost a decade, and his mentor was eager to push him out of the nest. After proddings and reassurances that he'd have plenty of time for field research, he agreed to go.

The assignment did not begin promisingly. Just before boarding ship in Amsterdam, Rolander suffered from an attack of fever, causing his new employer to delay the voyage for several months. But upon arrival in Surinam the following year, Rolander was a blaze of activity and discovery. He documented an abundance of novel rainforest life—anteaters, sloths, glow-worms, and magnificent butterflies—as well as one creature that had seemed mythical: the chameleon. He gleefully experimented with placing one in front of different-colored clothes, marveling at its ability to change patterns and colors at will. Rolander appeared to have years of productive work ahead of him in Surinam, with a seemingly endless supply of species to collect and describe.

But he was desperately unhappy, and increasingly on edge. His illness in Amsterdam left him worried about his health, fretting that the tropical climate would cause the fever to return. He despised the Dutch colonists, who he described as boorish, hard-drinking, and unspeakably cruel to the enslaved population. After just six months, in a fit of weariness and emotional exhaustion, Rolander impulsively decided to leave the Surinam rainforest on the next ship out. "I went to visit the forest one last time to examine its pleasantries and bid it farewell," he wrote in his journal. "I was not ignorant of the fact that the recesses of these forests held gifts from Nature; gifts that remained unseen to me at that point and would escape my last glimpses." The sudden retreat required a circuitous route through the Caribbean and across the Atlantic. By the time he arrived in the Dutch port of Texel in April of 1756, he had spent twenty months at sea and only four on land. Upon his return to Uppsala, Rolander was as penniless as Kahler before him.

Rolander was also in a fragile mental state. Upon his premature return, his attitude toward his mentor was strangely altered. Instead of paying obeisance and politely turning over his specimens, he kept them hidden away, demanding that Linnaeus either secure him an academic post or provide financial support as compensation. Unsociable and seemingly paranoid, he grew increasingly convinced of his collection's immense value. One specimen, he said, was a plant that produced a harvest of pearls. Another "might prolong human life, if not permanently, to an extraordinary length of time." Whether he was delusional or still suffering from the aftereffects of fever is undetermined, but no one disputes Linnaeus's reaction to his claims and demands.

The recent developments of the apostles—one death, two sudden returns with scant specimens—could not have come at a more stressful time. Linnaeus was grappling with an urgent crisis: Thousands of his fellow Swedes were starving to death. A series of crop failures had plunged large regions of the nation into famine, and the Swedish parliament was desperate for solutions. Linnaeus exhorted the starving to eat fir bark, nettles, acorns, Iceland moss, seaweed, burdock, polypody, bog myrtle, cherry tree resin, and thistles, but none of these could be harvested on a scale large enough to make a difference.

Linnaeus's patience was at an end. He went to Rolander's quarters and, finding him absent, broke in through a window. He emerged without the purportedly miraculous plants—Rolander was likely carrying them on his person—and with only a single specimen in hand. It was a plant he would name *sauvagesia*, a small creeping herb of minor medical significance (the natives of Surinam used it to treat diarrhea) but useless as a food stock.

Rolander soon discovered the burglary, but Linnaeus was not about to apologize. As a mark of his wrath he named only a single specimen after Rolander, a tiny beetle he dubbed *Aphanus rolandi*. In Greek, *aphanus* means obscure. The meaning was clear: Rolander was consigned to oblivion.

The former apostle moved to Stockholm, and by 1751 seemed recovered enough to be seriously considered for a teaching position there. Linnaeus used his influence to ensure that he was denied the job. Desperate, Rolander worked for a while as the gardener of Stockholm's Seraphimer Hospital, then drifted to Denmark, where he sold his specimens and his Surinam journals for a pittance (the miraculous plants turned out to be an ordinary lithosperm and a hibiscus). He died in 1793.

Linnaeus remained unrepentant of his treatment of Rolander. But the loss of Hasselqvist and Lofling, combined with Kahler's suffering, had begun to take a toll. "The death of many whom I have induced to travel have made my hair grey," he confessed to a friend. "And what have I gained? A few dried plants, with great anxiety, unrest, and care." He did not, however, refrain from anointing apostles.

Linnaeus's Homo nocturnus, Homo caudatus,
Homo sylvestris, *and* Homo troglodtyes, *1758*

EIGHTEEN

Governed by Laws, Governed by Whim

BUFFON'S CHIEF CRITIQUE OF THE LINNEAN SYSTEM—THAT IT
was inherently arbitrary—seemed to find validation in 1758, when the
system arbitrarily changed. That year, Linnaeus published the tenth
edition of his *Systema Naturae*, which listed 4,236 animal species and
completed the application of binomial names to all animals, comple-
menting his binomial naming of plants. All in all, it was "a work which
for a knowledge of nature has no equal," as Linnaeus congratulated
himself in his journal. Yet the effort had left him exhausted and depressed. "My hand is too weary to hold a pen," he confided in a letter
to a friend. "I am the child of misfortune. Had I a rope and English
courage I would long since have hanged myself. . . . I am old and grey
and worn out." He was, at the time, fifty years old.

In the first edition of *Systema Naturae*, Linnaeus had coined the term
Fauna as a companion to *Flora*. In the tenth, Linnaeus forged still more
language to encapsulate his concepts. Among the now-common
words that he coined was *Cactus*, for the order of prickly plants. He
adapted it from the Greek word *kaktos*, which at the time referred only

to a species of Spanish artichoke. *Lemur*, from the Latin word for "spirits of the dead," a reflection of the creature's nocturnal habits and unsettling stare. *Aphid*, for the genus of tiny insect: As he never explained the reason for this name, its etymology remains a mystery. Other names coined by Linnaeus in this edition did not remain solely scientific names but entered the common usage of modern language: *absinthe, azalea, amaryllis, tanager, boa constrictor*.

There were near-poetical acts of language encapsulated in the names themselves. Linnaeus, a lover of chocolate, named the cocoa plant *Theobroma cacao*, or "food of the Gods." But there was also gibberish. Linnaeus had originally used *Alcedo*, Latin for "kingfisher," for the genus of kingfishers. But as the category grew he decided to split *Alcedo* into two separate genera. To name them, he made two anagrams of the original term, labeling them *Lacedo* and *Dacelo*. The words were nonsensical but "sounded" right.

There were numerous reassignments of species, which retained their names but were shifted from one genus to another, evidence to Buffon that the system had been on shaky ground in the first place. Of the mongoose, he observed, "Monsieur Linnaeus made it first a badger, then he made it a ferret . . . others make it an otter, and others a rat; I cite these ideas . . . to put people on guard against these appellations that they call generic, and which are almost all false, or at least arbitrary, vague and equivocal."

But those were minor adjustments. Far more notable were the major revisions. Linnaeus now moved worms and insects into separate orders, and not all water dwellers were lumped into the category of "fish." The class of *Quadrupedia* (four-legged), which was now extended to include elephants, sea cows, whales, and all hoofed animals, was renamed *Mammalia*, from the Latin *mammalis*, or "pertaining to the breast."

A categorical relabeling was long overdue—whales and sea cows were indisputably not four-legged—but switching to lactation as the defining characteristic was a jarring move, and easily criticized. "A general character, such as the teat, taken to identify quadrupeds should at least belong to all quadrupeds," Buffon responded, pointing out that male horses do not have nipples (nor, for that matter, do male rats) and that the ability to produce milk was usually only a temporary condition in the lifecycle of a species. Identifying a newly discovered species as a mammal would be cumbersome—one would have to make sure they were observing a female specimen actively nurturing off-spring. But Linnaeus was only interested in using the number and position of teats as one of several grouping characteristics at his disposal. To seize on "some small relationship between the number of nipples or teeth of these animals or some slight resemblance in the form of their horns" struck Buffon as nothing less than "violence."

Yet the most consequential changes were Linnaeus's extensive renovations to the order *Anthropomorphia*. Retiring the tautology of "manlike," he renamed it *Primates*, or "of the first or highest rank." Buffon did not object to the implicit value judgment; he too had needed to tread lightly when including humans in the animal kingdom. "Man is a reasonable being, and animals are creatures without reason," he'd written in *Histoire Naturelle*, before gingerly piercing the barrier between the two: "Although the works of the Creator are in themselves all equally perfect, the animal is, according to our way of perceiving, the most complete work of Nature, and man is the masterpiece."

But within the order *Primates*, the genus *Homo* displayed dubious changes. Two years earlier, in the ninth edition of *Systema Naturae*, Linnaeus had at last applied binomial nomenclature to the human species, naming it *Homo diurnus*, or day-dwelling man. Now he renamed it *Homo sapiens*, or wise man (the description remained *know yourself*). Previously the genus *Homo* held only one species, but now Linnaeus felt compelled to give it company. While *Homo diurnus* had disappeared, there was now a *Homo nocturnus* (night-dwelling man), *Homo caudatus* (tailed man), *Homo troglodytes* (cave-dwelling man), *Homo sylvestris* (forest man), *Homo ferus* (wild man), and *Homo monstrosus*.

The crowded genus would last only for this edition. *Homo troglodytes* would be reclassified as *Simia satyrus*, the chimpanzee. *Homo sylvestris* would become *Pongo pygmaeus*, the orangutan. *Homo ferus* would be discontinued, as it represented not a separate species but incidents of humans raised in the wild. *Homo monstrosus* would likewise fade, as Linnaeus had used it to file away miscellaneous oddities, from the rumored giants of Patagonia to a reported "large-headed, horned" denizen of China. But *Homo nocturnus* and *Homo caudatus* were products of Linnaeus's own imagination.

Such errors were fleeting. Others were lasting, and profound. Linnaeus had originally said only that "man varies," sorting humanity into the vague, unspecified categories of White European, Red American, Tawny Asian, and Black African. Now he added deeply detailed—and profoundly prejudiced—descriptions:

Homo sapiens americanus
Red-colored, choleric, erect. Hair black, straight, thick; nostrils wide, face harsh; beard scanty. Stubborn, cheerful; free. Paints himself with the red lines of Daedalus.
Governed by customs.

Homo sapiens europaeus
Fair, sanguine, brawny. Hair yellow, flowing. Eyes blue. Gentle, acute, inventive. Covered by close vestments.
Governed by laws.

Homo sapiens asiaticus
Sallow, melancholic, strict. Hair blackening. Eyes dark. Severe, haughty, greedy. Covered by loose vestments.
Governed by opinions.

Homo sapiens afer
Hair black, contorted. Skin silken. Nose snub. Swollen lips. Women shameful of bosom [i.e., on display]; breasts long-lactating. Sly, slow, careless. Anoints self thickly.
Governed by whim.

Linnaeus sought to encode such sweeping characterizations as universal qualities, aspects of identity as real as the needles of an evergreen or the hooves of a horse. Later apologists have attempted to absolve Linnaeus of racism, pointing out that he did not specifically use the term *race*, nor did he declare one type superior to others. This is specious. In 1758, the word *race* was a fluidly applied collective term, demarcating any group one wished to consider as a whole—a usage that persists today in the term *the human race*. Defining one group as "governed by laws" and another as "governed by whim" could not be a more blatant assertion of superiority. And unlike *Homo nocturnus* and

Homo caudatus, these concoctions did not disappear. Their author would stand by them for the rest of his life.

Linnaeus believed many things that later admirers have chosen to ignore. He was convinced that epilepsy was caused by washing one's hair, that the liberal application of aquavit to a puppy's fur would cause it to remain a miniature dog, and that swallows slept through winter at the bottom of frozen lakes. These and other pronouncements, confidently stated but absent verification (Linnaeus, unlike Buffon, was not an experimenter), were at least confined to his minor writings. In the pages of his magnum opus, embedding equally unmoored conclusions was a spectacular act of ignorance masquerading as savantry.

It was also an act with enormous consequences. Humans' practice of objectifying other humans on the basis of physical characteristics is, of course, ancient, and the xenophobic prejudices encapsulated in what we now term *racism* date back to human prehistory. But the modern conception of *races*—along with the spurious pseudoscience of assigning innate characteristics to them—has a genealogy that can be traced directly to the pages of *Systema Naturae*.

. . .

Buffon's work was moving him in the opposite direction. As the ephemeral new members of the genus *Homo* demonstrated, Linnaeus's notions of what constituted a species were more vague than ever, but they still centered on physical appearance. In the first volume of *Histoire Naturelle*, Buffon had stated that while physical descriptions were important, "care must be taken against losing one's way in such a number of minor details . . . while considering too lightly the main and essential things." Physiological details varied with individuals, and "the history of an animal ought not to be the history of the individual, but that of the entire species," he cautioned. True natural history was "to depict things simply and clearly, without changing or oversimplifying them, and without adding anything to them from one's imagination."

He'd listed twelve factors he considered more important than appearance in defining a species. Now, after years of firsthand observation of his private menagerie, Buffon was more convinced than ever of the validity of that list:

Their generation (creation of live young)
The duration of pregnancy
Their nature of birth
Their number of young
Their care by their fathers and mothers
Their type of rearing
Their instinct
Their place of habitation
Their tricks
Their way of hunting
The services that they can render us
All the uses and commodities that we can obtain from them

Such multiple datapoints certainly helped him paint vivid verbal portraits in the pages of *Histoire Naturelle*. Yet taken as a whole, they fell short of usefully identifying a species; it was immensely impractical for field naturalists to standardize observation of all twelve. A breakthrough came when he realized that the first four qualities were interrelated: Generation, pregnancy, birth, and litter size could be seen as aspects of the same process, which he dubbed *reproduction*. It was in reproduction, then, that Buffon saw a practical species concept. "It is neither the number nor the collection of individuals who confirm the species," he wrote, "but the constant and uninterrupted succession of individuals who reproduce." Although *species* remained a "general and abstract word," it now seemed possible to approach a working definition. Not by examining a single specimen but a long chain of them, extending backward and forward in time.

> It is by comparing nature today to that of other times, and current Individuals to past Individuals, that we have taken a clear idea of what is called species; and the comparison of the number or likeness of Individuals, is only an accessory idea and often independent of the first. . . . The Donkey looks more like the Horse than the Barbet to the Greyhound, and yet the Barbet and the Greyhound are one and the same species, since they produce Individuals who can themselves produce other.

Were there not species that could interbreed? Yes, but only for a generation. Buffon had cited the horse and the donkey because, bred together, they produced a mule—a sturdy creature longer-lived than the horse and more tractable than the notoriously stubborn donkey. Yet mules are sterile: Each one must be born of a male donkey and a mare. An offspring of the opposite combination (female donkey and stallion) is possible, but rare. Called the hinny, it is equally sterile. But what of other animals? Were some capable of producing fertile hybrids, but simply did not do so because they had more compatible mates at hand?

The same year that Linnaeus published his tenth edition of *Systema Naturae*, Buffon published in the seventh volume of *Histoire Naturelle* a record of his ongoing experiments in cross-breeding, reporting on "a female wolf I kept three years, [who] although shut up alone in a large pen with a mastiff of the same age, was not able to accustom herself to living with him, nor to submit to him." He would spend years on similar experiments, enforcing the cohabitation of male and female cross-species pairs—dogs and foxes, hares and rabbits, pigs and peccaries, goats and sheep. It was only the last of these pairings that produced results: nine hybrid offspring, seven male and two female. Yet none of the goat-sheep could be bred to produce a subsequent generation. Like the mule, they were sterile.

Buffon would have had similar results had he tried interbreeding large cats. We now know that jaguars and lions can interbreed, despite the former being native to the Americas and the latter to Africa. Pumas, leopards, and tigers can also interbreed with both the jaguar and the lion, creating some very interestingly named hybrids: liger, jaguleps, pumapards. All are single-generation creatures, unable to reproduce.

Buffon had his working definition. The essence of species lay in reproduction, the ability of one generation to propagate another. He logically applied this measure to humanity: Since all ethnic groups seemed clearly capable of interbreeding with one another, they comprised a single species. "The dissimilarities are merely external, the alterations of nature but superficial," he concluded. "The Asian, European, and Negro all reproduce with equal ease with the American. There can be no greater proof that they are the issue of a single and identical stock than the facility with which they consolidate to the common stock." In sum, reproduction was proof that

there was originally but one species, which, after being multiplied and diffused over the whole surface of the earth, underwent diverse changes from the influence of the climate, food, mode of living, epidemical distempers, and the intermixture of individuals . . . that at first these alterations were less conspicuous, and confined to individuals; that afterwards, from continued action, they formed specific varieties; that these varieties have been perpetuated from generation to generation.

Buffon was not free of prejudicial assumptions. He considered it likely that the earliest humans were relatively light-skinned, concluding that "*blanc* appears to be the primitive color of Nature." But he would later amend this statement, hypothesizing that the earliest humans were dark-skinned Africans. And while the French word *blanc* is most commonly translated as *white* or *pale*, Buffon applied the term to a broad array of ethnicities. Inhabitants of the Barbary mountains in North Africa, of the northern provinces of Persia, and of the middle provinces of China—he described them all as *blanc*, making it clear he was referring to a far more diverse group than Linnaeus's *Homo sapiens europaeus*. Buffon's theory was that our species arose in a climate that the human animal finds most comfortable, namely "the most temperate climate [that] lies between the 40th and 50th degree of latitude."

To some extent this reflected a cultural bias: Those latitudes include Italy, Switzerland, and of course France. But Buffon was not pointing to those countries. He was demarcating a geophysical band stretching around the Earth—one that includes parts of North Africa and the Mediterranean Sea, Turkey, Ukraine, broad swaths of North America, most of Mongolia, the aforementioned middle provinces of China, and a generous portion of Japan. Indeed, Buffon considered Asia the likely origination point of human civilization, calling it "the ancient continent" and stating that Japanese and Chinese culture "is of a very ancient date. . . . Their early civilization may be ascribed to the fertility of the soil, the mildness of their climate, and the vicinity of the sea." In later pages of *Histoire Naturelle* he would make his point even clearer, writing that Europe "only much later received the light from the East."

Before the foundation of Rome, the happiest countries of this part of the world, such as Italy, France, and Germany, were still only peopled by men who were half-savages. . . . It is thus in the northern countries of Asia that the stem of human knowledge grew; and it was upon the trunk of the tree of science that was raised the throne of man's power.

Declaring just four varieties of humans (or subspecies; Linnaeus was never clear on the matter) was to Buffon an error of the highest magnitude. To affix inherent attributes to them was repugnant. Defining *Homo sapiens asiaticus* as "severe, haughty, greedy" was a grave disservice to our cultural forebears. *Homo sapiens afer* struck him as particularly ridiculous. Calling those of African descent "sly, slow, careless" served only to justify their exploitation, which Buffon adamantly opposed. "Their sufferings demand a tear," he wrote. "Are they not sufficiently wretched in being reduced to a state of slavery; in being obliged always to work without reaping the smallest fruits of their labour, without being abused, buffeted, and treated like brutes?"

> Humanity revolts at those oppressions. . . . How can men, in whom the smallest sentiment of humanity remains, adopt such maxims, and on such shallow foundations attempt to justify excesses to which nothing could ever have given birth but the most sordid avarice? But let us abandon those callous men, and return to our subject.

Linnaeus's coat of arms, with twinflowers and central egg

NINETEEN

A General Prototype

IN NOVEMBER OF 1761, TO COMMEMORATE THE TENTH YEAR of his reign, King Adolf Frederick of Sweden celebrated with several public acts of largesse. One of them was the bestowal of a knighthood upon Professor Linnaeus of Uppsala.

Linnaeus had already collected minor honors from the royal family. He'd been named to the Order of the Polar Star, a recent creation of the king to reward Swedish academics and civil servants, and along with several other physicians he shared the ceremonial title of *archiater*, or royal doctor (Linnaeus did not practice medicine, in court or elsewhere). But now Adolf Frederick wished to elevate him to genuine nobility as part of the *Riddarklassen*, the Class of Knights. It was a modest ennoblement, conferring no lands or seat in a deliberative body. But it did allow him to be addressed as Sir, to assume a genteel surname of his own creation, and to adopt a coat of arms.

The man who had named so many species particularly relished naming himself. After some deliberation, he shed Linnaeus for von Linné (although this text shall continue to refer to him as the former). The name was a curious hybrid. *Von*, a German prefix, was used by Swedish gentry, but only among those who could at least claim some German ancestry; Linnaeus could not. But *Linné*, a unique coinage, seemed to be at least vaguely French. Why combine allusions to two countries he had visited only fleetingly and whose languages he could not understand? No explanation is recorded. But it did have the effect of making him seem worldly, far removed from the stony fields of Småland province.

The newly minted Sir Carl von Linné also acquired property, using a cache of funds from what would prove to be his only profitable foray into scientific entrepreneurship. While studying mollusks, he'd hit upon a reliable method of seeding pearls in oysters, a method he presented to the Swedish government as a trade secret. It was not truly a trade secret—Asian countries had been cultivating pearls for centuries—and Sweden made no serious effort to enter the pearl industry. But by way of thanks a secret vote of Parliament awarded him 450 pounds, which he used to purchase Hammarby, a small farm nine miles southeast of Uppsala. As he began transforming it into a proper estate befitting a member of the *Riddarklassen*, he also addressed another privilege attached to his ennoblement: designing a heraldic coat of arms for the House of von Linné.

In his application to the king he proposed a shield with three fields of color, representing the three kingdoms (black for the mineral kingdom, red for the animal kingdom, and green for the plant kingdom), surmounted by a lone twinflower and emblazoned by a fertilized egg ("to betoken nature which is continued and perpetuated in an egg," as the application reads). For the family motto he proposed *Ex Ovo Omnia* (Everything Comes from the Egg), which struck Sweden's official heraldists as too eccentric; they rejected it and substituted *Famam Extendere Factis* (Fame Through Deeds).

The House of von Linné's coat of arms, however, retained the egg as a centerpiece. Linnaeus was uncertain of the mechanics of reproduction, but he was certain of the raw materials.

· · ·

Linnaeus genuinely believed in *ex ovo omnia*. He was a fervent ovist, which meant he held that the female egg contained everything needed to produce the next generation—the role of the male was to "excite" and trigger the process, not to contribute to it. He'd made this clear in the first edition of *Systema Naturae*, writing:

> If we observe God's works, it becomes more than sufficiently evident to everyone, that each living being is propagated from an egg and that every egg produces an offspring closely resembling the parent. Hence no new species are produced nowadays.

Ovism was just one theory. Aristotle, in *The Generation of Animals*, believed that the female's primary contribution was her internal supply of menstrual blood, which the male semen used as the material for frothing up an embryo. (This, he believed, explained why women stop menstruating when pregnant.) Yet Aristotle was at a loss to explain a crucial point: Why do children resemble their parents to various degrees? In 1651, William Harvey published *Exercitationes de Generatione Animalium*, or *On Animal Generation*, postulating that the female's entire body is "irradiated" by an "aura seminalis" that somehow began gestation; resemblance was a matter of the strength of that aura. Epicurus and others thought that the female must internally produce her own version of semen as well, and that their mixture begins the process of generation. These were different paths toward explaining why children can resemble their mothers as well as their fathers, but all were speculations.

It was not until 1669 that Jan Swammerdam, a Dutch scientist participating in early microscopic studies, formulated a theory that all animals come from an egg produced by a female of the same species. This was quite the leap, since at the time insects were widely believed to spontaneously generate, and mammalian eggs had never been observed. Still, Linnaeus and other ovists took their existence as an article of faith.

Animalculists, on the other hand, believed that it was the male contribution that mattered. And they had physical evidence to point to, since in 1677 the pioneering microscopist Antonie van Leeuwenhoek had discovered "animalcules" in the semen of humans and other

mammals, which we now call spermatozoa. The discovery led to
Leeuwenhoek's claim (again, not backed up by observational evidence) that "it is exclusively the male semen that forms the foetus and
that all the woman may contribute only serves to receive the semen
and feed it." In 1694, Nicolaas Hartsoeker went so far as to claim that
animalcules contained "homunculi," tiny but fully formed humans requiring the mother only for incubation.

Buffon considered ovist and animalculists equally ridiculous. Despite their different assumptions, each took as a given that life was
seeded—that it was contained and conveyed in a minuscule unit generally called the "germ," which the sexual act merely activated. But
this posed a logistical problem: Since the newly emerged lifeform also
gave rise to future generations, themselves germinated, did that mean
the germ itself contained even smaller germs? Did a chicken emerge
from the eggshell somehow containing the seeds of all chickens that
will descend from it in the future? Buffon described such theories of
reproduction as "that which suppose the thing already done." To believe in such a concept was "not only to admit that one does not know
how it happens, [but to] abandon the will to think about it." He believed there were no series of nesting minuscules but "an ever-active
organic matter, always given over to shaping itself, to making itself,
and to producing beings similar to those which house it."

What was this force, or process, or principle? To delve further into
the question he imported the British microscopist John Needham
and set up a laboratory in Montbard with the most sophisticated instruments then available. "If we do not succeed in explaining the
mechanism by which Nature accomplishes reproduction," he wrote,
"we shall at least arrive at something that has far more appearance of
truth than anything that has thus far been proposed." Together, Buffon and Needham examined the reproductive tracts of various animals drawn from Montbard's menagerie, and after several months of
study they announced a momentous discovery: They had observed a
female "spermatic fluid" in the ovaries of a dog.

They were, of course, wrong. It is unclear what they had actually
witnessed, but it was not female sperm. Their confusion is understandable: Even the best microscopes of the era were primitive by
modern standards (the true choreography of fertilization, the inter-

action of sperm and egg, would not be discovered until 1875). Still, the mistaken impression led Buffon in an important direction. Paired seminal fluids implied an intermingling, a joint contribution to the reproductive process by male and female. This made him lean in favor of epigenesis, the theory that the "recipe," so to speak, for creating a fetus was comprised of an interplay of instructions between the two genders. Epigenesis would explain why a child might have their father's nose and their mother's eyes, and why a sibling's features might be vice versa. It would also explain why some offspring were stillborn and others greatly deformed: The interplay of instructions had somehow gone wrong.

What was the nature of these instructions? While they might be effectively invisible, that was no bar to studying them as a force. After all, no one had actually "seen" electricity, but most people had observed means by which it could be processed, transmitted, and stored. Gravity was both omnipresent and unobservable, yet its laws could be discerned. "There is in Nature a general prototype in each species on which each individual is modeled," Buffon wrote, "but which seems, in realizing itself, to alter itself or perfect itself according to circumstances."

> So that, relative to certain qualities, this is an extraordinary appearing variation in the succession of these individuals, and at the same time a constancy which appears wonderful in the entire species. The first animal, the first horse, for example, has been the external mode of the moule intérieur, on which all horses which have been born, all those which now exist, and all those which will be born have been formed.

The *moule intérieur* was Buffon's term for nature's general prototype, the mechanism molding, or giving form to, organic matter and dictating reproduction. While its literal meaning is "internal mold," Buffon did not intend for it to be taken literally: A mold, by definition, cannot be on the inside of a form. Contemporary translators have concluded it might best be rendered in English as the "internal matrix."

This matrix was something he could only point to, not discover—the technology of the day was far too limited. But biologists would later recognize this as the first rough sketch of a working theory of reproduction. "If a physician were to attempt today to explain the functioning of the genes and the formation of the zygote in a highly simplified way to an intelligent child," science historians Otis Fellows and Stephen Milliken concluded two centuries later, "one totally ignorant of even the most basic concepts of chemistry, with no knowledge whatsoever of the structure of the cell, and with only the fuzziest notions of microorganisms, his explanation would inevitably bear a strong resemblance to Buffon's system."

Discovering the internal matrix, or even sounding its workings in greater detail, were tasks Buffon reluctantly left to future generations. Documenting the quadrupeds continued to consume every moment of his working hours. Published in 1760, the eighth volume of the *Histoire* covered sixteen species, from the guinea pig to the agouti. Especially vivid was his excursus on the beaver, a living specimen of which he'd exported, at great cost, from Canada to the Jardin du Roi. "One day he escaped and descended by a cellar stair into the caverns of the old quarries that are under the Jardin," Buffon wrote "He fled a long way, swimming in the rank water on the bottom of the caverns; nevertheless, as soon as he saw the light of the torches that I had brought there in search for him, he came."

> He is friendly without being affectionate or caressing, his requests are made through a little plaintive cry or a few gestures. . . . He gnaws away at everything that comes in his way, cushions, furniture, or trees. The barrel in which he was shipped had to be reinforced with sheets of tin.

Buffon wrote on. Volume nine, devoted to the great cats (lion, tiger, panther, cougar, leopard), the hyena, the civet, and the black wolf, arrived in 1761. Volume ten (peccaries, opossums, pangolins, marmosets) followed in 1762.

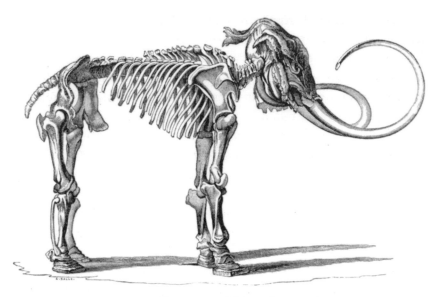

The first fully reconstructed skeleton of a mammoth

TWENTY

Breaking the Lens

LINNAEUS SEMI-RETIRED IN 1763, SUCCESSFULLY PETITION-
ing to be excused from his university teaching duties—although, in
the tradition of Uppsala professors before him, he continued to retain
both the title and the salary. The move was intended to anoint as successor his son, Carl Junior, now twenty-two years old, who was appointed adjunct professor of botany with the understanding that he'd
become a full professor after his father's death. There was grumbling
about the arrangement. It was an act of blatant nepotism, perhaps
excusable for a lesser academic figure but not for someone at the center of a circle of brilliant disciples. The junior Linnaeus had been
groomed for such a succession since childhood, but it was strikingly
clear that the elevation was at best premature. He had taken no exams,
defended no thesis, and obtained no degree, nor had he demonstrated
any passion for the profession. As Linnaeus himself privately admitted, his son "never helped me in botany, and has no love for it." The

witticism whispered around the school was that young Carl, a convivial soul who enjoyed female companionship, was "inquiring less after Flora than after the nymphs."

There was an ample supply of true botanical zeal, even brilliance, in the von Linné family, but it didn't belong to Carl the Younger. While Linnaeus had taken great pains to educate his only son, he'd actively discouraged the intellectual development of his four daughters, to a degree unusual even by the misogynistic standards of the day. A university professor's daughter was at least expected to refine her character by learning French, mastering etiquette, and acquiring the smattering of knowledge necessary for polite conversation. Linnaeus forbade his daughters, Elisabeth Christina, Lovisa, Sara Christina, and Sophia, from even attending school. The girls were destined for domesticity, he concluded, and as such their instruction need not extend beyond the household. "Under those disadvantages, the education of the children of Linnaeus could not but be of an inferior description," wrote one eyewitness to their upbringing. "The young ladies, his daughters, are all good-tempered, but rough children of nature, and deprived of those external accomplishments which they might have derived from a better education."

Carl and Elisabeth Christina von Linné

Yet their household was on the grounds of a botanical garden. Despite a childhood circumscribed by the woodstove and the washtub, the eldest daughter, Elisabeth Christina, privately nurtured her curiosity about the garden and her father's work. She overheard his conversations with the naturalists who came to call, and made acquaintances of some of the younger ones. At the age of nineteen, she wrote a paper on the curious optical phenomenon of *Tropaeolum majus*, or Indian cress; she'd noticed that at twilight, under the right circumstances, the cress seemed to send out luminous flashes, not unlike small lightning bursts. Someone other than Linnaeus read the paper, and thought highly enough of it to submit it to the Swedish Royal Academy of Sciences. They duly published *Om Indiska Krassens Blickande*, earning Elisabeth Christina her father's ire and a place in the history books as Sweden's first female botanist. The mystery of sparking cresses, now known as the Elizabeth Linnaeus Phenomenon, was solved in 1914, when Professor F.A.W. Thomas of Germany proved it was not residual phosphorus (as some believed) but an optical illusion caused by the anatomy of the human eye. At twilight our vision shifts color receptors, causing for a brief moment the perception of brightness in the particular colors of *Tropaeolum majus*'s petals.

Elisabeth also fell in love with one of her father's students, Daniel Solander (not to be confused with Daniel Rolander, the shunned apostle of Surinam), who returned her affections and proposed marriage. Linnaeus rejected the notion out of hand. The thwarted suitor became one of his apostles instead, departing for England and signing on for a naval expedition, but Elisabeth Christina continued to hold sway in his heart. Solander wrote to Linnaeus from Rio de Janeiro, requesting to be remembered to "your eldest daughter, whom I had hoped would make me happy."

Elisabeth Christina was instead married off quickly to Major Carl Fredrik Bergencrantz, a Swedish soldier. The match was a miserable one. Bergencrantz abused her so severely that she fled the marriage two years later, returning to the Linnaeus family household with two young children in tow (named Carl and Sara, after her parents). Traumatized and broken in health by the ordeal, she would die at the age of thirty-nine. Neither of her children lived to adulthood.

. . .

Johann Gustav Acrel was just seventeen when he arrived at Uppsala in 1758, eager to study medicine at the feet of the great professor Linnaeus, but obligatory military service interrupted his education two years later. After completing his tour of duty in 1765, Acrel returned to Uppsala, only to find his old professor visibly diminished.

Acrel had been aware of Linnaeus's health issues, remembering that "at fifty he began to shuffle his feet forward in place of lifting them." But his mentor's infirmities were now clearly mental as well as physical. As a younger man, Linnaeus had been celebrated for lecturing extemporaneously for hours on end, without once referring to notes. Now he seemed hardly able to stand for such durations, much less retain the thread of his thoughts, and he taught astride what he called his study horse, a lectern with an integrated seat designed to be straddled like a saddle. At his side stood dutiful students, feeding him the notes from which he read. "Even when I was with him, he could not remember the names of his dearest friends and relatives," one of those students recalled. "I recollect an occasion when he was terribly embarrassed because, having written a letter to his father-in-law, he could not remember his name." At the time, Linnaeus was fifty-six.

Buffon, also fifty-six, remained vigorous and seemingly at the height of his powers, a fact he attributed to his continued regimen of sparse diet, physical exercise, and regular hours, the latter still enforced by his faithful manservant. "I owe Joseph three or four volumes of *Histoire Naturelle*," he quipped. Now at the pinnacle of his fame, he was presently serving as both the permanent treasurer of the Académie des Sciences and the director of the Académie Française. Other naturalists began to wonder why he did not simply compromise with Linnaeus, consolidate their work, and occupy a point at the pinnacle of their profession. "How miserably Monsieur Buffon falls foul of him every now and then!" the American botanist Alexander Garden wrote of Linnaeus and his most prominent opponent, attributing Buffon's use of indigenous species names to pure spite.

> His affect of retention of the Iroquois and Brasil names, merely out of opposition of Linnaeus, is ridiculous and puerile. I admire his general dissertations very much; and on many animals he is full, clear, decisive and satisfactory. He writes in an easy, agreeable, lively style, and is often truly a painter. . . . I am vexed at him for snarling so at Linnaeus. Plague on it, why cannot they agree?

It is understandable that Garden, an ardent follower of Linnaeus (who would name the flowering plant *gardenia* in his honor), wished to frame the conflict as a reconcilable squabble, but in truth the gap between the two transcended mere catalogue matters. Linnaeus remained convinced that life on Earth was essentially unchanged from the date of God's creation. "We count so many species as there were in the beginning," Linnaeus had written in the introduction to the first edition of *Systema Naturae*. Since life presented a conveniently static target, both had no doubt it could ultimately be catalogued. In a 1763 lecture, Linnaeus confidently announced "the whole sum of the species of living creatures will amount to 40,000," breaking it down thusly:

20,000 species of vegetables
3,000 of worms
12,000 of insects
200 of amphibious animals
2,600 of fishes
2,000 of birds
200 of quadrupeds

Even forty thousand species struck Linnaeus as an unnecessary complexity. To his students, he wondered out loud why God, "who has created everything in the most simple and wise way," had not created man as worms and "the globe as a cheese, which we worms could have gnawed." Since Nature existed solely for the benefit of humanity, surely matters could have been streamlined. Still, he felt certain of his estimate: The Bible, after all, stated that Adam had known the names of all the animals, and even the first man would have had difficulty fitting more than several thousand names in his head. There was also, of course, the great filter of the Ark and its built-in limitations.

Deciding on quantity did not, however, solve the conundrum of diversity: Why were some forms of life clearly suited for their environment? Why, for instance, were polar bears wonderfully adapted for the arctic, and camels comfortable in the desert? Linnaeus theorized that the Garden of Eden was, at the time of Creation, the only landmass on Earth: an island located somewhere around the equator, surrounded by water. It was a special island, crafted to contain every single climate and habitat found today: rainforests, swamps, dry deserts, and snowy peaks. God had populated each section with suitable organisms. Because they were confined to the island that population was concentrated, and Adam was able to wander at will and get to know them all. Only afterward did the waters recede, revealing more landmasses, and life migrated away.

. . .

In contrast, Buffon was beginning to suspect that Earth's inventory held far more than forty thousand species—and that the planet's history held more species still. Like Linnaeus, he'd begun his work conceiving of life as a static subject, writing in the introduction to *Histoire Naturelle* that "we shall find, through all of Nature, that what can be, is." But he was now revisiting that notion. The Cabinet du Roi held numerous fossils that appeared to be bones of extremely large animals. While the term *dinosaur* would not be coined until 1824, these and a number of other large-scale fossil discoveries seemed to point to species that no longer existed—or that, if they did, were inexplicably good at hiding.

Take the case of the Cornwall Femur, discovered in 1676 in a limestone quarry in Cornwall, England. While some speculated it belonged to a biblical-era giant, Linnaeus was adamant that it was an interestingly shaped rock, just as all so-called fossils were simply curious rock formations. He scoffed at those who yielded to the "temptation" of classifying them; as he'd explained in the mineral schema of his *Systema Naturae,* such "petrifications" were purely part of the mineral kingdom, shaped into animal forms either by accident or unknown divine purpose.

Buffon could not guess that that Cornwall Femur would one day be identified as belonging to a megalosaurus, but he believed it to be no more the bone of a human giant than the similarly labeled "Giant's Bone" remnant on display in the Cabinet (which would prove to belong to a giraffe). He was cataloguing newer shipments of multiple bones from a recently discovered creature, called an *incognitum* or *mammoth*, which appeared to resemble an elephant. The first of these curiosities arrived at the Jardin in 1738, after being discovered by French soldiers in what is now Kentucky and taken by barge down the river to New Orleans.

Kentucky being very far away from Africa and Asia, where elephants made their home, Buffon concluded that they were of a different species entirely, and that no living specimens remained: They were extinguished, or extinct. "The prodigious mammoth no longer exists anywhere," he wrote. "This species was certainly the first, the largest, the strongest of the quadrupeds: Since this species has disappeared, how many others, smaller, weaker, and less noteworthy, must have perished without having left us evidence or information on their past existence."

Asserting extinction as fact was controversy enough—again, orthodoxy held that the death of an entire species would imply a flaw, an error, in God's perfect creation. But Buffon pressed on, insulating what he knew to be dangerous ideas by making them islands of text in a sea of general observations about ospreys and otters. As he put it in a private letter to a friend, "One can slip ideas into a quarto volume which would cause a public outcry if they appeared in a pamphlet."

Having addressed the disappearance of species, he turned to the far more incendiary subject of their appearance. The sheer diversity of life outraced his pen, and even his imagination; Linnaeus's imaginary island could not account for it, nor could any variation on the theme of a singular act of creation. Just as species passed *from* existence, he concluded that they must come *into* existence as well, throughout the expanse of time. In 1753, the fourth volume of *Histoire Naturelle* had contained Buffon's observation that while humans and horses were greatly dissimilar from each other in outward appearances, the horse's hoof contained the same inventory of bones as the human hand. This was a "hidden resemblance" that evaded Linnaeus's systematics entirely, and which to Buffon "seems to indicate that in creating these animals the supreme Being wished to employ one idea, and to vary it at the same time in all possible ways."

Buffon thought that these resemblances were more than coincidences. If his postulated shaping internal matrix (*moule intérieur*) could account for individual variations within species, might it not also be responsible for even larger, inheritable changes? Such changes might accumulate, to the degree that they added up to an entirely new species. In the eleventh volume of *Histoire Naturelle*, published in 1764, he began to consider a distinction between *genuine* species and *derived* ones.

It is here we begin to enter a thicket of significant linguistic differences between Buffon's era and ours. He did not mean "genuine" to imply that some species were more real than others, but instead deployed it in its original sense of "natural, not acquired" (it is from the Latin *genuinus*, or innate). Genuine species were ones whose internal shaping matrix had remained relatively unchanged, allowing them to reproduce along original lines. Minor differences within the matrix could account for variety within species—hair color, eye color, et cetera—but it was also possible that these small differences could accrue over time. At some point, those accrued differences could add up to a "derived" species, bearing certain resemblances but distinctively apart. Where was the point of demarcation, when the genuine and the derived were separate species? When they could no longer interbreed.

Having established the *how* of species change, Buffon then tackled the *why*. Species, he believed, changed not by whim but through adaptation to their environment. "If we consider each species in the different climates which it inhabits," he wrote, "we shall find perceptible varieties as regards size and form: they all derive an impress to a greater or less extent from the climate in which they live." He attributed this to

> that older [process] existing from time immemorial, which seems to have occurred in each family . . . the elephant, the rhinoceros, the hippopotamus, the giraffe form genera or simple species. . . . All the others seem to make up families in which one ordinarily notices a principal and common ancestor, from which different stems seem to have come.

Linnaeus had freely used the term *genus*, and its plural *genera* (Latin for "family" and "families"), but he and other naturalists of his stripe had never intended to imply that such groupings were *literally* related, sharing descent from a common ancestor. Family resemblances were only resemblances, the Creator's version of theme and variations. But Buffon began to treat the term as literally accurate: the branches of a family tree. Furthermore, he found the audacity to take this line of reasoning to the very root of the tree, musing that perhaps even "genuine" species were an illusion—that all species could be traced back to a common ancestor.

> If the point were once gained that among animals and vegetables there had been, I do not say several species, but even a single one, which had been produced in the course of direct descent from another species; if for example it could be once shown that the ass was but a biological departure from the horse—then there is no further limit to be set to the power of nature, and we should not be wrong in supposing that she knew how to draw through time all other organized forms from one primordial type.

If that passage strikes the reader as a succinct definition of evolution, that's because it is—in fact, later English translations render the phrase "to draw through time" (*tirer a travers le temps*) simply as "to evolve." But in 1765 the term *evolution* had no such meaning. It meant only "to unroll," as in the unrolling of a scroll.

It was also a very dangerous idea, as Buffon well knew. He promptly and theatrically disclaimed it, writing in the very next line:

> But no! It is certain from revelation that all animals have alike been favored with the grace of an act of direct creation and that the first pair of every species issued full formed from the hands of the Creator.

Trusting this blatant rhetorical flourish to appease the censors, Buffon experimented with giving his concept a name—"drawing through time" was a tad unwieldy. Here the linguistic thicket deepens. Buffon tried out several terms, including *dénaturation* and *dégradation*, but the one he ended up using the most was *dégénération*, most likely because it seemed to underscore that species change was a force apart from *generation* itself. As the biologist Stephen Jay Gould would later make clear, this *dégénération* was

> not intending the modern sense of deterioration, but the literal meaning of departure from an initial form . . . at the creation (*de* means "from," and *genus* refers to our original stock).

Nor was the word original to Buffon: He was adapting Leibniz, who used it in the same context as early as 1718. The difference between *dégénération* and *degeneration* is profound, but it is difficult to avoid superimposing one meaning on top of another. A less confusing and more accurate translation is *exogeneration*, employing the Greek *exō* ("outside of") to capture the original sense of the term.

It is tempting to simply substitute the word *evolution*, as others have, but that is perhaps over-freighting Buffon. He floated the concept but did not delve into the theoretical principles by which it might operate. It is enough to say that it represents Buffon's most forceful attempt to break the lens of fixity. Buffonian exogeneration was a broad-scale assertion that life transformed itself, appearing in new guises while disappearing in others across a vast expanse of years. "These changes are only made slowly and imperceptibly," he reminded the reader. "Nature's great workman is Time."

> He marches ever with an even pace, and does nothing by leaps and bounds, but by degrees, gradations, and succession he does all things; and the changes which he works—at first imperceptible— become little by little perceptible, and show themselves eventually in results about which there can be no mistake.

Linnaeus's final portrait

TWENTY-ONE

My Cold Years

FREED FROM HIS FORMAL TEACHING DUTIES, LINNAEUS MARshaled sufficient strength and focus to complete a new edition of *Systema Naturae* in 1765. It would be his last. He labeled it the twelfth edition, despite it being only the fifth one prepared by Linnaeus himself—he had the habit of counting other authorized versions, published outside of Sweden, as editions in their own right. The work had grown from the 12-page original to a shelf-swallowing compendium of 2,400 pages, each in the folio format of fifteen inches tall and one foot wide. The animal kingdom, which readers of the first edition could take in at a glance, now commanded 1,504 pages. The plant kingdom, still adhering to Linnaeus's admittedly artificial sexual system, occupied 878 pages of the subvolumes *Species Plantarum* and *Genera Plantarum*. The mineral kingdom, now a neglected afterthought, weighed in at a relatively vestigial 118 pages. The multiple volumes were massive enough to warrant printing in stages, spread out over a three-year period.

The edition included the final tweaks to the Linnean hierarchy and a few new coinages that would enter general use. He included sponges for the first time, creating the class *Zoophyta*, which means simply "a plant-like animal." It was a full-circle gesture: Historically the term had been used to refer to some of the mythical creatures he'd consigned to the now-disappeared order *Paradoxa*, including the barnacle tree and the vegetable lamb. Linnaeus debuted the term *larva* to refer to immature insects, a word that in Latin meant both "ghost" and "mask." Did he intend it in the sense that a larval form "masks" the adult, or in the sense that the pale, wriggling form was reminiscent of a ghost? He never explained.

· · ·

Another lasting consequence of the final edition of *Systema Naturae* was its near-complete avoidance of microscopic life. The dictum that "classification and name-giving will be the foundation of our science" enforced a prejudice against analyzing the invisible, or even the barely visible. If something could not be seen well enough to be described and taxonomized, it scarcely existed at all.

The notion that independent organisms existed beneath the level of the visible—and that such organisms interact, often consequentially, with larger life—was an ancient concept. In 37 B.C., the scholar Marcus Terentiaus Varro wrote in *Res Rustica* that "there are bred certain minute creatures which cannot be seen by the eyes, which float in the air and enter the body through the mouth and nose and cause serious diseases." These were confirmed in the 1670s, when a Dutch fabric seller, looking to improve upon the magnifying glasses he used to examine cloth, constructed a crude single-lens microscope and entered the invisible world. The fabric seller (the aforementioned Antonie van Leeuwenhoek) looked with wonder upon the structural components of complex life—the flow of blood vessels in capillaries, the contractions of muscle fibers. But he also noticed what he called *dierkins* ("small animals"), minute lifeforms that seemed to exist on their own.

This was startling enough, but further microscopic investigations by his English contemporary Robert Hooke unveiled what seemed like a paradox. Beginning with a cross-section of cork and moving on to more complex specimens, Hooke discovered discrete units he called "cells," from the Latin *cella*, or small room. Even simple organisms seemed to have countless thousands of these cells, while Leeuwenhoek's *dierkins* (savants were now calling them "animalcules") were clearly comprised of a single cell.

If life required the interplay of multiple cells, how could a single cell be alive? Yet unicellular organisms clearly existed, and as such needed to be accounted for in the overall tally of life. Buffon, an enthusiastic microscopist who purchased the most advanced instruments of his time, announced that he would add a volume to *Histoire Naturelle* specifically dedicated to them (it was never realized). He also suspected that they were not all animals, or plants for that matter, but instances of what he had predicted in his earliest critiques of Linnaeus: "We will find a great number of indeterminate species and in-between objects which we do not know where to place and which necessarily upset the project for a general system." It would be another century before the term *animalcule* was replaced with the coinage *microbe*, reflecting the truth of his observation. Although he did not anoint the cell as the primary building block of all organisms (that realization came in 1838, with the advent of even better microscopes), Buffon postulated strongly in that direction.

> We are led to conclude that there exists in nature an infinity of organic particles, of the same substances with organized beings . . . as the accumulation perhaps of millions of cubes are necessary to the formation of a single grain of sea salt that is perceptible by our senses, an equal number of similar organic particles are requisite to produce one of those numberless germs contained in an elm.

He further proposed that studying the simplest lifeforms could be a key to understanding life as a whole:

> That these small organized beings [entities] are composed of living organic particles, which are common both to animals and veg-

etables, and are their primary and incorruptible elements; that an assemblage of these particles constitutes an animal or a plant; and, consequently, that reproduction or generation is nothing but a change of form, effected solely by the addition of similar particles; and the death, or resolution of organized bodies, is only a separation of the same particles.

Linnaeus, meanwhile, wanted nothing to do with the microscopic. He shunted them into the kingdom *Animalia*, the order *Vermes* (worms), a genus he named *Chaos*, and for good measure into a single species epithet that bore the same name: *Chaos chaos*.

As made clear by other entries, his powers of description were rapidly failing. He had doubled the number of listed birds, and added 102 new species of insects and crustaceans. Yet many were so vaguely demarcated that subsequent generations of scientists have failed to determine which species, exactly, Linnaeus was talking about. These have collectively been consigned the label *nomen dubium*, dubious names. Despite their quasi-existence, the names he gave them cannot be re-used.

Linnaeus also fell victim to at least one hoax. When he received two specimens of a brightly patterned butterfly with predominantly yellow wings, purportedly discovered by one William Charlton, Linnaeus named the species *Papilio ecclipsis* and included it in the twelfth edition, as well as in a supplementary work entitled *Centuria Insectorum Rariorum*. But both specimens were relatively crude forgeries, created by daubing black and blue inks onto *Gonepteryx rhamni*, the common brimstone butterfly. The deception was obvious enough to

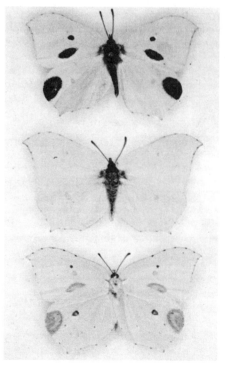

Papilio rhamni *from Linnaeus's collection (center),* with re-creations of the forged Papilio ecclipsis

later examiners: In 1795, a curator indignantly stamped both specimens to pieces.

One change in the text went practically unnoticed. For the first
time, Linnaeus omitted the introductory passage asserting that "hence
no new species are produced nowadays." In the subvolume *Genera plantarum* he appeared to soften his position on the fixity of species, vaguely
acknowledging that contemporary plants may be variations on Edenic
originals. The strongest framing of this line of thought survives in a
handwritten note, scribbled in a manuscript margin.

> First earth little, for only one plant of each species. . . . God created
> classes, from their mixture orders, from the orders genera, from
> the genera species.

It was a soft, vague speculation, one that he would pursue no further. But in an unpublished aside written at the same time, Linnaeus
allowed himself a brief unguarded moment. He would have been willing to believe in species change, he confessed, "if Holy Scripture
would allow."

. . .

In December of 1772, Linnaeus moved from semi-retirement to full
retirement, although he still planned to lecture privately. This required a ceremony in Uppsala Cathedral, capped by a speech by the
departing professor. He made a few remarks and then abruptly cut
himself short, declaring, "But this cold season, this cold cathedral, my
cold years and your patience, gentlemen, which is beginning to cool,
command me to desist." It was his last public address.

"Many things have happened to warn me that my time is nearly up
and that my fate will be a stroke," Linnaeus confided to a friend. "I get
giddy, especially when I bend down, and stumble like a drunken
man. . . . Perhaps God will take me in time to spare me the misery that
must inevitably lie ahead."

Linnaeus at last gave up on a project that dated back to 1747, when
he'd instructed his first apostle, Christopher Tarnstrom, to bring back
a living specimen of *Camillia sinsesis*, the black tea plant that formed a
cornerstone of export economies in China and India. Tarnstrom had

died before reaching China, but Linnaeus had persisted. Between 1745 and 1751, he'd dispatched five apostles to the East, each with instructions to bring back a tea plant. None of them succeeded. In 1757, Linnaeus ecstatically received what were billed as two living tea plants from the Swedish East India Company. He tended them carefully, but when they flowered in spring he realized they were not tea bushes, and not even native to China. (Linnaeus denounced this as "the treachery of the Chinese," and an intentional substitution.)

The quest had continued. In 1760, exactly ten tea seeds arrived from England. They were far from fresh, and failed to grow. In 1761, the Swedish East India Company informed Linnaeus they'd transported a live tea bush as far as Germany—where it had been nibbled to death by the ship's rats. Finally, in 1763, the company wrote Linnaeus that a third batch of tea plants was on the way. "Is it possible?" Linnaeus replied. "For all that is holy and famous in the world, treat them with the most tender care."

The plants were dead when they arrived. Linnaeus was inconsolable, but he cheered up when he learned that the ship's captain was still hanging on to ten additional plants. These, carefully transported to Uppsala by the captain's wife in a wagon, arrived alive. Two of them survived for two years, allowing him to cautiously pluck a few leaves and later claim that "tea was first seen away from China in the Uppsala Garden." But the plants did not survive the next Swedish winter. Linnaeus turned to devising "substitutes" for tea: sloe berries, bog myrtle, even his beloved *Linnea borealis*. None caught on.

Another goal remained elusive: a truly natural botanical system to replace his sexual one. He tried several new organizational metaphors, and wrestled with finding new patterns in the relationships between the reproductive parts of flowers—the calyx, corolla, pericarp, pistil, seed, and stamen. But "notwithstanding the great extent of his exertions, those productions remained fragments," one student reported. "Many plants still are left, to which he could not assign a place." Neither could he improve upon his classifications of animals. "Linnaeus himself was very sensible that his system of the animal reign was not built upon so safe a foundation as his botany," another student wrote, "and that his generical characters were far more tottering and more undefined."

. . .

In what would prove to be his last burst of ambition, Linnaeus worked on creating a systematics of medicine, hoping to build upon his 1759 publication *Genera Morborum* (Families of Diseases). *Morborum* had been an attempted taxonomy of illness, cataloguing 11 classes, 37 orders, and 325 species of maladies. Those causing fevers in the blood and brain belonged to the class of *Febriles*; those affecting temperament fell into the class of *Morbi*, further subdivided by manifestation. Depression, mania, and delirium were lumped into the species *Mentales idealis*. Hypochondria was *Mentales imaginarii*. Rabies, excessive thirst, bulimia, and pronounced sexuality were all *Mentalis pathetici*.

With that squared away, in 1766 he published *Clavis Medicinae Duplex* (The Two Keys of Medicine), a thirty-page treatise intended to link such classifications with their treatments. The "two keys" referred to what Linnaeus considered a crowning insight. "Diseases are CURED by diseases, opposites by opposites," he wrote (emphasis his). "The main principle is that of the antithesis of opposites." Therefore a patient diagnosed with *Mentalis pathetici* due to an unseemly amount of sexual desire should be treated with "rank-smelling" items—lizard orchid, stinking goosefoot, toadflax—on the basis that people seldom wish to copulate while in the throes of nausea. Conversely, a nauseous patient should be infused with "ambrosiacs" such as ambergris, musk, and civet.

It is difficult to find an opposite for every disease, which is likely one of the reasons *Clavis Medicinae Duplex* is only thirty pages long. Nevertheless, a rave review soon appeared in the Swedish journal *Larda Tidningar*, praising "our learned and industrious author."

> We may justifiably assert that no one who has studied medicine, pharmacy or surgery can do with out it; indeed that it cannot be but of use and pleasure to the most learned medical men.

The review was anonymously written by Linnaeus himself.

It is around this time that Linnaeus began compiling his strangest book, a semi-mystical work entitled *Nemesis Divina*. In its pages he struggled to settle old scores, to divine meaning out of life, and, characteristically, to impose a system on philosophy itself. The manuscript, unpublished during his lifetime, is largely a collection of aphorisms intermixed with anecdotes that seek to demonstrate a symmetry in action and retribution—a taxonomy of morality, so to speak. These are punctuated with interludes of nihilism.

> This life? This world? Wars, pestilences, conflagrations, banditry, thieves, poisons, shipwrecks, diseases, injuries, extreme weather, tyrants, bitter desires for love objects, slavery, universal taxes, lusts, envies, every manner of enemy. The war of all against all.

The passages are reflective of a strained mind, seeking to find patterns everywhere. In one of the anecdotes he states that his old opponent Siegesbeck had the misfortune of witnessing his son's suicide, implying that this was karmic retribution of some sort. He also strained to see portents in the departures of his ill-fated apostles, noting that both Lofling and another apostle stuttered upon bidding him farewell. The manuscript is disjointed, reaching for unities never quite grasped. But it is painfully earnest in its tone, and as such a window into Linnaeus's mind at the time.

> Daily we die; the passage of any hour you choose subtracts a portion of life. You have already died a considerable portion of your death.

> Everyone unfortunate would have been lucky, had they only died sooner.

> Death is nature's law.

· · ·

Death is nature's law. By now Linnaeus was weary of grieving his apostles. There would be seventeen, all told. Half of them would never return home. These included:

Carl Fredrik Adler, whose medical training allowed him to sign on as a ship's doctor on vessels bound for East Asia. He got as far as Java, skirting pirates and amassing what he tantalizingly reported as a large bounty of specimens, but died there in 1761 while preparing to return home. Hoping Adler had shipped off his collection before dying, Linnaeus spent the next several years waiting fruitlessly for its arrival.

Johan Peter Falck, the apostle who traveled the farthest on land. In 1760, Linnaeus had assigned him to Siberia and other far-eastern regions of Russia, a near-blank botanical slate at the time. But it was a long journey there, made longer still by Falck's need to pay his own way. It took him until 1763 to reach St. Petersburg, and until 1768 to move farther east. Passing through Kazan in March of 1774—after traveling for fourteen years over an increasingly bleak landscape— Falck exchanged pleasantries with other travelers, wandered away from the encampment, and died by suicide. As his papers revealed, he'd failed to discover a single new species. His rambling field notes include elaborate descriptions of *Avena sesquitertia*, ordinary wheat.

Andreas Berlin, a Swedish farmer's son who'd impressed Linnaeus with an imaginative thesis on the usefulness of mosses. Dispatched to the tiny Banana Islands off West Africa in 1773, an area with the highest concentration of malaria in the world, he managed to send only a few specimens to Linnaeus before dying—predictably, of malaria—on the Iles de Los in Guinea.

At least two apostles were still afield, their fates unknown. It wasn't hard to understand why Jacob Wallenberg, a prominent naturalist and author, had jokingly implored Linnaeus not to make him an apostle:

> I must kneel for his majesty of the kingdom of plants, duke over crocodiles and mermaids, etc., and lord of the quadrupeds, birds, and insects, our great knight Linnaeus, asking most humbly to be free from these stony excursions.

• • •

Shunning social occasions and neglecting his appearance, Linnaeus drifted into eccentricity. One visitor could not contain his disappointment upon meeting the great man, describing him as a "somewhat aged, not large man with dusty shoes and stockings, markedly unshaven and dressed in an old green coat from which dangles a medal." The latter was his decoration as a Knight of the Polar Star. His insistence on affixing it to an increasingly threadbare garment was one of the reasons people began to call him *gubbe* behind his back. It is a Swedish term, meaning both "old man" and "blunder."

One eccentricity concerned Buffon: Linnaeus began to believe that he had decisively vanquished his old rival, forcing him to endure a public humiliation. In 1774 he wrote that "Buffon, the antagonist of Linnaeus, was compelled, like it or not, to have the plants in the Jardin du Roi arranged after [my] system, since they have been so arranged by the Kings of France and England and in most Gardens in Europe." Nothing could be further from the truth—Buffon was not forced to replant the Jardin, and he had always included Linnean names on the plant labels anyway. In a letter written that same year, Linnaeus dismissed Buffon's work as "without pretty figures; wordy descriptions; oratorice; beautiful ornate French; much anatomy with skeletons; without any method; criticizes everyone, but forgets to criticize himself, although he himself has erred the most. Hater of all methods."

Yet Linnaeus also offered begrudging praise. "Buffon did not extend the boundaries of science," he wrote, "but he knew how to make it popular; and that too is a way of serving it to advantage."

• • •

Another tendency began to emerge as Linnaeus entered his sixty-first year, one that could not be written off as a harmless quirk. His private tutoring routine usually began by welcoming students on the ground floor of his Botanical Garden home, then accompanying them upstairs to the chamber he maintained as a lecture space and small museum. But at the start of one session he ushered in his pupil, then stopped suddenly in his tracks. "Aha! Is it you sitting there, Carl?" he spoke in a clear, high voice, acknowledging a presence in the room. "Sit in peace, I will not disturb you."

The room was empty.

"Sir, whom are you addressing?" the startled student asked.

Linnaeus pointed to the table and his usual chair. "I sometimes think that I am sitting there, working," he replied, as if it were an everyday occurrence. He began his lecture, studiously avoiding the area where his doppelgänger presumably remained.

It was the first recorded episode of Linnaeus's increasing bouts of autoscopy, a persistent visual hallucination and conviction that he now shared his existence with a second version of himself. Modern medicine would likely classify it as a monothematic delusion known as Subjective Double Syndrome, a psychological/neurological condition observed in patients undergoing great stress, a spectrum of psychoses, or organic brain disease. Linnaeus was fortunate in that his doppelgänger seemed to be benign, and busily at work; he did not mind it much. Other sufferers report that their doubles are sleeping with their spouses, hovering in wait to assume their lives. Others report that the assumption has already taken place—that the imagined double is now the real self, that they are either someone else or no one at all.

"Linné limps, can hardly walk, talks confusedly, can scarcely write," Linnaeus wrote of himself in early 1776. Those were among the last intelligible words to emerge from his pen. By spring he'd lost most of his powers of speech, and by summer exhibited only a tenuous awareness of his surroundings. "He ventured to go a few steps from his chair without help, but with extreme difficulty," recalled Anders Sparrman, one of his last students. "If anyone takes him into the garden, he is pleased to look at the plants, but cannot recognize any. He laughs at almost everything, but sometimes weeps, can speak only three or four words, but listens to all."

"He had forgotten his own name," reported Adam Azfelius, another student, "and mostly seems to be unconscious of others' absence and presence."

Yet there were still flashes of acumen. "For a few short periods, here and there, his power of thought returned," Azfelius wrote, "as when he found lying near him some books of botanical or zoological contents, even his own." The man who once trumpeted that God had "let him see more of his created works than any mortal man" was reduced to admiring *Systema Naturae* without realizing it was his. "He would turn the leaves with evident pleasure, and let it be understood that he would think himself happy if he could have been the author of such useful works."

By 1776, he had entered into a silence from which he would not emerge. Although he occasionally rallied enough to enjoy a few pleasures—smoking a pipe, drinking a tankard of beer—he lingered in his private twilight for another year and a half. No family members were present when he died, at eight in the morning of January 10, 1778. Only the student John Rotheram and Samuel Duse, his youngest daughter's fiancé, were by his side.

Sir Carl von Linné, born Carl Linnaeus, Archiater of Hammarby, *Riddarklassen* and Knight of the Polar Star, was interred in Uppsala Cathedral twelve days later. In an evening ceremony, the fieldhands from his country estate carried torches to illuminate the funeral cortege. Each of the pallbearers wore a medal bearing Linnaeus's portrait, drawn from a supply Linnaeus himself had commissioned. The body on the pall was not so smartly adorned. "Put me in the coffin unshaved, unwashed, unclad, enveloped with a sheet," Linnaeus had directed in his funeral instructions. "And close the coffin immediately, so that no one may see my wretchedness."

As one of the attendees remembered,

> It was a dark and still evening, the darkness only dispersed by the torches and lanterns carried by the mourners, the slow progression of the procession whose silence was only disturbed by the multitude of people assembled in the streets and the great bell's majestic tolling, were the only sounds to be heard.

"It makes no difference to me how big my tomb is after my death," Linnaeus had written in his *Nemesis Divina.* "Everywhere death comes in one size." But three decades earlier he'd purchased a large and fairly prominent burial space, beneath the cathedral's organ loft. With characteristic immodesty, he had also ordered that a bronze medallion inscribed with *Princeps Botanicorum* (The Prince of Botany) be affixed to the tomb, along with the dates of his birth and death.

The medallion was not cast. His tomb remained unmarked for another twenty-one years.

A bust of Buffon in his seventies

TWENTY-TWO

The Price of Time

WHILE BUFFON SURELY NOTED THE PASSING OF HIS RIVAL, AL-most none of his private writings during this period survive. He was still in the habit of burning all his papers once he was through with them, and he had much to burn. At the age of seventy-one, he was busier than ever. Eleven years earlier, after the considerable invest-ment of nearly half a million livres, he'd opened an ironworks in Buf-fon, his namesake hamlet near Montbard. Serving as both a foundry and a research center into metallurgy, the ironworks had grown to employ more than four hundred workers, casting cannon for the French army and navy alongside experimental items, such as the or-nate iron fencing Buffon installed in the Jardin. As one of the first attempts at private large-scale industrial manufacturing in France, the Buffon ironworks were a template for factories to come. They were also quite profitable. Buffon poured some of the profits into acquiring adjacent properties and expanding the Jardin, trusting that the king would eventually compensate him for the expenditures. (He would be disappointed.)

He'd also entered into his most prolific phase as a writer. In just three years he'd produced five more volumes of *Histoire Naturelle*, bringing their total number to fifteen. From that he at last turned away from the subject of quadrupeds, commencing his survey of birds. He began by constructing an aviary in Montbard, this time not in the Parc Buffon but adjacent to his manor. The location was a gift to his wife, who enjoyed the soundscape of birdsong.

Buffon was also continuing to boost the careers of younger natural historians. One notable patronage began in September of 1764, when a man named Philibert Commerson arrived at the gates of the Jardin, humbly inquiring after a job he'd been offered seven years earlier. Born to a prosperous middle-class family, Commerson was a physician who'd never bothered to practice, having attended the Montpellier medical school in southern France chiefly to take botany classes. He'd made an early name for himself in botanical circles, but not an entirely positive one—he had a reputation for brilliance but also for flouting authority. Instead of asking for permission to take cuttings from the university botanical gardens, he'd simply helped himself. When caught and banned from the grounds, he'd climbed over the fence at night and continued adding to his collection.

Still, his research had earned the attention of Linnaeus, who commissioned him to write a study on Mediterranean fish, and Jussieu, who offered him a place at the Jardin du Roi. He'd quickly fulfilled the commission but turned down the job, preferring instead to botanize in the French and Swiss countryside.

The next seven years did not go well. Commerson's family began to withdraw their financial support. He suffered multiple injuries during his field studies—falls from rock-climbing, near-drownings in torrential streams, and ultimately a severe dog bite that left him bedridden for three months (he would walk with difficulty for the rest of his life). In want and in pain, Commerson agreed to his family's hastily arranged match with a woman seven years his senior. They attempted to start a family. She died in childbirth, at the age of forty-one.

Now thirty-six years old, Commerson quietly departed for Paris, accompanied by a single family servant. He rented a second-floor apartment on the Rue des Boulangers, a few blocks away from the Jardin, where Jussieu informed him that, regrettably, the offered position had long since been filled. But Buffon welcomed him nevertheless, giving him the run of the Jardin, introducing him into the social circles of Parisian savantry, and sponsoring him two years later, when the botanical opportunity of a lifetime arose.

The opportunity began with the unhappiness of Louis Antoine de Bougainville, a decommissioned French army colonel who loathed being remembered for the lowest point of his military career: surrendering Montreal to the British during the Seven Years' War. After that war, after a defeated France had ceded almost all of its territory in North America, Bougainville used his own funds to help 150 colonists relocate from French Canada to a group of islands off the coast of South America, in a settlement he named Port Saint Louis.

It was an expensive achievement, and quickly dismantled. After returning to France, Bougainville learned that the nascent colony was seen as a threat by Spain, who considered it a prime location to stage attacks against Spanish holdings on the mainland. King Louis XIV, aware that the British probably agreed, decided to sell the island chain to Spain rather than risk the costs of having to defend it. Bougainville was ordered to sail back to the archipelago (which Britain would indeed later claim as the Falkland Islands) and hand it over to Spain.

Bougainville did not particularly relish adding another surrender to his record. He proposed making the voyage grander in scale, transforming it from yet another French concession into a triumphal act. No French ship had ever circumnavigated the globe. What if he sailed to South America, turned over the islands, and just kept going? It would be a prideful achievement, but more important a covert scouting mission for heretofore-unknown lands that France could claim as territories. To mask this intent, Bougainville suggested making it a scientific expedition.

The king, receptive to the suggestion, reached out to the Jardin to provide a suitably intrepid naturalist. Buffon steered the appointment to Commerson, who eagerly accepted, asking only permission to bring along Jean Baret, his longtime servant and assistant. There were no objections. The expedition, consisting of two ships, the *Etoíle* and the *Boudeuse*, departed on November 15, 1766.

. . .

On board ship and at ports of call, Jean Baret struck his fellow expeditionaries as exactly the sort of assistant the slightly built, sickly Commerson needed. In his mid-twenties, sturdily built, and strong, Baret earned the nickname "Commerson's beast of burden" and the admiration of Bougainville, who wrote of him "carrying, even on those laborious excursions, provisions, arms and portfolios of plants with a courage and strength." Working in tandem, Baret and Commerson collected or documented so many specimens that Commerson was struck by the implications on Linnean taxonomy. "What presumption to lay down the law as to the number of plants and their characters in spite of all the discoveries which are yet to be made," he wrote. "Linnaeus only proposes some 7,000 to 8,000. . . . I venture to say however that I have already made by my own hand 25,000, and I am not afraid to declare that there exist at least four or five times as many species on the whole world's surface."

In April of 1768, after successful specimen-collection stops in Brazil and Patagonia, Bougainville's two ships crossed the Pacific, reached the Society Islands, and dropped anchor in Tahiti. They were only the second group of Europeans to visit the Polynesian archipelago: The British ship HMS *Dolphin* had passed through just ten months earlier. Upon learning that the British had not staked a claim, Bougainville promptly claimed the islands for France, naming the region New Cytheria.

When the expedition arrived, one of the first Tahitians to come aboard the *Etoíle* was a young man named Autourou. Instead of being impressed by the formal uniforms of Bougainville and his retinue, Autourou saw Baret among the crew assembled on deck and began circling in fascination, saying "Aiene, aiene." The next day, when Commerson and Baret were onshore collecting specimens, they found themselves surrounded by Tahitians looking at the assistant with wonder and again repeating "Aiene." One of the strongest among them stepped forward, picked up Baret, and began to carry him away. An officer of the *Etoíle* drew his sword and pursued, thwarting the abduction and compelling the expeditionaries to puzzle out what was going on.

Aiene, they soon gathered, meant "female." Unaware of the clothing and behavioral conventions that Europeans associated with gender, the Tahitians had immediately recognized what had passed unnoticed by Bougainville and company during the months of voyage: Commerson's valet was not Jean but Jeanne. "With tears in her eyes [she] admitted that she was a girl," Bougainville wrote, "that she had misled her master by appearing before him in men's clothing at Rochefort at the time of boarding. . . . When she came on board she knew that it was a question of circumnavigating the world and this voyage had excited her curiosity. She will be the only one of her sex [to do so] and I admire her determination."

Philibert Commerson and Jeanne Baret

The statement that she had "misled her master" was a lie, and soon unraveled; Commerson and Baret had been lovers for years. They'd even had a child together not long after arriving in Paris, a son they named Jean-Pierre, who'd died a few months later. The ruse had been planned from the beginning by them both. The double deception— of Baret's gender switch and Commerson's pretended ignorance— roused the anger of their fellow expeditionaries. It marked Commerson as a criminal: It was illegal for a woman to sail aboard a French vessel. To the relief of Bougainville, they disembarked shortly thereafter on the island of Mauritius, where the two at last lived openly as a couple. Commerson happily continued to acquire specimens both on Mauritius and the island of Madagascar, until his health began to fail. Baret tended to him until he died in 1773, at the age of forty-five. Two years later, Baret returned to France, in the process becoming the first woman to circumnavigate the globe.

Baret brought with her Commerson's collection of specimens, which by now contained thousands of species entirely novel to European savantry. Buffon could have summarily confiscated the collection in the name of the king, but instead he insisted on treating Baret respectfully, entering into negotiations for the collection's sale. They settled on a price that allowed her to live comfortably until her death in 1807.

Buffon assigned the collection to the Jardin du Roi, conveying it into the care of another protégé: Antoine-Laurent de Jussieu, the twenty-nine-year-old nephew of Bernard de Jussieu, who'd arrived from Lyon as a teenager and taken a medical degree before installing himself in the Jardin. While living with and caring for the aging Bernard, he'd been surprised to learn that the elder Jussieu had given up on waiting for Linnaeus to develop a natural system of botany and had begun quietly developing one himself. Bernard had made significant progress, going so far as to lay out a garden at Versailles according to his new taxonomy, but with characteristic modesty had refused to publish even a brief description of his principles. The garden was eventually plowed over, and the project languished until 1772, when his nephew Antoine-Laurent decided to take it up. Unlike his friend Adanson, the younger Jussieu was more deferential than defiant in his opposition, admitting only to "not being wholly Linnaean" and calling Linnaeus himself "a great man." Yet he was convinced that botany deserved an alternative to Linnean systematics, which, he wrote, "seem to keep science away from its true goal." He was

Antoine-Laurent de Jussieu

still working on a Jussieu System when his uncle Bernard died in 1777, just two months prior to Linnaeus's passing. The Commerson-Baret collection came into his hands soon afterward, and his progress accelerated.

. . .

That same year, Buffon released his fourth and fifth volumes on birds, and a fifth volume of "suppléments" to earlier portions of *Histoire Naturelle*. One of these supplements was in truth a direct revisitation of Earth's deep history—the same subject he'd used to begin the series, and the subject that had drawn the angry censure of theologians and the Sorbonne. In an essay entitled *The Epochs of Nature*, he divided the history of the Earth into seven "epochs," clearly intended as counterparts for the seven days of biblical creation. He did not specify the time period for each epoch, as he could not—no technology existed for accurately measuring geological time. But the essay's radical claim remained clear: Earth's history far exceeded the brief span enumerated in the Old Testament.

Buffon began by assuming the role of his inevitable critics. "How could you reconcile, one might say, such a great age that you give to matter, with the sacred traditions, which provide for the world only some six or eight thousand years?" he wrote. "To contradict them, is that not to lack respect to God, who had the goodness to reveal them to us?" On the contrary, he argued, "The more I have penetrated into the heart of Nature, the more I have admired and profoundly respected its Creator. But a blind respect would be superstition."

> Let us try to hear reasonably the first facts that the divine Interpreter has transmitted to us on the subject of creation; let us collect with care the rays that have escaped from that first light; far from obscuring the truth, they can only add a new degree of luminosity and of splendor.

The first epoch, he wrote, began with the creation of the solar system. He theorized that the Earth and other planets were accreted masses of matter ejected from the sun at approximately the same time, since all "circle around the sun in the same direction, and nearly all in the same plane"—a theory that, in broad strokes, remains predominant today. Earth's second epoch began when it solidified, coalescing into a dense, rough spheroid. In the third, Earth became an ocean planet ("for a long time the seas covered it entirely, with the possible exception of some very high ground"); the fourth began when the waters retreated and "the lands raised above the waters became covered with great trees and vegetations of all kinds," Buffon wrote. "The worldwide sea was filled everywhere with fish and shellfish."

Then life arrived on land. The fifth epoch dawned "when the Elephants and the other animals of the South lived in the north," and the sixth was marked by the separation of the continents. The seventh and current epoch was the age of humans, or more specifically "when the power of man has assisted that of Nature."

The Epochs of Nature represented Buffon's final attempt to break the lens of fixity, and as such it was a calculated provocation. He anticipated a backlash, which came in November of 1779 when the Sorbonne once again formally denounced him and began censure proceedings. The theology faculty formed a committee to take up the question, but this time Buffon had the tacit support of the king himself. Louis XVI trod lightly amid the academic indignation, quietly requesting only that the committee "proceed with circumspection," but this had the effect of stopping them in their tracks. The committee never issued a report. The Sorbonne nevertheless prepared a new retraction for Buffon to sign, one very similar to the retraction he'd put his signature to twenty-eight years earlier. Again, he gladly signed it, and again promised to incorporate the retraction in the next volume of *Histoire Naturelle*.

This time he did not keep his promise. In fact, he went so far as to republish *The Epochs of Nature* as a separate volume, the better to reach the widest possible audience. Wary of pressing the point further, the Sorbonne published his retraction themselves in a brochure (written in Latin) in 1780. "When the Sorbonne picked petty quarrels with me, I had no difficulty in giving it all the satisfaction that it could desire," he confessed to a friend. "It was only a mockery, but men were foolish enough to be contented with it."

• • •

"No man has known the price of time better than the Count de Buffon," one relative observed. "No man has employed, more constantly or with more uniform purpose, all the moments of his life." Yet in the spring of 1782, the formidably busy Buffon stopped in his tracks. A special visitor had arrived at the Jardin, and Buffon wished to give him his full attention.

The visitor was Carl von Linné the Younger, now forty-two years old and well into his tenure as his father's successor. Buffon had never met the senior Linnaeus in person, but here was an unprecedented opportunity to defuse the old perceptions that he had been the father's nemesis and perhaps forge a form of peace. He welcomed the son effusively, insisting on being his personal tour guide and directing that they wander undisturbed, that "on that day he would be spoke to by none but him." As assistants rushed to clear the Jardin paths and the rooms of the Cabinet, their tour began.

The occasion was the professor's first trip out of Sweden in an official capacity, a circuit of courtesy calls to prominent naturalists in Denmark, the Netherlands, and now France; London would be next. It was a junket, but a welcome respite for von Linné. His life back in Uppsala bordered on the miserable.

A taller, thinner version of his father, von Linné was generally well liked as a person and respected (albeit grudgingly by some) for gamely doing his best to fulfill familial expectations. But as a teacher, he could not step to the podium without revealing his limitations. "His delivery was fluent, but mixed with a certain cold indifference," one lecture audience member reported. "It appeared as if his exertions were rather a performance of the duties of his station, than a real zeal flowing from a natural fondness of his science."

Even before his father's death, he'd made a botch of Uppsala's academic politics. In 1777 he petitioned the king of Sweden to assume the last of his father's duties by becoming a full professor. Owing to his father's worsening condition the petition had been granted, but by appealing directly to the king he'd gone over the head of the university's chancellor, which angered both the chancellor and his colleagues, who called him behind his back a "wretched boy" and a "lazy loon in a superlative degree." The task of simply maintaining his father's collection was exhausting: He was fighting off rats, wood mice, moths, and molds until he complained he was "as tired as a day laborer." But worst of all was his domestic situation. While never warm to her husband, the Widow von Linné positively hated her bachelor son. "It was singular that the lady of Linnaeus should have had so particular an aversion," one visitor to the household wrote. "He could not have had a greater enemy in the world than his own mother. . . . Her only son lived under the most slavish restraint and in continual fear of her." Humiliated on almost all fronts, he considered his role the hollowest of legacies.

"Poh! My father's successor," he privately grumbled to a friend. "I would rather be anything else. I would rather be a soldier." Still he plodded on, albeit slowly. Despite his announced intention to update his father's *Systema Naturae* and its sub-catalogues, he had thus far only completed a *Supplementum*, a single volume intended to augment both *Genera Plantarum* and *Species Plantarum*.

For Buffon, getting to know Professor von Linné carried at least a pang of sadness. For all the indications that the younger man was struggling to walk in his father's footsteps, he was at least an undisputed successor. Buffon was attempting to construct a similar legacy for his own son. It was a battle he'd already half-lost, and he seemed on the verge of losing it completely.

On that afternoon in 1782, Buffon had been a widower for thirteen years. His wife, Marie-Françoise, had been severely injured in a riding accident in Paris, suffering a disfiguring jaw injury that kept her immobilized and in pain for almost three years. When she died at the age of thirty-seven, the grief-stricken Buffon formally mourned for the next two years; he would never remarry. Instead he devoted himself to raising their only child, Georges-Louie-Marie Buffon, universally known as Buffonet.

As Linnaeus had done with Carl the Younger, Buffon sought to anoint Buffonet as his successor from the start. Linnaeus had enrolled his son at Uppsala University at the age of nine; Buffon had succeeded in convincing King Louis XV to promise that Buffonet would inherit his position as intendant of the Jardin, even though the boy was only five years old at the time. The attempt at a dynasty only lasted two years. In February of 1771, Buffon had been struck down by an attack of dysentery so virulent he was expected to die at any moment. The

king's assurances had been for sometime in the distant future; Buffonet at the time was still only seven. Under these circumstances
other ambitious men began jockeying for Buffon's job, and the king
began to listen. In between bouts of agony, Buffon learned that the
arrangement was suspended. His successor would not be his son but
the Count d'Angvillier, a rank amateur in matters of natural history.

Buffon did not die. He rallied back to health, much to the embarrassment of the king, who felt exposed to criticism at court for having
jumped the gun. To save face he made Buffon a count, creating the
title by declaring that Buffon's holdings in Burgundy comprised a
county in themselves. As an additional honor, the king awarded the
newly ennobled Comte de Buffon the status of royal attendant, allowing him to enter the royal bedchamber instead of requesting formal
audience. The king, however, did not retract the future appointment
of the Count d'Angvillier. Buffon was now an aristocrat, but it had
come at a bitter price.

He continued to raise his son as the prodigy of natural history he
wished him to be. While Buffonet might not be destined to run the
Jardin, he could still carry on the work of the *Histoire Naturelle*. Yet despite twelve years of private tutors and his father's personal instruction, Buffonet displayed neither the intellect nor the inclination to
engage in the subjects at hand. He pranked his tutor by spraying ink
across his shirt. Sent off on a tour of European capitals, the boy dined
privately with Catherine the Great of Russia, an admirer of the senior
Buffon. The occasion prompted Catherine to observe "the strange
irony by which men of genius seem fated to father sons who are virtual imbeciles." Now eighteen, the handsome but shallow Buffonet
was likely to be found anywhere but in the Jardin.

Buffon and Linnaeus the Younger walked the earthen avenues
formed by the spaces between plant beds, and past ubiquitous identification labels that disproved the rumors that Buffon had suppressed
Linnean botanical names (they were prominently displayed, alongside Tournefortian ones). As they entered the Cabinet du Roi, the
visitor could not help but notice the building's centerpiece, a feature
his host found personally embarrassing. It was a heroic statue of Buffon himself.

To Buffon's dismay, the statue had been the second of the king's

expiatory gestures for reneging on his promise to Buffonet. It was ordered as a surprise, with the commission given to Augustin Pajou, one
of the foremost sculptors of the day, at the significant cost of fifteen
thousand livres. The secret project took six years to complete, and was
installed while Buffon was away at Montbard. "It would have given
me greater pleasure if it were installed only after my death," he confided. "I have always thought that a wise man must rather fear envy
than attach importance to glory." As he'd written in *Histoire Naturelle*'s
fourth volume:

> Glory . . . becomes no more than an object without appeal for
> those who have come close to it, and a vain and deceitful phantom
> for others who have remained at a distance from it. Laziness takes
> its place, and seems to offer easier paths and more solid benefits to
> everyone, but disgust precedes it and boredom follows it. Bore-
> dom, that sad tyrant of all things that think, against which wisdom
> can do less than madness.

Buffon was not immune to flattery, but there was a line between
flattery and glorification. Though confident of his historical importance, he was also well attuned to public relations. He sensed that the
statue would be seen as a sop to his ego, even though he had not commissioned it, posed for it, or even authorized its existence. When he
finally saw it in person, his suspicions were confirmed.

The Buffon it depicted was nearly nine feet tall, heroically muscu-
lar, and, in the words of one contemporary, "absolutely naked, enveloped only by a drapery to conceal from view those parts which

Detail of statue

modesty orders one to hide." The figure holds a stylus in his right hand, seemingly poised to write upon a tablet propped upon a globe of the Earth. The overall pose evokes Moses receiving the Commandments. Beneath the globe sponges grow, representing the vegetable kingdom. The animal kingdom is represented by a lion, either dead or insensible, a serpent winding through his mane, and a sheepdog. A cluster of crystals, representing the mineral kingdom, grow by the lion's head. In a touch later described as an "unfortunate marriage of actual and symbolic," the sheepdog is earnestly licking the toes of Buffon's bare left foot.

In the third volume of *Histoire Naturelle*, Buffon had written a description of the (male) human animal:

> He carries himself straight and tall, his attitude is that of command, his head looks to heaven and presents an august face on which is printed the character of his dignity. The image of the soul is painted there by his physiognomy, the excellence of his nature breaks through his material organs and animates the features of his face with a divine fire.

Augustin Pajou, the sculptor, seems to have intended to bring this passage to life, with Buffon as the embodiment. But the idealized image was already proving to be more of an embarrassment than even Buffon had imagined. The original inscription at the base had been *Naturam amplectitur omnen*, "He embraces all of Nature," but persistently reoccurring graffiti added *Qui trop embrasse mal étreint*, "He who embraces all, grasps poorly." At the time of Linnaeus the Younger's visit, the inscription had recently been removed and a new one substituted: *Majestati Naturae par Ingenium*, "A genius to match the majesty of Nature." Buffon doubtless did not linger long in the statue's vestibule, instead hurrying into the Cabinet itself, where he and his guest whiled away several companionable hours.

Professor von Linné departed in a hail of pleasantries, pleased to have been treated with far more respect than he'd been accustomed to back in Uppsala. But he could not help being awed. Much expanded by Buffon's methodical acquisitions of adjacent properties (often using his own money), the Jardin now extended over 110 acres. The

Uppsala botanical garden comprised 11 acres. Its only staff was a part-time gardener; Buffon employed more than a hundred. His cabinet occupied a single small outbuilding, and it certainly bore no towering statue of his father.

The two never saw each other again. Von Linné the Younger returned from his tour nine months later with a case of jaundice, contracted in London, and died the following year. On November 30, 1783, his body joined his father's under the floor of Uppsala Cathedral. After the funeral oration, attendees bore witness as the family coat of arms was broken into pieces, in accordance with Swedish tradition. Since there were no male survivors to assume the *Riddarklassen* title, the House of von Linné ceased to exist.

There were, however, female survivors. As Linnaeus Junior had never married, his inheritance reverted to his mother. This included the entire collection of specimens, which Linnaeus Senior had considered his most valuable bequest. "Let no naturalist steal a single plant," he'd written in his will. "Invaluable as they are, they will increase in value as time goes on. They are the greatest the world has ever seen." Linnaeus Junior had taken this injunction seriously. Not long after his father's funeral he'd turned down a wealthy Englishman's offer of twelve thousand pounds for the collection, a decision that may have explained some of his mother's animosity toward him.

Faced with the costs of ushering her two remaining daughters into respectable marriages, Sara-Lisa von Linné was unsentimental about selling off the family's chief asset. She reached out to the English collector, but he was no longer interested. The best price she could muster was from another Englishman, an amateur naturalist who offered slightly more than one thousand pounds. It was a fraction of the original offer, but the widow could pry no higher price from any other interested parties. The deal was done, and on September 17, 1784, the British brig *Appearance* sailed from Stockholm, carrying twenty-six crates from Uppsala in its cargo hold.

• • •

It was time for Buffon to grapple with his own mortality. Outwardly, he was still a fine physical specimen. A visitor to Montbard in 1785 described him as "looking at age seventy-eight like many do at fifty-six or fifty-eight . . . a man of large stature, with a very happy physiog-nomy, brown eyes, black eyebrows, a thick head of hair, white as snow." But unknown to all but a few intimates, his famously rigorous routine was now being interrupted by excruciating attacks of abdominal and lower back pain, and agonizing urination. Buffon was suffering from nephrolithiasis, more commonly known as kidney stones. His solitary habits had thus far allowed him to conceal his attacks, but they were increasing in frequency, often bringing his work to a standstill.

As to the work itself, no end loomed even on the distant horizon. To Buffon it was patently obvious that even under the best of circum-stances, he'd never live to see the completion of the *Histoire*. It was time to choose a successor.

He'd abandoned hoping that it might be Buffonet, now twenty-one. Buffon had at last let his son become a soldier, purchasing for him an officer's commission in the elite Gardes Française. He'd also facili-tated Buffonet's marriage to the daughter of the Marquise de Cepoy, a match that came with a generous dowry. Buffon added an extrava-gant stipend of twenty thousand livres a year and wished the couple well.

Georges-Louie-Marie Buffon, known as Buffonet

Yet Buffon, like Linnaeus, had a far more capable offspring. Visi-tors to Montbard often found themselves attempting to avoid staring at Monsieur Lucas, Buffon's young assistant, for two reasons. For one, he was exceptionally handsome. "He was remarkably tall and had features of great regularity," as one guest described him. But moreover "the distinction of his form and the beauty of his face made him look, very strongly, like the native son of Buffon."

The suggestion that the resemblance might be coincidental was pure politesse. It was unspoken but long accepted that Monsieur Lucas was in fact Buffon's son, the result of a liaison with a village woman. Unlike his half-brother, he was an adept manager, a gifted horticulturalist, and a man of the outdoors. Buffon trusted Lucas to manage the Montbard estate for several years before relocating him to the Jardin, where Lucas occupied an apartment in the Cabinet building. There, Lucas became the de facto intendant while Buffon was away, holding the purse strings and executing his father's plans. But Buffon could not publicly acknowledge him as his son, nor train him as his successor.

Who, then, would carry on the work of *Histoire Naturelle*?

It would not be Michel Adanson. While he had once annoyed Buffon by openly lobbying to become his successor, by now he was enmired in a project of his own. In 1775, Adanson had presented to the Académie des Sciences a plan to extend his complexist taxonomy across twenty-seven volumes of "a Natural Method comprising all known beings," followed by an eight-volume continuation of *The Natural History of Senegal*, then "a Natural History Course, a Universal Vocabulary of Natural History to serve as a basis for the universal order, then of a collection of 40,000 figures of 40,000 species of known beings, and finally a Collection of 34,000 species of beings preserved in my office."

This was, as everyone but Adanson himself recognized, monumentally overambitious: The task would be greater than the lifelong toils of Buffon and Linnaeus combined. Buffon had attempted to make him see reason, counseling Adanson to dial back his ambition, but he could not be persuaded. The academy, naturally, voted not to fund the project. Adanson had vowed to undertake it nonetheless, at his own expense. In the eight years since, his dogged insistence on being the vast work's sole author had taken him away from the Jardin. When his manuscripts and specimens overflowed his rooms at the Jussieu house,

he'd left it for larger quarters in a cheaper, more distant neighborhood. With no income other than a meager pension Buffon had arranged for him through the academy, he attempted to make ends meet
by privately teaching botany.

Students came rarely, and rarely lasted long. He was still welcome
at the Jardin—Buffon made sure of that—but the crush of work and
the indignity of encroaching poverty had turned him into something
of a recluse. Overwhelmed by sheer volume, his effort to develop a
complexist taxonomy came to a standstill. The task of constructing an
alternative to Linnean systematics was now the sole purview of
Antoine-Laurent de Jussieu, years into the project of building upon
his late uncle's work. Unwilling to interrupt that work, Buffon did not
consider the younger Jussieu a candidate to succeed him.

The winning candidate would be Bernard-Germain-Étienne de La
Ville-sur-Illon, the Comte de Lacépède, born in Gascony in 1756.
Buffon could see similarities between the young count and himself:
Lacépède had inherited his fortune (and in his case, a title) from a
childless great-uncle on his mother's side. Like Buffon, he grew up
dividing his time between a rural estate and his province's capital. A

Bernard-Germain-Étienne de La Ville-sur-Illon,
Comte de Lacépède

recent entry to the circle of Jardinistes, Lacépède was only twenty-seven, and he had grown up obsessively reading *Histoire Naturelle*.

In 1784, Buffon appointed Lacépède as curator and assistant demonstrator in the Jardin, on the understanding that the young man would study to become his literary successor as well, practicing to write in Buffon's distinctive style. This was important: Buffon not only wanted Lacépède's volumes to blend seamlessly with his, he believed that style itself was critical to *Histoire Naturelle*'s success. "Style is the only passport to posterity," Buffon had written.

> It is not range of information, nor mastery of some little known branch of science, nor yet novelty of matter that will ensure immortality. Works that can claim all this will yet die if they are conversant about trivial objects only, or written without taste, genius and true nobility of mind.

Most important, Lacépède shared his core philosophy: It took discipline to perceive nature in all its complexities. The urge to tame such complexities through oversimplifying, by seeing demarcations when none existed, was a human failing, and ultimately a human blindness. Lacépède understood the distinction between the mask and the veil—between imposing order and patiently waiting for nature to reveal what it would. With any luck, Lacépède would pass this understanding on to his own successor, and that successor would do the same in turn. It was now apparent that the *Histoire Naturelle* might best be considered a multi-generational project, extending well into the nineteenth century if not beyond.

Buffon had made his peace with that realization, with knowing that his life's work would comprise not a monument in itself but a foundation for others to build upon. After four decades of unstinting discipline he'd failed to achieve his goal, yet the effort left him more awestruck than disappointed. "Far from becoming discouraged, the philosopher should applaud nature," he wrote, returning to his metaphor of the veil.

> Even when she appears miserly of herself or overly mysterious, and should feel pleased that as he lifts one part of her veil, she al-

lows him to glimpse an immense number of other objects, all worthy of investigation. For what we already know should allow us to judge of what we will be able to know; the human mind has no frontiers, it extends proportionately as the universe displays itself; man, then, can and must attempt all, and he needs only time in order to know all.

Buffon's own time, he knew, was running out.

. . .

In May of 1784, while walking the streets of Philadelphia, Thomas Jefferson noticed an unusually large panther skin hanging from the door of a hatter's shop. The forty-one-year-old Jefferson bought it on the spot, determining, as he put it, "to carry it to France, to convince Monsieur Buffon of his mistake with this animal, which he had confounded with the Cougar." He was in town waiting for a ship to take him to the Continent, where he was about to succeed Benjamin Franklin and become the American republic's second ambassador to France. Yet to him, a posting to France meant entering the domain of the great Buffon, and he most earnestly wished to make his acquaintance.

For one, his predecessor was one of Buffon's dearest friends. Franklin and Buffon had grown so close that other Frenchmen, wanting Buffon's attention, often first approached the American ambassador. But also among his many other interests, Jefferson considered himself something of a naturalist. In fact he was working on a book on the subject at the time: *Notes on the State of Virginia*, the only book Jefferson would publish. As soon as the first printing of two hundred copies emerged from a Paris press, he bundled a copy with the panther skin and sent both on to Buffon.

In January of 1786, he received an acknowledgment:

Monsieur De Buffon offers his thanks to Mr. Jefferson for the animal skin that he was kind enough to send him. Would his health have permitted he would have conveyed his thanks in person, but as he cannot travel, he hopes that Mr. Jefferson would enjoy coming . . . to enjoy a meal in the garden.

> This cougar differs from the one that was supplied by a Monsieur Coliunou, in that the body is shorter . . . it also has a shorter tail, and so it seems to fall in the middle between Coliunou's cougar and the South American variety.

By referring to it as "the animal skin" and later "this cougar," Buffon was pointedly refuting the reason Jefferson had sent it in the first place: He did not believe it to be a panther skin, as he did not believe panthers existed in North America. (He was correct; even the so-called Florida panther is actually a cougar.) Nevertheless, the invitation was too tempting to turn down. It was a long journey from Paris to Montbard, particularly for a single evening, but Jefferson was curious to see both the great man and the park that had become his natural element. He gladly accepted.

"It was Buffon's practice to remain in his study till dinner time, and receive no visitors under any pretense," Jefferson observed, "but his house was open and his grounds, and a servant showed them very civilly." Whiling away his time wandering through the gardens, he caught a glimpse of his host deep in thought, pacing down another path. He knew better than to acknowledge Buffon's presence, as that was standard protocol. Their formal meeting came that evening, when Buffon entered the dining room in his characteristic full finery. "I was introduced to him as Mr. Jefferson, who, in some notes on Virginia, had combated some of his opinions," the ambassador reported. "He proved himself then, as he always did, a man of extraordinary powers in conversation."

The two men were amiable but did not exactly mesh. Buffon, who abhorred slavery, was all too aware that Jefferson was a slave owner. Buffon had also read *Notes on the State of Virginia*, and was quite aware that while Jefferson had praised him as "the best informed of any naturalist who has ever written" in its pages, he'd also spent a good bit of the book directly rebutting one passage from *Histoire Naturelle*, in which Buffon had written "the animals common both to the old and new world are smaller in the latter, and that those peculiar to the new are on a smaller scale."

To Buffon this had seemed a reasonable observation. The large cats of Africa and Asia—lions and tigers, and yes, panthers—were larger than the reported large cats of the Americas. The European bison is larger than the American one. There were no American equivalents of the giraffe, nor of the elephant.

Ah, but there was an American elephant, Jefferson contended. The *incognitum* or *mammoth*, whose fossilized bones were on display at the Jardin: Those had been unearthed in Kentucky. True, no living mammoths had yet been captured, but the American wilderness was large, and largely unexplored. Surely they would eventually turn up. Jefferson did not believe in extinction, not because of religious orthodoxy but because it struck him as too extreme for the subtleties of nature. (Fourteen years later, then-president Jefferson would dispatch the Lewis and Clark Expedition with standing instructions to shoot and preserve portions of any living mammoths they might encounter.)

Buffon was in no mood to explain his conception of deep time. "Instead of entering into an argument," the American recalled, "he took down his last work, presented it to me, and said, 'When Mr. Jefferson shall have read this, he will be perfectly satisfied that I am right.'" Jefferson does not record the volume. Since at the time Buffon was publishing *Histoire Naturelle*'s survey of minerals, it seems likely "his last work" refers to the most recently published supplementary volume, which includes the following passage:

> This communication of elephants, from one continent to another, must have taken place through the northern lands of Asia neighboring America.... One can thus presume, on some basis, that the eastern seas beyond and above Kamchatka are not very deep. And, one has already seen that they are littered with a great number of islands.

In other words, Buffon envisioned a time when the separation of continents was not quite complete. The Old and New Worlds, he speculated, had once been connected by a land bridge stretching from the Kamchatka peninsula, of which only the Aleutian Islands remained. Numerous species, including humans, had likely migrated across this land bridge. Over vast expanses of time—enough for the land bridge to disappear—the migrated species adapted to the new climates of the Americas. The *incognitum*, however, could not. (Buffon was right, but only in the broader sense; it now seems likely that mastodons could not "adapt" to the increased hunting of human population growth during the Pleistocene.)

Still contentious, Jefferson told Buffon that he had conflated two different species into one, the American deer and the European red deer. "I attempted . . . to convince him of his error. . . . I told him our deer had horns two feet long; he replied with warmth, that if I could produce a single specimen, with horns [antlers] one foot long, he would give up the question." He also found the count "absolutely unacquainted" with another North American animal, the moose. Buffon thought it was a kind of reindeer. Jefferson assured him they were a completely different species, and that "the reindeer could walk under the belly of our moose." That North America could produce an animal so large was, of course, difficult for Buffon to believe. Jefferson pointed out that Pehr Kalm, Linnaeus's American apostle, had mentioned the moose's enormous size, but this did not quite convince Buffon; Kalm had been known to exaggerate.

Jefferson had to leave it at that. It was exactly 9:00 P.M. Buffon, sticking to the same schedule he'd maintained for more than four decades, thanked his guest and retired for the evening.

Jefferson left Montbard determined to provide Buffon with indisputable proof that North America was indeed home to mammals of impressive size. On January 7, 1786, he commissioned John Sullivan, an American friend, to obtain a freshly preserved moose. "It would be an acquisition here, more precious than you can imagine," Jefferson wrote. "I will pray you send me the skin, skeleton and horns just as you can get them."

In the winter of March 1787, under heavy snow, Sullivan dispatched a hunting party of twenty men to Vermont, where they shot a magnificent specimen: a moose standing seven feet tall, with antlers towering four feet higher still. It took two weeks to deliver the carcass to Boston, by which time it was rapidly decomposing. Sullivan did his best to render the bones from the flesh, but found it "a very troublesome affair" to assemble a complete skeleton. Unable to salvage the antlers, he procured another pair to "be fixed on at pleasure." The crated-up remains were placed on a ship and transported to Le Havre-de-Grace in France, where they subsequently went missing. Jefferson called it "a proper catastrophe," mourning that "this chapter of natural history will still remain a blank." To his elation the boxes of moose eventually turned up, along with a staggering bill—the equivalent of about thirty thousand dollars today. Jefferson paid up, declaring the entire project "an infinitude of trouble." Still, he took great pride in delivering the moose to the Jardin du Roi and felt vindicated when Buffon agreed to have it "stuffed and placed on his legs in the King's Cabinet." As he reported to Daniel Webster, the giant moose "convinced Mr. Buffon. He promised in his next volume to set these things right."

. . .

There would be no such volume. Buffon had already published what would be his last entry on a living creature, the stub-billed penguin ("a kind of goose which does not fly") in 1783. The next volumes were devoted to minerals, a forum he used to again assert that "petrifications and fossils" comprised the remains of living species, many of which were now extinct.

> This operation of Nature is the great means which she has used, and which she still uses, to preserve forever the footprints of perishable beings; it is indeed by these petrifications that we recognize her oldest productions, and that we have an idea of these species now annihilated, whose existence preceded that of all beings now living or vegetable; they are the only monuments of the first ages in the world.

He also took a few last jabs at systematists. "Nothing has more retarded the progress of the sciences than this logomachia, this creation of new words," he wrote in an essay on sulfur. "Such is the defect in all methodological nomenclatures," he noted in his entry on salt. "They vanish, their inadequacy exposed, as soon as one tries to apply them to the real objects of nature."

On March 28, 1787, Buffon's final volume was approved for publication by the royal censor. It was a treatise on magnets, capped off with 361 pages of meticulously compiled tables documenting the declination of compass needles observed by mariners at various locations throughout history. *Histoire Naturelle, Générale et Particulière* now stood at thirty-five volumes—three introductory ones on general subjects, twelve on mammals, nine on birds, five on minerals, and six miscellaneous ones he entitled *Suppléments*. It was only a fraction of his original plan, which had been to cover "the whole extent of Nature, and the entire Kingdom of Creation" in fifteen volumes. He never touched upon amphibians, fish, mollusks, or insects. The realm of plants, for which he'd originally allocated three volumes that would bring him in sharpest contrast with Linnaeus, still awaited a viable plan of organization.

It was a staggering achievement nonetheless. Even incomplete and excluding the Cabinet descriptions (outsourced and credited to his assistant Daubenton), the *Histoire* comprised the most extensive nonfiction work to emerge from the pen of a single author, rivaling in size and scope encyclopedias compiled by hundreds of contributors.

· · ·

Histoire Naturelle had upset far more factions than systematists and the Sorbonne. There were other ideologies, other worldviews, that Buffon had offended, particularly when he found himself unable to mask his condemnation. His excoriation of slavery (*How can men, in whose breasts a single spark of humanity remains unextinguished, adopt such detestable maxims?*) was difficult to ignore, particularly in the pages of a bestselling book at a time when the international slave trade was at its height. His musings that civilization had first arisen in Asia were well outside the mainstream worldview, which held that civilization was a torch passed directly from the Greeks to the Romans to contemporary Europe. He'd courted controversy in his *Discours sur la Nature des Animaux*, a section of the fourth volume of *Histoire Naturelle*, where he addressed the notion that animals had no souls. Centuries of observation, he pointed out, made it clear that animals could experience "all that is best in love," including bonding, maternal and paternal connections, and even self-sacrifice. He could not, however, conclude directly that this proved animals had souls, since that would bring down the wrath of Catholic theologians, who assured the faithful that animals had nothing of the kind. This skirted the censors, but it made for unsettling reading: the idea that un-souled creatures were nevertheless experiencing romantic attachments. The discourse offended prudes as well as theologians, since Buffon made it clear that "all that is best in love" also included sexual attraction and gratification, an elevation of sexuality that ran against the notion of consigning it to base and shameful urges. After reading *Discours sur la Nature des Animaux,* the king's mistress Madame Pompadour showed her disapproval by wordlessly swiping Buffon in the face with her fan.

Misogynists were also displeased. Elsewhere in *Histoire Naturelle,* Buffon unleashed a scathing attack on how men had used the concept of "virginity" to control women. "A virtue existing solely in purity of heart," he observed, "has been metamorphosed into a physical object, in which most men think themselves deeply interested."

> This notion, accordingly, has given rise to many absurd opinions, customs, ceremonies, and superstitions; it has even given authority to pains and punishments, to the most illicit abuses, and to practices which mock humanity.... I have little hope of being able to eradicate the ridiculous prejudices which have been formulated on this subject.

This contention, that virginity was a "moral being" and not a "physical object," meant that a woman lost her virginity when she first chose to have sex, not when it was forced upon her. He also pointed out that the so-called proof of virginity, an intact hymen, was elusive—when anatomists could find this membrane at all, he pointed out, its configuration varied so widely that it could often not be disturbed by intercourse at all. "The symptoms of virginity," he concluded, "are not only uncertain, but imaginary."

Buffon had even attempted to dismantle a near-universal belief: that the Earth's abundance was unlimited. Writing of what we now call biomass in his essay "*De la Terre Végétale,*" he wrote, "Nature's productive capacity is so great that the quantity of this vegetal humus would continue to augment everywhere, if we did not despoil and impoverish the earth by our planned exploitations of it, which are almost always immoderate." The idea that humans could "despoil and impoverish" a God-given natural cornucopia was almost as blasphemous as the notion that it had not been created in a single act.

But each of the thousands of pages of *Histoire Naturelle*, controversial or not, was the direct production of Buffon's unstinting application of the complexist worldview, the willingness to peer patiently through the veil of Nature rather than imposing a mask. "The true and only science is the knowledge of facts," he'd declared in his youth. The facts had taken him in directions where he practically stood alone. He could only hope that Lacépède would summon similar discipline, and courage.

. . .

After forty-seven years of clockwork scheduling, Buffon's springtime return to Paris seemed as predictable as the arrival of the season itself. But in early 1788, the staff of the Jardin du Roi received an unprecedented notice: The intendant had broken routine, wrapping up his Montbard affairs and preparing the transfer of his household far sooner than usual. It was still winter when a welcoming party of demonstrators, gardeners, and assorted Jardinistes assembled in the Jardin's esplanade, watching an unfamiliar carriage pass through the gates and halt before the Cabinet du Roi. Why was the Count not arriving in his usual style, in his personal coach? That question faded at the sight of the man who emerged and stepped down haltingly, under an assistant's hovering care.

Buffon was eighty-one years old. Advanced nephrolithiasis compelled him to walk with a deliberate gait, his customary composure occasionally broken by a wince. The carriage had been borrowed, in the hopes that its more robust suspension would better shield him from the jarring of ruts and potholes.

His condition, Buffon knew, was in its terminal phase. Unable to be removed by surgery, the kidney stones had accumulated to the point at which the kidneys themselves were beginning to shut down. The pain dominated every waking hour, and there were far too many waking hours. "Sleep had abandoned him for the last three years of his life," his grand-nephew would later recount, and yet "the uselessness of remedies did not seem to surprise or irritate him."

> He saw his approaching destruction . . . with courage and resignation, without being surprised and without complaining; he calmly

dictated his last dispositions. The movements of the soul did not
disturb the freedom of the spirit.

"What kindness! You have come to watch me die," Buffon said a
few days later, grasping the hands of the woman arriving at his bedside. "What a spectacle for a sensitive soul!" It was a morbid jest, and
one that no longer unnerved the visitor. Buffon had been welcoming
her in this fashion for several mornings now, as if she hadn't already
spent the previous day sitting quietly by his side. She was Suzanne
Curchod Necker, an elegant fifty-one-year-old who moved in the
highest circles of Parisian society. Her husband, the former director-
general of the Royal Treasury, was a wealthy banker. She and Buffon
had loved each other for more than fifteen years.

Suzanne Curchod Necker

Theirs was what the French call *une amitié amoureuse*, an intimacy
that is nonsexual but nonetheless passionate, conducted in plain sight
and with the tolerance of significant others. "I love you and I will love

you all my life," he'd written, in one of the stream of letters they exchanged. "I love and respect you beyond all that I love."

His words were well chosen. Respect, not ardency, had won her affection. The founder and formidable center of one of Paris's most prestigious literary salons, Madame Necker had little time for those admirers who praised only her beauty and social graces while failing to acknowledge her keen, far-ranging intelligence. This had never been the case with Buffon: The great man cultivated her acquaintance with a refreshing lack of condescension, treating her not as an object of flirtation but as an intellect fully capable of sparring with his own. He sent Necker drafts of his own work to "submit to her judgement, asking at the same time that she may be indulgent and frank with him." They argued over religion (she was a Calvinist, and hoped to convert him), exchanged ideas and confidences, and grew close.

Their relationship was conducted entirely in the open—her husband Jacques and Buffon often corresponded about "our Suzanne"—but it was indisputably one of chaste intimacy. Despite the passionate language that passed between them, neither was interested in a sexual relationship. Madame Necker was aware that her husband's political career made him vulnerable to scandal. Buffon, still devastated by his wife's passing ("Her death has left an incurable wound in my heart," he confessed) was elated to pursue intimacy on an entirely different plane. "Ah, God!" he wrote to Necker. "It is not a sentiment without fire; it is rather a warming of the soul, and emotion, a movement sweeter and also quite as strong as that of any other passion. It is enjoyment without violence, a happiness more than a pleasure. It is a communication of existence more pure, and yet more real than the sentiment of love. The union of souls is a penetration, that of bodies only a contact."

It is clear that Buffon saw her not only as an equal but in many respects a superior. "But for the union of souls, is it not necessary that they be both upon the same level, and can I flatter myself that mine will ever raise itself as high as yours?" he asked in a letter. "I think so sometimes because I wish it, because you are my model, because I love and respect you more than I have ever loved any one."

Such feelings were reciprocated. "When I see him," she wrote of Buffon, "my heart deceives me in two ways directly contrary to each other. I believe I am admiring him for the first time, and I think that I have loved him all my life."

Over the years he'd confessed to her about the indignities of aging, of fading eyesight and a trembling hand that increasingly struggled to hold a pen. He'd shared his heartbreak over Buffonet's unsuitability as a successor, and the unvarnished truth about his present suffering. It was Necker who, having immediately grasped the significance of his plans to leave Montbard early, sent her most luxurious carriage to minimize the agony of the journey. Now he was refusing all food and medicine, and she was here to hold his hand.

"I feel myself dying," Buffon said. "You are still a charmer to me, when every other charm has gone from me."

She remained by his bedside for five more days, until the morning when he surprised everyone by asking them to leave. He wished to make one last tour of the Jardin, in as solitary a fashion as possible. On an April afternoon, "at a time when a warmer sun was gilding the new shoots," Necker and other friends retreated to a distance. "We could see an old man," she wrote, "supported by two servants, wrapped in warm furs, walking for a moment in the alley trees crossing the garden." The furs had been a gift from Catherine the Great of Russia.

The Jardin he teetered across was significantly different from the one he'd entered in 1749. He'd expanded it 450 percent, increasing the original stock of two thousand plants to more than sixty thousand. The former Cabinet of Drugs, renamed the Cabinet du Roi, now occupied not a few crowded rooms but a stately building. Buffon had tripled the staff, and doubled the courses of study available, at no cost, to anyone who wished to undertake them. He had turned the former apothecary garden into the largest institution of natural history in the world.

Passing summarily by the monumental statue of himself, donated by a guilty king, he took in a view of the open-air structure he considered his true monument. The highest point in the garden had been topped with a flagpole when Buffon took command of the Jardin, but in 1786 he'd made it the site of a gazebo he called the Gloriette. Cast at Buffon's own eponymous ironworks in Burgundy, it was the first all-iron structure in Paris. It seemed to defy gravity. The slender columns were designed as a series of spear-like projections, holding to earth a cupola that was not a weighty dome but a delicate latticework that seemed woven like a basket, its own arc being pulled higher into the sky by a smaller structure surmounted by a weather vane and armillary sphere. Unlike other cupolas, which were designed to convey solidity and stolidity, the Gloriette was like the gondola of a balloon, about to take flight. It marked noon by means of an unusual mechanism: a magnified sunbeam burned through a single strand of hair holding back a counterweight. When released, a hammer in the shape of a globe struck a gong, allowing passersby to adjust their timepieces and notifying workers it was time to replace the hair. The uppermost part bore the inscription *Horas non numero nisi serenas*, "I count only the bright hours."

Buffon's Gloriette, at the highest point in the Jardin

The Gloriette's elegant blend of simplicity and complexity would not be fully appreciated until decades later, when mathematicians noticed that the airy, net-like roof drew its solidity not only from solid iron casting but from the pattern of the lattice. It is a one-sheeted hyperboloid, considered the most complicated of quadratic surfaces yet discovered. Maximizing the ratio of structural integrity to surface area, one-sheeted hyperboloids would not enter the lexicon of architecture until the twentieth century, when they became the underlying formula for the cooling towers of power plants. The Gloriette, like Buffon, was ahead of its time.

The Jardin was thriving, and would endure. So too would *Histoire Naturelle*. To Buffon's mild annoyance, his handpicked successor the Count de Lacépède was so eager to carry the torch that he'd already published a volume of his own, a work on egg-bearing quadrupeds and snakes. Buffon would have preferred that Lacépède wait until his passing, but at least he could enjoy his protégé's praise in the introduction ("a general and immense natural work, of which his genius conceived the vast whole in such a sublime manner"), and read enough to assure himself that Lacépède's voice on the page reasonably resembled his own, and that their philosophies remained in alignment.

Buffon was not a religious man, but "when I become dangerously ill and feel my end approaching, I will not hesitate to send for the sacraments," he'd confided to a friend. "One owes it to the public cult." He believed that a family priest was at his side during the evening of April 14, but he was hallucinating: It was Madame Necker who heard his last confession. "Dear Ignatius," he said to the imagined cleric, "you have known me for more than forty years; you know what my behavior has always been. I did good when I could and I have nothing to be ashamed of. I declare that I die in the religion in which I was born." He drank three teaspoons of Alicante wine, closed his eyes, and died at forty minutes past midnight.

•••

In accordance with his wishes, Buffon became his final specimen. During an autopsy conducted the following day, the stones in his urinary tract were extracted and counted. There were fifty-seven, enough to overflow an adult's cupped hand—an astonishing amount of internal shrapnel, representing an unfathomable quantity of pain. Buffon's heart was excised and drained of blood, then given per his instructions to Barthélemy Faujas de Saint-Fond, a geologist who had assisted him in his research on mineralogy. After a longer procedure requiring the delicate deployment of a bone saw, the autopsists removed Buffon's brain. They measured it, assessed it as "of a slightly larger size than that of ordinary brains," and presented it to Buffon's son.

Buffonet did not want it. He was all too aware of how his own intellect had disappointed his father, and the prospect of a lifetime tending to the relic of his father's brain appears to have been less than appealing. Instead, he proposed a switch: He would keep his father's heart, while the geologist could depart with the brain. As Saint-Fond was compelled to discreetly mention, the heart was intended for another. Buffonet understood, and saw an opportunity to clear himself of the matter altogether.

Saint-Fond departed with both the heart and the brain. He placed the latter in a crystal urn and attached an engraved label reading *Cerebellum of Buffon, conserved by the method of the Egyptians.* (It would remain in his family for the next seventy-nine years.) Buffon's heart, encased in gold and crystal, made its way to the private quarters of Madame Necker, the chaste love of his last years and comfort of his final moments.

"Buffon died Wednesday," the periodical *Literary Correspondence* reported. "He just closed the door on the most beautiful century that could ever honor France." The next morning, twenty thousand Parisians gathered to witness the funeral procession.

. . .

While she treasured the reliquary of Buffon's heart, Madame Necker decided that her truest earthly memento of Buffon was not an object but a place. "I go every week to the Jardin du Roi, to pay homage to the living remains of a friend who is dear to me," she wrote to an acquaintance. "Sometimes still his great soul rises from the midst of his ashes to commune with me. Nothing, however, can compensate me for his usual society."

Her weekly visitations with Buffon's great soul did not last long. Four months later King Louis XVI, desperate to stave off an impending financial crisis, appointed her husband Jacques Necker the Chief Minister of France. The following January, on her husband's advice, the king sought popular support for fiscal reforms by reviving the ancient, nearly forgotten concept of Estates General, an advisory assemblage of representatives from the nobility, the clergy, and commoners.

When the assembly provided not compliance but a list of their own grievances, the frustrated king fired his chief minister.

It was July 11, 1789. Historians point to various incidents as the inciting moment of the French Revolution, but the dismissal of Jacques Necker looms large among them. He was a popular figure, respected for his openness about the country's near-bankruptcy and willingness to consult the populace. In Paris, street protests broke out the next day. They continued, and escalated. Two days later, protestors stormed the Bastille.

PART III
God's Registrar

Men argue. Nature acts.

—VOLTAIRE

Dedication of bust of Linnaeus in the Jardin, August 23, 1790

TWENTY-THREE

Germinal, Floreal, Thermidor, Messidor

ON THE EVENING OF AUGUST 23, 1790, A LITTLE OVER A YEAR into what was now unquestionably a revolution, a torchlit procession marched through the gates of the Jardin du Roi.

They were revolutionaries (or at least they billed themselves as such), but not revolutionaries of the stripe that had overrun the Bastille prison, killed its warden, and paraded his head impaled on a pike. These styled themselves as revolutionaries of the intellect, and they carried not a head but a bust, crafted of stucco but painted to resemble bronze. A bust of Linnaeus.

The forty-year dispute of Linnaeus and Buffon had not ended with their deaths. Systematists and complexists continued to clash, winning different degrees of influence in various countries. In Italy, Linnean systematics had been banned outright until 1773, when Clement

XIII relaxed a papal proscription enacted fifteen years earlier. The British were mixed on the subject. The Germans, still registering the objections of Haller and Siegesbeck, had given it a tepid reception. In France, where Buffon and his circle had held sway for decades, Linnean thought had been driven to the fringes. But in the topsy-turvy year of 1790, the fringes marched steadily toward the center.

. . .

Since the fall of the Bastille—only a mile away, directly across the Seine—the staff and students of the Jardin had kept a low profile, anxiously awaiting the seemingly inevitable moment when the revolution addressed the matter of their existence. It was unobtrusive as far as institutions went, a gated patch of green on the edge of the city, but it was still the Garden of the King, containing the Cabinet of the King. The royal association could not be denied.

As Buffon lay dying in 1788, his official successor d'Angvillier had decided to pass the appointment on to his brother, the Marquis de la Billaderie. The marquis was a military man, a grand marshal in the king's army, with no experience in natural history and no inclination to acquire any. He did not move into Buffon's house on the premises, routinely ignored requests from the staff, and even before the revolution hadn't bothered with one of the most important aspects of Buffon's old job, which was advocating for the Jardin's financial allocations at court. With far weightier matters to deal with, neither Louis XVI (at the time still clinging to the throne, as a constitutional monarch) nor the new National Assembly had taken notice of the Jardin's rapidly dwindling operational funds—or, for that matter, of the Jardin at all. The Jardinistes tightened their belts and quietly coped, hoping to wait out the tide of unrest.

But others saw unrest as opportunity. One of them was Auguste Broussonet, permanent secretary of the Royal Agricultural Society and a private, but ardent, Linnean. In a published speech calculated to inject natural history into the roiling national discourse, he launched an attack on Buffon and his complexist legacy. Conceding that Buffon was "gifted with a lively and penetrating mind, with a vast and fertile imagination," he nevertheless concluded that the late intendant "seemed to bear his eyes upon this rich confusion of things only to paint them in this very confusion, and to throw variety upon the great spectacle." Linnaeus, on the other hand (Broussonet concluded) was "filled with a genius no less vast, no less ardent, but which he submitted to observation," who "saw in this mixture of beings only an apparent disorder that it was necessary to dispel, to render more useful natural knowledge."

This was a tactical strike. Broussonet was one of the leaders of a clandestine group calling themselves the Société de Linné de Paris. Formed in December of 1787 (while Buffon was still alive), the SLP met in secret—to escape official attention, and to mask obvious acts of betrayal—in hopes of taking advantage of Buffon's expected passing. When no advantage materialized, the secret meetings tapered away; after only one year, the society was effectively defunct. But the growing air of change had led to a revitalization, and now the group was at least sixteen members strong.

Broussonet, a provincial originally from the Mediterranean city of Montpellier, occupied a peripheral position in Parisian savantry, an elite that had long held Buffon at its very center. But he was motivated by more than a desire for petty vengeance. As a physician turned agriculturalist, Broussonet was steeped in the practical concerns of natural history, not its abstractions. For those who did not have the luxury of engaging in pure research, whose chief concerns were identifying and utilizing individual species, there was simply no substitute for the sexual system, artificial as it was. To dismiss Linnaeus as distortive and flawed was to place the life's work of savants like himself in a similar light. Broussonet was sincere in wanting to banish "the old errors and its prejudices" (as another SLP member put it), "to better see the Nature that Buffon had painted in broad strokes . . . to return to France the importance that it must have in the science of natural history."

Other members had different motivations. Another leader was André Thouin, who'd been a lifetime beneficiary of Buffon's largesse. Buffon had hired Thouin's father as an ordinary gardener, then promoted him to the post of head gardener; when the father died, Buffon gave the job to André, despite the fact that he was just seventeen at the time. Over the years Buffon considered Thouin one of his most trusted associates, often leaving him in charge of the Jardin when decamping for Montbard. But even before Buffon's passing, Thouin had

decided to throw in his lot with the Linneans, knowing his future options would be limited. There would be a changing of the guard, and
the informal authority of a second-generation gardener with no formal education seemed shaky ground on which to stand.

Having tested the winds with Broussonet's speech, in August of
1790 the Société de Linné de Paris decided to go public, albeit in a
low-key manner: They would install a bust of Linnaeus in a forested
area just outside the village of Saint-Germain-en-Laye, a half-hour's
carriage ride from Paris. The ceremony backfired. The locals, seeing a
group of well-dressed men carrying a large wooden crate as well as
several smaller tin boxes (intended to gather botanical specimens
along the way), grew suspicious and formed a crowd, driving them
away. Bolder, or at least less ambiguous, action seemed called for, so
the SLP changed tactics. Instead of scurrying around on the outskirts
of Paris, they petitioned the National Assembly for the right to place
the bust inside the Jardin itself.

In carefully worded language, Broussonet and company repack-
aged Linnaeus as a scientific revolutionary, a true man of the people,
asking only to humbly erect a "plain stone monument" to a man they
referred to not as Sir Carl von Linné but Charles Linné, democratiz-
ing the name by stripping away the "Sir," and emphasizing its French-
ness by eliminating the German prefix *von*. Placing such a tribute in
the middle of the so-called Jardin du Roi would clearly strike a blow
against the *ancien régime*. The growing SLP had sixty-three signatures
to affix to the petition. They sent it to the National Assembly and
waited.

They did not have to wait long. Despite the fact that the National
Assembly had no clear jurisdiction over the Jardin du Roi, they approved the petition. On the evening of August 23, 1790, a torchlit
procession marched through the Jardin gates and gathered just below
the promontory of the Gloriette, at the base of a large cedar tree.
They quickly erected the pedestal, which was painted to resemble
marble, and placed upon it the bust of Charles Linné.

Speeches were delivered. The society's members, as well as new-
comers beginning to style themselves Linneans, wrote their names on
bits of paper and placed them in a vase at the base of the monument.
The names were set afire, illuminating the column's poetical inscrip-

tion: *In this beautiful garden, Linnaeus transports my eyes to paradise. But what a strange destiny.*

Étienne Geoffroy Saint-Hilaire

Watching this spectacle with undisguised disdain was a nineteen-year-old named Étienne Geoffroy Saint-Hilaire, usually referred to by the surname Geoffroy. "I myself witnessed the extravagant tumult of zealots gathering at a designated time in the gardens embellished by Buffon to insult the memory of this great man," he later recalled. "The bust of the foreigner, religiously placed under the shade of the great cedar of Lebanon, was . . . the glorification of the memory of Linnaeus. However, it was less a question of honoring such a great reputation than of annihilating the development of the school of Buffon."

Geoffroy was a recent graduate of a local Catholic school, where he'd developed an interest in mineralogy and been pointed to the Jardin by one of the teaching priests. He considered himself merely a beginner in "the school of Buffon," and like the other Jardinistes could only look on in silence as the SLP began to treat the Jardin as their own, taking over the auditorium to conduct their meetings and initiate new members. As their numbers grew into the hundreds, the organization changed its name: No longer the Société de Linné de Paris, they became the Société d'Histoire Naturelle. This was not to dis-

tance themselves from Carl Linné but to eliminate any distance whatsoever. As far as they were concerned, Linnaeus *was* natural history.
An amateur naturalist named Grandmaison (yet another of the Société's co-founders) took the opportunity to denounce "Buffon, whose
writings seduced by the magic of style and which fixated on him the
attention of his fellow citizens and all of Europe."

> It cannot be denied that he has delayed the progress of true knowl-
> edge in Natural History, by the contempt he manifested and in-
> spired, of systems and methods without which this science can
> only offer confusion, can only be an inextricable labyrinth.

• • •

Despite the occupation and victorious rhetoric, there was still no official word on the Jardin's fate. De la Billaderie resigned as intendant
on July 1, 1791. His replacement, Jacques-Henri Bernardin de Saint-
Pierre, was even less qualified to lead the institution. Saint-Pierre was
an author whose novel *Paul et Virginie* was a bestseller at the time, but
his sole connection to natural history had been his friendship with the
late philosopher Jean-Jacques Rousseau, who had notably lavished
praise on both Buffon and Linnaeus. In 1770, while visiting Buffon at
Montbard, Rousseau had ostentatiously knelt to kiss the threshold; he
was, as he explained, about to enter holy ground. But the following
year he'd also written fawningly to Linnaeus, imploring him to "receive with health, Monsieur, the homage of one of the most ignorant
but most zealous of your disciples. . . . Monsieur, continue to offer and
interpret for men the book of nature. . . . I read you, I study you, I
meditate on you, and I honor you and your companions with all of my
heart."

Proving as much a fence-straddler as his late friend Rousseau,
Saint-Pierre kept a low profile and did nothing. The Jardin entered a
limbo. Young Geoffroy resumed his studies in mineralogy until the
days following August 10, 1792, when the storming of the Tuileries
Palace led to the imprisonment of the king, the establishment of a
new government called the National Convention, and a descent from
revolution into reactionary chaos.

• • •

One of the National Convention's first acts was to round up and begin executing "non-juring" priests: those who refused to take a vow of obedience to France over the church. Geoffroy, discovering that most of the faculty of his former school had been arrested, implored the Jardinistes to help in their release. When their influence was successful in freeing only a few, the young man vowed to rescue the rest himself. On the morning of September 2, he dressed as a prison commissioner and walked right into the open-air holding pens of Saint Firmin prison, where he urged the priests to follow him to freedom.

To his astonishment, they refused to go. Likely unaware of the immediate danger they were in, they thanked Geoffroy but chose to remain in solidarity with the other captives. He retreated with only one priest in tow, pondering what to do next. He knew the executions were just beginning—in fact, that day marked the beginning of the September Massacres, a four-day killing spree that would claim the lives of half the prison population of Paris.

Waiting until nightfall, he procured a ladder and scaled one of Saint Firmin's parapet walls, then crouched at the top. He stayed there for eight hours, listening to the carnage below and knowing full well his ladder might be spotted at any minute. At last he recognized the priests by moonlight and gestured that they climb toward him. He pulled twelve of them over the wall and down to safety in a nearby lumberyard. Just before scrambling down the ladder himself, Geoffroy felt a bullet rip through his coat. He'd been so occupied with the rescue he hadn't noticed the sun rising, fully exposing him to the guards.

· · ·

One month later Jacques and Suzanne Necker fled Paris, in their haste leaving behind the preserved heart of Georges-Louis Leclerc de Buffon. They were captured at the Swiss border, and detained for several days while the local commune debated the question: Were they heroes of the revolution, or enemies? Jacques had, after all, set much in motion with his defiant release of the royal budget in 1781, had worked with the National Convention to broker a peaceful transition of power from the monarchy, and had done his best to keep the French economy from collapsing, at one point personally donating most of his fortune into the national treasury. They were allowed to cross over on September 11, but only as the revolution moved to confiscate their remaining assets.

Suzanne never recovered from the shock of their fall and flight. She became a recluse, giving in to fits of hypochondria and a morbid obsession that compelled her to perpetually write and rewrite her will and funeral arrangements. She died in exile in 1794, at the age of fifty-four, leaving behind instructions for her body to be embalmed, propped up in her bed, and exhibited to visitors for a period of four months. Jacques lived until 1802, trying in vain to recover at least some of the fortune he'd sacrificed for France.

King Louis XVI, stripped of his titles and put to trial as Citoyen Louis Capet, was led to the guillotine on January 21, 1793. He attempted to address the crowd, shouting out "I die innocent!" before a drumroll, ordered by an impatient revolutionary general, drowned out his final words. Revolutionaries sang "La Marseillaise" and dipped handkerchiefs in his blood, then moved on to dismantling his legacy. By February they were petitioning the National Convention to "suppress" (i.e., demolish) the Jardin, branding it "an annex of the king's palace." A subcommittee was ordered to decide the issue, but at the same time the Convention rolled out the Revolutionary Tribunal, a new, more radical court aimed at meting out justice. The tribunal opened on March 10, beginning what would come to be known as the Reign of Terror.

The Jardin's future looked dismal, as did the survival odds of several Jardinistes. Aware that his aristocratic title put his life in danger, Count Lacépède immediately fled Paris to lie low in his native Leuville, resigning his post as sub-keeper and assistant demonstrator of the cabinet of natural history. Bernardin de Saint-Pierre, the novelist turned intendant, approached Geoffroy with a startling proposition: Would he agree to be appointed as Lacépède's successor?

Geoffroy was staggeringly unqualified, and he knew it. He was an unaccomplished twenty-one-year-old student, with zero experience in research or teaching, but he understood the subtext of the offer. Saint-Pierre and the Jardinistes knew he'd risked his life during the September Massacres to save the priests that had been his teachers. They hoped that if need be, he'd rescue them as well.

He accepted the post, and the unspoken charge of protecting his new colleagues, with the understanding that his duties would not include continuing the *Histoire Naturelle* (that matter would hinge on Lacépède's survival). And not a moment too soon: Within days of his appointment the Reign of Terror was in full swing, with the Convention's innocuously named Committee of Public Safety effectively seizing power and summarily killing off its enemies, real or imagined. Within a year, an estimated seventeen thousand citizens would be marched to the executioner, with some ten thousand more dying in prison before they could meet a similar fate. While Geoffroy could do his best to keep the Jardinistes from joining their ranks, he was powerless to rescue the Jardin itself from a suppression that now seemed to border on the inevitable.

That rescue came from the unlikeliest of sources: fifty-two-year-old Jean-Baptiste Pierre Antoine de Monet, the Jardin's humble herbarium keeper, who also happened to be a knight.

. . .

Born in the Picardy region of northeastern France, the herbarium keeper was the youngest of eleven children in a family that, while poor, passed down to him the last vestige of an aristocratic past. He'd inherited the minor title of Chevalier de Lamarck (a *chevalier* was roughly equivalent to knighthood), which gave him sufficient social cachet to join the French army's officer corps. He did so, under the name of Lamarck, and gained a reputation for able leadership and courage under fire, but then came a humiliating and painful incident. In 1763, during a deployment to Monaco, a fellow officer impulsively lifted the diminutive Lamarck by the head. This bit of horseplay severely injured his neck, requiring surgery and a long recuperation, after which Lamarck decided not to return to the ranks. At the age of thirty-two he drifted to Paris, worked briefly as a bank clerk, and was thinking of becoming a professional musician when he discovered the Jardin. An introduction to Buffon soon followed.

There was nothing in his background that fitted him for a career in natural history, but Buffon, who had himself assumed control of the Jardin without formal qualifications, took a liking to the battle-weary veteran. He hired him as a tutor for Buffonet, and in 1777 was instrumental in arranging for state-sponsored publication of La-

Jean-Baptiste Pierre Antoine de Monet, Chevalier de Lamarck

marck's first book, *Flore Française*—a book decidedly complexist in tone, in which the author took Linnaeus to task for "giving orders to nature, forcing her to deploy her productions like a general his army, by brigades, by regiments, by battalions, by companies, etc." The book sold well, and in 1779 Lamarck was accepted into the Académie des Sciences.

But over time, he and Buffon drifted apart. His term as Buffonet's tutor had been humiliating—even years later, he shuddered at the memory of the brat splashing him with ink, ruining his clothes. More important, he and Buffon could not agree on a crucial point. While Buffon's notions of exogenesis depicted life as fluid and changing, Lamarck continued to believe in the fixity of species: The mechanisms of life struck him as far too intricate to be anything other than a singular act of Creation. Relations between the two men remained cordial, but the intendant's patronage ended with awarding him the post of keeper of the Jardin's herbarium. It was a low-paying job, but Lamarck kept it as a lifeline to continue his own work.

His book *Flore Française* had marked him as an anti-Linnean, which he unabashedly was. But that was an unfortunate reputation to have when the new intendant was allowing free rein to Linneans. Convinced he would soon be out of a job, Lamarck had rapidly drafted a pamphlet to distribute to the National Assembly, in which he argued for his livelihood and for the utility of the Jardin du Roi. Cannily,

Lamarck entitled the pamphlet *Memoir on the Cabinet of Natural History, and Particularly on the Jardin des Plantes.*

Not *Jardin du Roi*, but *Jardin des Plantes.* The King's Garden, he implied, was no more: It was merely a Garden of Plants. He signed himself not Chevalier de Lamarck but simply Citizen Lamarck.

The Cabinet had always been open to the public, he noted. No admission had ever been charged to stroll through the Jardin, and even the most advanced courses of instruction had always been free. Citizen Lamarck humbly suggested that the informal role of depository be formalized—that the Cabinet of Natural History, enlarged by the people's generosity, become the people's Musée de Histoire Naturelle, or Museum of Natural History, wreathed in the people's greenery of the Garden of Plants. It would no longer be led by a sole intendant or other political appointee, but managed under the more egalitarian leadership of six "professor-administrators."

The name was a clear capitulation to the Linneans, who had changed their organization's name to the Société de Histoire Naturelle. Lamarck anticipated that they would dominate the council of professor-administrators, but he hoped there would be room for himself.

The pamphlet disappeared into the National Assembly. Three years passed. The National Assembly itself disappeared, to be replaced by the National Convention. In June of 1793, the government had weightier matters to contend with—in addition to the Reign of Terror, it was fighting simultaneous wars against Austria, Prussia, Great Britain, and Spain—but nevertheless a decision on the Jardin finally emerged. Somehow, yet another Convention committee (Public Instruction and Finances) had happened across a copy of Lamarck's pamphlet, decided it made a splendid template for reorganization, and endorsed it with the sole modification of increasing the number of proposed professor-administrators from six to twelve. The Convention passed the plan without objection.

Lamarck's pamphlet had shielded the institution from the worst of the turmoil of the revolution, and given it a new name. Since his proposal had treated the Jardin as a subset of the Cabinet and not the other way around, the term *cabinet* was upgraded to *museum* and given precedence. The Jardin du Roi was now officially the Muséum National d'Histoire Naturelle, avec les Jardin des Plantes.

This inversion underscored the institution's utility. As throngs of revolutionaries stormed the homes of the elite to confiscate their jewelry and other valuables, they were also carrying off the contents of that formerly fashionable status symbol, the private cabinet of curiosities. Some looters, upon realizing most specimens held little inherent value, chucked them into a trash-heap or the Seine, but others discarded them at the Jardin. The paths leading to the Cabinet/Museum were regularly littered with abandoned collections of stuffed birds, pinned butterflies, fragments of fossils, and miscellaneous objects floating in jars.

· · ·

The task of dividing up the organization into twelve departments was left to the denizens of the Jardin themselves. The plant kingdom was split between two chairs—one in Horticulture, awarded to André Thouin, and one in Botany, given to René Desfontaines, a physician and botanical hobbyist whose only real credential was being a cofounder of the SLP. The chair of anatomy was split into Human and Vertebrate departments as well (although humans were obviously vertebrates), and the animal kingdom was split into Higher and Lower departments.

For Higher Zoology, the obvious candidate was Lacépède, but he had fled Paris. The task of recruiting a new candidate was taken up by Louis Daubenton, the Montbard native Buffon had imported to write *Histoire Naturelle*'s supplemental inventories of the Cabinet and supervise its anatomically precise illustrations. At seventy-seven years of age, Daubenton had no hopes of filling the role himself, but he desperately wished it on to someone with Buffonian sensibilities, to counterbalance the ruling majority of Linnean systematists. He paid a visit to Geoffroy, who once again found himself implored to take an office for which he was utterly unqualified. "I will undertake the responsibility for your inexperience," Daubenton reassured him. "I have a father's authority over you. Take this professorship, and let us one day say that you have made zoology a French science." Geoffroy accepted the chair.

That left Lamarck himself, a man the Linneans actively disliked but who had clearly saved the day. He was awarded the professorship of

Lower Animals, specifically "of insects, of worms, and microscopic animals"—the vaguest of all possible categories, covering the species that Linnaeus had first lumped into *Vermes*, then *Vermes* and *Insecta*.

He accepted. He was a botanist, not a zoologist, and it was meant to be a humiliating posting, but it was better than his previous role as keeper of the herbarium. Young Geoffroy, realizing that Lamarck now faced a learning curve even greater than his own, formed a life-long respect for the older man. As he recalled, "The law of 1793 had prescribed that all parts of the natural sciences should be equally taught."

> The insects, shells, and an infinity of organisms—a portion of creation still almost unknown—remained to be treated. . . . This task, so great, and which would tend to lead him into numberless researches; this friendless, unthankful task he accepted.

While the assignment was intended to exile Lamarck to an obscure corner of the animal kingdom, in truth it placed him at its center. Arthropods and mollusks ("insects, shells") happen to encompass approximately nine-tenths of all animal species alive on Earth. This compelled abundance, as well as his forced transfer from the kingdom *Plantae* to the kingdom *Animalia*, would have momentous consequences.

• • •

The transformation of the Jardin was taking place against a backdrop of even more radical changes to French savantry. In August of 1793, the Académie des Sciences and the Académie Française were abolished, along with all other royal academies. The plaster bust of Linnaeus disappeared from the Jardin—by some accounts it was mistaken for a bust of an ancient French king and smashed to bits. But the posthumous reinvention of Linnaeus as Charles Linné, revolutionary hero, continued unabated that autumn, when the National Convention turned toward revolutionizing time itself.

One morning, French citizens awoke to learn that the date was not the 25th of October 1793, but the Fourth of Brumaire, Year Two. The newly proclaimed *calendrier républicain français* retroactively declared the day after the Convention's own founding (September 22, 1792) as the first day of Year One; subsequent years would begin on the autumn equinox in Paris. Time was now reckoned in decimals. Weeks were ten days long, and the hours of each day redivided into units of ten, each consisting of one hundred minutes of one hundred seconds apiece. Months were now three weeks long—an arrangement that required four or five extra days to sync up with the astronomical year, which were swept into a loose period called "complementary days." It was, like so much of the revolution, produced by committee. One committee member, the actor and poet Fabré d'Englantine, had been given the task of finding new names for the months, stripping them of any imperialist or religious connotations. After paying a visit to André Thouin at the Jardin, he'd emerged with notes cribbed from Linnaeus's 1756 *Calendarium Florae*, an eighteen-page pamphlet describing the year in botanical terms.

D'Englantine unleashed his poetic sensibilities on the assignment, dubbing the February-ish month Ventose, or "time of the winds." The October-ish month became Brumaire, or "time of the mists." But four of the names were cribbed directly from *Calendarium Florae*: Germinal, Floreal, Thermidor, Messidor. In France, Charles Linné was now immortalized as the man who'd named much of the calendar.

With everything to learn about zoology, Étienne Geoffroy Saint-Hilaire decided to build a zoo. Since the Convention had banned private ownership of exotic animals, revolutionaries were rounding up traveling circuses and delivering them to the Jardin along with their keepers, making them Geoffroy's problem; thus far he'd found himself the caretaker of a tiger, a panther, a leopard, a white bear, two eagles, and several mandrills. He was supposed to dismiss the keepers and have the animals killed, stuffed, and put on display, but he had other ideas. Scrambling to find the funds, he began constructing enclaves in the lower part of the Jardin and hiring the former circus men as zookeepers. The zoo opened on December 11, 1794, and quickly became one of the Jardin's most popular features. As peaceful crowds began to stream through the gates in search of idle entertainment, Lamarck and Geoffroy cautiously resumed the institution's oldest tradition: free lectures on natural history, open to any and all.

. . .

Yet these were still lethal times. Georges-Louie-Marie Buffon, known as Buffonet before becoming the second Count de Buffon, had been an early supporter of the revolution, serving as a colonel in the revolutionary army's fifty-eighth infantry. But it was difficult to shake off the fact that he had inherited his father's title, and he was stigmatized further by being associated with the royal family, a connection that came strongly against his will.

Seven years earlier, one of the ailing Buffon's last parental tasks had been breaking disturbing news to his son: The Duke of Orleans, cousin to the king, had taken Buffonet's wife as his mistress. He counseled his son to move clear of the affair without confronting the duke. Buffonet had heeded his words, resigning his commission in the royal army and separating from his wife with as little fuss as possible. But the revolution was extirpating as much of the ruling class as possible, and eventually being the estranged husband of the mistress of the king's cousin was enough to cost him his life. Robespierre accused the twenty-nine-year-old of taking part in a conspiracy "to destroy the unity and indivisibility of the republic," and the Committee of Public Safety handed down a verdict of death. The execution was carried out on 22 Messidor, Year Two—July 10, 1794.

In his final hours, the once frivolous Buffonet appears to have summoned a dignity reminiscent of his father's stoic end. Eyewitnesses report that he walked calmly to the guillotine, and from the scaffold addressed the crowd with a declaration of pride. "Citizens, my name is Buffon!" he shouted, before the blade fell.

His body did not join those of his father and mother in the Montbard family crypt. That small structure, attached to a chapel on the edge of the Parc Buffon, had recently been desecrated by revolutionaries seeking lead for their bullets. They pulled down the coffins from their niches, emptying them and stripping them for their metal linings. Buffon's body, so recently interred it was barely skeletonized, was left strewn on the chapel floor, intermingled with the bones of his wife.

· · ·

In July of 1795, a young man named Georges Cuvier arrived at the Jardin gates, bearing a letter of invitation. "Come and fill the place of a Linnaeus here," the twenty-three-year-old Geoffroy had written to the twenty-five-year-old Cuvier. "Come and be another legislator of natural history."

The son of a lieutenant in the Swiss Guards, Georges Cuvier was born in 1769 in Montbéliard, a town near the Swiss border that was culturally French but territorially German, having been purchased from France by the duchy of Württemberg in 1536. An intelligent child with limited local educational opportunities, Cuvier spent his formative years entranced by Buffon's *Histoire Naturelle*, reading every volume he could get his hands on and committing his favorite passages to memory. He also found an influential role model in Buffon himself: a sophisticated savant ranging freely over a broad range of topics with a storyteller's flair, elevating natural history while at the same time popularizing it. He would later pattern his ambitions along similar lines.

To lift himself out of his circumstances, he took advantage of Montbéliard's technically foreign status, traveling to study in the duchy's capital of Stuttgart despite not knowing how to speak German. He picked it up so rapidly that he was soon winning prizes for declamation, earning a reputation as a brilliant student in whom "the driest chronological facts, once arranged in his memory, were never forgotten." But his funds ran out in 1788, and the nineteen-year-old was obliged to return to France and find what employment he could. By 1790 he was living in rural Normandy, working as tutor to a nobleman's son. The only intellectual life he could pursue consisted of occasional visits to a nearby village, where an informal group of farmers, burghers, and other citizens met to discuss agriculture.

At one of these meetings, Cuvier began paying special attention to another attendee, a physician from a nearby field hospital who seemed to know quite a lot about specialized topics such as grain diseases and the decay of wheat. There was something about him—his cadences, the specifics of his erudition—that struck Cuvier as familiar. From the depths of his prodigious memory he summoned up entries he'd read in an encyclopedia, and recognized the man as their author. He was Henri-Alexandre Tessier, the director of Louis XVI's royal farm at Rambouillet, famous for having introduced the Merino sheep to France. Excited to discover such an eminence in their midst, Cuvier

Georges Cuvier

cornered the man and began asking him questions, addressing him by name in the process.

Tessier was horrified. He'd bolted from Paris to escape the revolution, and was living under an assumed name in a rural village specifically to elude being recognized. "I am known, then," he blurted, "and consequently lost."

"Lost!" Cuvier replied. "No; you are henceforth the object of our most anxious care."

Relieved, the former royal agronomist grew increasingly impressed with the young man. "I have just found a pearl in the dunghill of Normandy," he wrote to a friend back in Paris. Word of the remarkable young man soon reached Geoffroy. Lonely for a colleague his own age, he arranged for an assistant's post in the department of animal anatomy and extended his invitation.

. . .

The two young naturalists hit it off immediately, forming a deep friendship that quickly took on the contours of a partnership. "Geoffroy and Cuvier knew no jealousy," one account reads. "Geoffroy had a position; he shared it with his friend; he had books and collections; they were open to his rival; he had lodging in the Museum; they were shared by his new brother." Their growing air of collaboration discomfited some Jardinistes, who gently reminded Geoffroy that he occupied their highest rank, while the new arrival was a low-ranking assistant. Far from being another Linnaeus, Cuvier was an inexperienced savant with dues to pay. Geoffroy did not give it much attention.

A peace of sorts began to settle over France after 13 Vendénmiaire, Year Four (October 5, 1795), when royalist forces marching into Paris were met with a rain of fire from artillery concealed in the winding city streets. The surprise barrage scattered the last vestiges of armed resistance to the revolution, and brought fame to the artillery commander, a young brigadier general named Napoleon Bonaparte. The monarchist cause was ended, but so too was the anti-monarchist bloodbath: Robespierre, chief architect of the Reign of Terror and the man who consigned Buffonet to the guillotine, had died under the same blade. As extremism gave way to the practical business of running France—a country now externally at war with several other nations—antipathies began to fade.

Count Lacépède emerged from hiding and returned to the Jardin, not minding that the reorganization had left him with no formal office. It gave him more time to resume working on the *Histoire Naturelle*. Lamarck attempted to open a new phase in his career, petitioning the Convention for funds to write a proposed *Systeme de la Nature*, which he described as a French-language "work analogous to the *Systema Naturae* of Linnaeus." Resigned to the triumph of the Linnean system, he no longer wished to replace it but instead to bring it up to date. But he made the error of describing Linnaeus's last published edition as "filled with gross mistakes, omissions of double and triple occurrence, and errors in synonymy."

The request was not approved. In the posthumous flowering of reputation that science historians would later call the *néolinnéisme*, Linnaeus remained a darling of populist culture, his lack of high-flown language and easily grasped classification system championed as a model of the democratization of thought. A poem of the period by René-Richard Louis Castel encapsulated the pervading spirit:

> Obtain, O Linné, this immortality!
> You came, the order appeared. A lively light
> Reflects suddenly the whole of nature

The deep dark bed of the various minerals
Agile child of the air and the inhabitants of the waters
The plants in the spring zephyr revived.
You knew all when you lived. You made all known to us.

Even though Lamarck had broken with Buffon by disbelieving in species change, his connections to the leader of the pre-revolutionary Jardin remained a stigma. "It is well known that Buffon, who did not understand the Linnaean system, nor chose to give himself any trouble to do so, had frequently censured Linnaeus," reads a 1794 account of the Société Linnéenne insurgency, underscoring what would become an enduring mythos: that Buffon's opposition to Linnean systematics arose not from an opposing complexist philosophy but from pigheaded ignorance. Linnaeus, on the other hand, "belongs to that small number of luminaries, who made a fresh epoch in the annals of literary greatness, raised their merit beyond the limits of their age, and rendered imperishable the splendor of their name."

• • •

By the time he'd risen to the rank of general, Napoleon Bonaparte fancied himself not only a soldier but a savant. His initial military success was founded on his considerable skills in deploying artillery, skills in turn based on his strong command of mathematics. When he was elected to the Académie des Sciences in 1796, it was not a courtesy title. When the Directory—the currently ruling junta—placed him at the head of a mysterious deployment in 1798, his preparations included an unusual move: He established a special Commission des Sciences et des Arts and invited 167 members of the savantry community to join, Geoffroy among them. Once they had assembled, the twenty-eight-year-old general made them a proposal. They could join him on his mission, on the condition that they depart in ignorance of his military briefing. They would board French warships for an expedition of unspecified duration, to a classified location.

The fact of the offer itself was a revelation. Until now, rumor had held that the troops and ships quietly massing in French ports were part of an invasion of England, which would not seem to require a special corps of zoologists, astronomers, chemists, and geographical

engineers. To twenty-five-year-old Geoffroy, this plunge into the unknown was too intriguing to turn down: Assured that his position at
the Jardin would be waiting upon his return, he signed on. So did all
but 16 of the 166 other savants, many of them swayed by the political
optics of refusing to come to France's aid in its time of need. As far as
Geoffroy was concerned, it was a shame that his friend Cuvier, the assistant animal anatomist, had not been deemed important enough to
warrant an invitation.

On May 19, 1798, Geoffroy sailed with Bonaparte on his flagship
l'Orient as part of a four-hundred-ship flotilla. Only when *l'Orient* was
out of the sight of land did the general inform him of their destination. France was invading Egypt.

. . .

Why Egypt? The tactic had been Napoleon's idea. He'd convinced the
Directory that the region was ripe for the taking—at the time it was a
much-neglected province of the fading Ottoman Empire, which
would be hard-pressed to defend it. It would give France a strategic
presence to disrupt British trade routes to both India and the East
Indies, and the Directory was especially intrigued by the idea of link-
ing the Mediterranean and the Red Sea with a canal (an idea that
much later became the Suez Canal). It would also be a cultural coup,
claiming the region generally regarded by Europeans as the cradle of
civilization. Which was why Napoleon went to the trouble of recruiting his corps of 151 savants: They provided the rationale that the
French were there not only to conquer but to discover and preserve as
well. The French fleet swept through the Mediterranean, conquering
Malta on the way and landing at Alexandria on the first day of July.

From a military standpoint, the invasion was not a success. Attacks
by the British fleet under Admiral Nelson resulted in France ceding
almost complete control of the Mediterranean Sea to the British. On
land, French control of Alexandria and Cairo wavered after Napoleon
failed to convince Egyptians that he was there to liberate them from
the Ottomans. The Egyptians' open rebellion led to the loss of more
than thirty thousand troops—half casualties of battle, half victims of
disease. In August of 1799, Napoleon abandoned the expedition and
returned to France, ostensibly to gather reinforcements but in reality

to take advantage of deteriorating political conditions. Within a month of landing, Napoleon had enlisted as a conspirator in a planned coup, staged his own coup *within* the coup, and emerged as First Consul of France with near-dictatorial powers. "The revolution is over," he proclaimed, ending an era. "I am the revolution," he added, beginning another.

Egyptian ibis

TWENTY-FOUR

Transformism and Catastrophism

TWO YEARS AND TWO MONTHS LATER, ÉTIENNE GEOFFROY Saint-Hilaire made his own return from Egypt. It was January of 1802, and the now twenty-nine-year-old Geoffroy was among the last of the 151 enlisted savants trickling back from the remnants of the expedition. In the wake of their leader's abandonment he'd made the best of the situation, conducting fieldwork in between troop movements and sending a steady stream of specimens back to the Jardin. It had been a heady, fascinating time, particularly when the troops melted away and he was left to his own devices, possibly forgotten but in a landscape deeply touched by time.

He found a hero's welcome awaiting him in Paris, and a rapidly transforming city. Napoleon had spun his unsuccessful military foray into a triumph of national pride, by presenting Egyptian artifacts as cultural coups. Egyptian statuary and votive objects were prominently on display at the Louvre (soon to be renamed the Musée Napoleon), and a tide of Egyptomania had swept fashionable France. Geoffroy walked through streets populated with women wearing turbans, past townhouses adorned with sphinxes and scarabs in a new architectural style called *retour d'Egypte* (return from Egypt). Other former members of the expedition were already hard at work adapting their sketches and field reports, which First Counsel Napoleon was eager to publish in lavish form, at public expense.

Within the Jardin gates, much had happened in Geoffroy's absence. He resumed his old position to effusive praise, but behind the scenes the balance of power was beginning to shift. In a series of deft moves, his friend Georges Cuvier had taken advantage of several factors—the lassitude of senior Jardinistes, the absence of 151 prominent members of the savantry community, and the general tide of revolutionary reform—to meld politics and natural history into a career-making trajectory. He'd finessed his way into temporarily occupying the Museum's chair of comparative anatomy, a position he would soon occupy permanently. He'd also secured a professorship at the Collège de France, a new institute reconstituted from the pre-revolutionary Collège Impérial. Combined, these roles gave him a solid base from which to consolidate influence and launch himself as a public figure.

Cuvier was, in fact, becoming famous, lecturing to crowded audiences that often seemed more interested in the speaker than the subject at hand. "He was then at his maturity, and might pass for a handsome man," one contemporary wrote. "His shock of red hair was now cut and trimmed in Parisian style. His dress was that of the fashion of the day, not without a little affectation." Geoffroy took in all of this with equanimity. Cuvier was still his friend. Geoffroy had never been much for grandstanding in the lecture hall anyway; his ambitions ran toward expanding his zoo and analyzing his trove of Egyptian specimens.

But the most impressive achievements in the Jardin belonged to Lamarck. Now fifty-eight years old, the professor of insects, worms, and microscopic animals had burst into the most productive phase of his career. Resigned to working within the Linnean hierarchy, he proceeded to expand it. Linnaeus "carried such great weight among naturalists that no one dared to change this monstrous class of worms," he wrote, while ushering *Vermes* into a new class for which he invented the term *invertebrates*. He moved crustaceans out of the order *Insectae*—a lobster, he concluded, was not a bug. Nor were spiders and other arachnids, which he ushered into an order of their own. Over his lifetime he would become second only to Linnaeus in the giving of names, assigning classifications to at least 1,700 species.

Another notable innovation had arisen from Lamarck's forced move from botany to zoology. He'd come to realize that "natural history," while acceptable as a category of savantry, was too vague a term for the discipline now taking on the recognizable contours of a science. Under natural history, botany and zoology had wandered along separate tracks. Having negotiated those tracks, Lamarck had coined a new term for a single science that encompassed both. He called it *biology*, the unified study of living things.

Under Linnean dominance, natural history focused on differences. Lamarck's biology focused on commonalities, even across the so-called kingdoms of animal and vegetable. Linnean natural history had a mystery at its core: The mechanisms and principles by which it operated were simply aspects of divine intent. Lamarck's biology was a discipline of dismantling, examining, and questioning.

Lamarck was not done surprising his young colleague Geoffroy. He'd parted ways with Buffon because he could not bring himself to believe his patron's theories on the fluidity of life, clinging instead to the fixity of species. But faced with the complexities of cataloguing nine-tenths of all animal life, he'd undergone a radical change in thinking. Such profusion was not the result of a single act but a process. Species were not fixed.

This had opened up new vistas in his work, in which he'd begun to expand upon the Buffonian concept of exogenesis. In the spring of 1800, during a series of public lectures on his freshly coined invertebrates, he began to follow through on the implications of his change of perspective. Buffon's musings on species change were general and conceptual, but Lamarck produced a fleshed-out theory centered on the environment, which he characterized as "the conditions of life." As these conditions changed, he envisioned species changing with them, in a two-part process.

The first process he called "the law of use and disuse." If an organism relied on specific aspects of its physiology, those aspects would be strengthened across the span of subsequent generations. If other aspects were no longer used, physiologies would conserve metabolic resources by rendering them vestigial, or by abandoning them entirely. Lamarck cited the genus *Spalax*, the blind mole rat, so adapted to life underground that while it still had eyes they no longer functioned: They were minuscule, and covered entirely by a layer of skin.

The second process was "the law of inherited characteristics." Once acquired, physical characteristics could be passed on to new generations—to which the first law would still apply, producing further changes. Thus life moved both laterally, along *l'influence des circonstances*, or the adaptive force, and vertically, via *le pouvoir de la vie*, the complexifying force.

Lamarck believed that this interplay of adaptation and sophistication, of use and disuse, accounted for species change. Upon Geoffroy's return, he was preparing for publication *Research on the Organization of Living Bodies*, the book that would debut his theory of what he called *transformisme*, or transformism.

. . .

Cuvier, however, had grand theories of his own. As acting chair of comparative anatomy, he had free rein over the increasing number of fossil specimens acquired by the museum. Their preponderance had convinced him that, in one regard at least, Buffon had been right: Extinction was a fact. As savants pieced together more and more fossil species, Cuvier came to the conclusion that across the geological scale of time, numerous lifeforms had died out, and new ones had emerged to take their place.

This did not, however, mean he believed in species change. Far from it—by his reckoning, every single species was the result of a unique act of creation. Citing what he called the "correlation of parts," Cuvier marveled at the intricacies of both fossil and living anatomies: how elegantly each bone coordinated with the next, how effectively they interrelated to constitute a functioning whole. Everything fit together as tightly and efficiently as the clockworks of a master craftsman, revealing an intelligent design so admirably suited for its environment that it was impossible to imagine incremental change.

How could one species possibly transform into another? Any modification to the design would upset the correlation of parts, causing it to no longer function. Cuvier thought Buffon's exogenesis, and even more so Lamarck's newer, more refined theory of transformism, were as ridiculous as saying a grandfather clock could, through the gradual replacement of parts, become a pocket watch while continuing to keep time.

How then, to account for the emergence of new life in the wake of extinction? Championing a doctrine that would come to be known as catastrophism, he argued that the Earth had undergone several periods of cataclysmic change, one of which had been the Great Flood referenced in the Bible. Some species survived these catastrophes, enduring unchanged into the present day. Others perished, and remained extinct. A third group of new species were simply divinely called into existence, in the wake of each catastrophe. The actions described in the book of Genesis had indeed taken place; it's just that there had also been repeat performances.

Lamarck's transformism and Cuvier's catastrophism arose at roughly the same time, and in diametric opposition to each other. Geoffroy's return provided the fuel for their conflict to erupt into the public eye.

. . .

In Egypt, Geoffroy had discovered that human mummies were vastly outnumbered by mummified animals—cats, crocodiles, dogs, jackals, snakes, and especially birds. One site, the catacombs of Tuna el-Gebel, held an estimated four million mummified birds, which curiously all seemed to be of the same species. Arriving with numerous examples from the Tuna el-Gebel trove, Geoffroy inadvertently laid the groundwork for what became known as the Sacred Ibis Debates.

The debates began in 1802, when Cuvier, Lamarck, and Lacépède jointly presented the mummies to the Académie des Sciences. After carefully unwrapping and examining several specimens, all three agreed that they were approximately three thousand years old, and that, in the words of Lacépède, "these animals are perfectly similar to those of today." Lacépède did not get further into the matter, choosing instead to return to his work on *Histoire Naturelle*. But Lamarck and

Cuvier drew very different conclusions, framing a public contention that continued for years.

Upon measuring the skeleton, Cuvier discovered that the mysterious bird species was not, as assumed at the time, the yellow-billed stork (*Tantalus ibis*)—in fact, it was not a stork at all, having a curved beak instead of a straight one. Digging around in the archives of the Jardin, he found specimens of an as-yet-unclassified bird that seemed a far better match. He named this species *Numenius ibis*, or sacred ibis (it would later be reclassified *Threskiornis aethiopicus*).

Comparing the mummified bones with the ones pulled from the shelf, Cuvier concluded that "we certainly do not observe more differences between these creatures and those which we see today than between human mummies and today's human skeletons." This he offered as definitive proof that Buffon and Lamarck were wrong. The categories of species were firmly fixed after all.

It was a glaring instance of circular logic. He had reconstructed an old skeleton, sought out a newer one that resembled it, then declared—on the basis of that resemblance—that they were one and the same species. And since his newly christened *Numenius ibis* hadn't changed in three thousand years, Cuvier concluded that therefore it would never do so in the future.

In his own address to the Académie, Lamarck replied that he'd have been surprised if the sacred ibis *had* changed. For one, the process of transformism could very well play out over far longer timeframes, but more important, the climate and conditions of Egypt had not changed appreciably since the specimens were mummified. Adaptation was the engine of transformism: It stood to reason that once a species adapted to an environment, change would be halted until the environment itself changed.

To people still struggling with the idea that the Earth was more than six thousand years old, the idea that three millennia were an eyeblink was difficult to grasp. A general consensus arose that Cuvier had won the debate, and decisively. As the late-nineteenth-century science historian William Locy concluded of Cuvier: "It is undeniable that his position of hostility in reference to the speculation of Lamarck retarded the progress of science for nearly half a century."

The debates sent Lamarck's reputation into eclipse, while firmly

Geoffroy and Cuvier in later life

establishing his opponent as one of the foremost figures in French science (a term increasingly replacing savantry). Capitalizing on his celebrity and heightened social status—he would eventually be ennobled as Baron Cuvier—the man praised for demolishing the theory of species change now tackled another concept he considered erroneous: the idea that all human beings were created equal.

The spurious pseudoscience surrounding the codification of humans into distinct categories began with Linnaeus, but it had taken on momentum with the 1795 publication of the third edition of *De Generis Humani Varietate Nativa* (On the Natural Variety of Mankind), by the German naturalist Johann Friedrich Blumenbach. In previous editions, Blumenbach had only slightly tweaked the categories of *Systema Naturae*, writing, "Linnaeus allotted four classes of inhabitants to the four quarters of the globe respectively. . . . I have followed Linnaeus in the number, but I have defined my varieties by other boundaries."

In his third edition, Blumenbach felt emboldened to make more fundamental changes. He replaced Linnean terminology (*Americanus rubescens*, etc.) with coinages of his own, deciding that "five principal varieties of mankind may be reckoned . . . Caucasian, Mongolian, Ethiopian, American and Malay." While he himself labeled these only "principal varieties," he cited others using the term *race* to the same purpose. Race it would be.

Blumenbach's categories did not neatly overlap those of Linnaeus. There were no Asians, only Mongolians. All Africans, including Egyptians, were Ethiopians (the region now known as Ethiopia was then more commonly referred to as Abyssinia). The Malay category included Australian aborigines and Polynesians. The Caucasians included Arabs, Jews, Armenians, Hindus, Persians, and nearly all Europeans (he excluded Lapps and Finns). In other words, each category included a wide variety of skin colors, which was exactly Blumenbach's aim. He was trying to remove Linnaeus's color-coding while at the same time maintaining the convenience of broad categories.

But *Caucasian* was more than a replacement for Linnaeus's *Europeanus albus*. While Linnaeus's categories were rife with prejudices, they were not prioritized; none was explicitly placed above the others. Blumenbach, however, did not hesitate to declare racial superiority. He was a collector of skulls, and in his opinion the prettiest specimen (in terms of pleasing proportions) in his collection was that of a female from the Caucasus, a mountain range between the Black and Caspian seas. Guided purely by personal aesthetics, Blumenbach wrote, "I have allotted the first place to the Caucasian," explaining, "I have taken the name of this variety from Mount Caucasus . . . because its neighborhood, and especially its southern slope, produces the most beautiful race of men. . . . In that region, if anywhere, it seems we ought with the greatest probability to place the autochthones of mankind."

Autochthones meant original forms. In examining his skull collection, Blumenbach imagined a line of descent from the Garden of Eden. Guided solely by personal aesthetics (of bones at that), he concluded that Adam and Eve must have been Caucasians. As the naturalist Robert Gordon Latham wrote in 1850, "Never has a single head done more harm to science than was done in the way of posthumous mischief by the head of this well-shaped female from Georgia." By assigning labels and declaring an "original" form of humanity, Blumenbach effectively combusted modern racism into existence.

Cuvier stood ready to amplify Blumenbach's aesthetic musings into outright assertions of racial supremacy. To him, Africans were "the most degraded of human races, whose form approaches that of the beast and whose intelligence is nowhere great enough to arrive at regular government." Meanwhile, "the Caucasian [race], to which we belong, is distinguished by the beauty of the oval which forms the head, and it is this one which has given rise to the most civilized nation—to those which have generally held the rest in subjection."

The phrase *to which we belong*, assuming an exclusively Caucasian readership, was an especially fatuous touch. For all his prejudices, Cuvier at least gingerly conceded to Buffon's definition of a species. "Since the union of any of its members produces individuals capable of propagation," he wrote, "the human species would appear to be single."

Other naturalists were not so certain. Charles White, the author of the 1799 work *Account of the Regular Gradations in Man*, took Buffon to task, arguing that surely the Creator could bestow separate species with the ability to interbreed. "There are but two ways of accounting for this great diversity in the human frame and condition," White concluded. "To suppose that the diversity, great as it is, might be produced from one pair, by the slow operation of natural causes. . . . Or to suppose that different species were originally created with those distinctive marks which they still retain."

White firmly believed in the latter. He was an adherent of *polygenism*, the doctrine that humans looked different because they were fundamentally different—God had created them at different times, and for different reasons. Reproduction was immaterial: Races were not generalized groupings but distinctly separate species.

Polygenism's roots were ancient. Paracelsus, Walter Raleigh, and Giordano Bruno all postulated various scenarios that took separate creations as a given. But polygenism as pseudoscience dates back to 1655, when a French theologian named Isaac de la Peyrere published *Men Before Adam*, a tract in which he argued that the truth of humanity lay in chapter five, verse thirteen of the New Testament's Epistle to the Romans: *Sin is not charged against anyone's account where there is no law.* To de la Peyrere, the story of Adam and Eve accounted for the origins of the Jewish people, not for humanity as a whole. Since Adam sinned, it followed that there was law in the Garden of Eden, and if there was law, there must have been what he termed "pre-Adamites," preceding populations that created such laws.

The theology was contortive, and controversial. *Men Before Adam* was burned in Paris, and de la Peyrere briefly imprisoned. But the ideas espoused gained a broad and persistent following, particularly

among those who added their own interpretation. The author's original intent had merely been to resolve some of the perplexing incon-
sistencies in the Bible—where, for instance, did Cain's wife come
from?—but polygenism dovetailed neatly with a rising interest in jus-
tifying conquest, colonial exploitation, and slavery. Under pre-
Adamite theory, those subjugated were divinely ordained to be
separate, and therefore unequal. "Our wise men have said that man is
the image of God," wrote Voltaire, one of the more prominent polygenesists. "Here is a pleasant image of the Eternal Being: a flat black
nose with no intelligence!" Polygenesists held particular scorn for
Buffon, whose theory of divergence from common origin struck them
as preposterous. "Doth M. Buffon think it sufficient to say dryly, that
such varieties may possibly be the effect of climate, or other accidental
causes?" wrote the Scottish Lord Kames in 1774. "We are put off with
mere suppositions and possibilities."

. . .

Lamarck had survived the French Revolution by maintaining a low
profile, and in the aftermath of the Sacred Ibis Debates he continued
to do so. He was reluctant to call attention to his domestic arrangements in Buffon's former residence, which were unconventional even
by Parisian standards. For fifteen years he'd lived with Rosalie Dela-
porte, his mistress and the mother of his six children, consenting to
marry her only on her deathbed in 1792. He remarried two or three
more times (the record is unclear), had two more children, and struggled to maintain his large family on what remained a minuscule stipend. Lamarck's last attempt to maintain a public profile came in
1809, when he requested an appointment to present a copy of his
masterwork *Philosophie Zoologique* to then-emperor Napoleon. He was
rejected, as Napoleon mistakenly thought it was a book of meteorol-
ogy, "which Napoleon considered unworthy of a member of his Académie des Sciences," one chronicle records. "The elderly zoologist
was reduced to tears." By 1810 he had begun to retreat even further
from society, this time to hide a secret: He was slowly but inexorably
losing his eyesight. Over the next eight years, he would go completely
blind. Relying on his daughters to help him read and write, he became
a virtual hermit in the Jardin des Plantes.

The aged, blind Lamarck

Still, with Geoffroy's encouragement, he continued to develop his thoughts on transformism. In his final book, *Analytic System of Positive Knowledge About Man*, published in 1820, he argued that diversity arose from different points of digression along a single path. "Reptiles . . . build a branching sequence, with one branch leading from turtles to platypuses to the diverse group of birds, which the other seems to direct itself, via lizards, towards the mammals," he wrote. "The birds then . . . build a richly variable branching series, with one branch ending in birds of prey."

It was a description of life's abundance as a reflection of unity, a series of branches on a common tree. It also bore striking parallels to Buffon's decades-earlier observation of Nature: "We should not be wrong in supposing that she knew how to draw through time all other organized forms from one primordial type."

The other protégés of Buffon began to fade away. In the postrevolutionary era, Count Lacépède had published two more volumes of *Histoire Naturelle*, one on cetaceans and one on fishes, in which he remained faithful to Buffon's complexism. "Why not proclaim an important truth? The species is like the genus, the order, and the class: It is basically an abstraction of the mind, a collective idea," he wrote, echoing his predecessor's cautions. Like Buffon he acknowledged the practicality of working in categories, calling them "necessary for apprehending, comparing, knowing, instructing" but warning that human convenience was not the same as physical reality. "Nature has only created beings that resemble each other and beings that differ," Lacépède wrote in 1804. All further demarcations were constructs of imagination, entities of reason.

He left off there. In the era of *néolinnéisme*, when even the calendar bore Linnaeus's stamp, it seemed pointless to continue to caution against species, genus, order, and class. Quietly, with little public notice, the writing of *Histoire Naturelle* came to an end. Lacépède drifted off to launch a brilliant second career in politics, becoming president of the French senate and a minister of state under Napoleon. But he remained loyal to the mentor who had created and bequeathed to him "a general and immense natural work." He died in 1825, at the age of sixty-eight. "I shall see Buffon again" were his last words.

• • •

Michel Adanson, the former Jardiniste who'd sought to create a natural system all on his own, died in abject poverty in 1806, his small pension from the Académie des Sciences having disappeared when the Académie itself disappeared during the revolution. His massive work, which he stubbornly refused to amend during his lifetime, went unfinished and unpublished. To note his passing, Cuvier delivered a bitingly dismissive eulogy. "Monsieur Adanson devoted himself to his great work," he said. "Henceforth his ideas were no longer fed or improved by those of any other. His genius now wrought upon its own foundations only, and these foundations underwent no further renovation."

But Adanson's work would not go entirely in vain. After nearly two decades of patient toil, Antoine-Laurent de Jussieu had at last published the first volume of his *Genera Plantarum*, a functional natural method for the classification of plants. While it retained Linnaeus's binomial nomenclature, *Genera Plantarum* applied them differently. Unlike the Linnean sexual system, which picked only one characteristic (the arrangement of stamens and pistils), Jussieu's system took multiple characteristics into account. But it prioritized some over

Michel Adanson

others, layering them in three categories: uniform characteristics, almost uniform characteristics, and semi-uniform characteristics.

Genera Plantarum arranged thousands of flowers into fifteen classes, which in turn encompassed one hundred "family" sub-groupings. Of the hundred families, thirty-eight were directly borrowed from Adanson's unfinished taxonomy, a debt Jussieu freely acknowledged.

Despite his reputation as a zealous guardian of his own work, Adanson had appreciated the tribute. He'd made a single request for his funeral, asking that his grave be adorned with a wreath of thirty-eight flowers. Each represented one of his contributions to the new Jussieu system.

Genera Plantarum, however, seemed destined to make almost no mark in botanical savantry. It had been published on the unfortunate date of August 4, 1789, coincidentally the day of a pivotal event. On that date the National Assembly abolished France's feudal system, stripping away the rights of nobility and setting the revolution on its radical, reactionary path. A new book by a little-known Jardiniste attracted little notice. Jussieu abandoned botany, taking part in Paris's municipal government and rising to the directorship of the city's hospitals. He would eventually return to the Jardin and work on a new edition of *Genera Plantarum*, but failed to find enough interest to justify publication. Only the introduction to a second edition would see print, and that only after his death in 1836.

. . .

Adanson's was not the last grave Cuvier would dance upon. When Lamarck died in 1829 at the age of eighty-five, Cuvier wrote a eulogy but did not deliver it in person—diplomatically so, as it contained not a scintilla of praise. In Lamarck, he concluded, "too great indulgence of a lively imagination had led to results of a more questionable kind." He went on to attack Lamarck's "attachments to systems so little in accordance with the ideas which prevailed in science." Such attachments "were not calculated to recommend him to those who had the power of dispensing favours," Cuvier concluded, all but pointing at himself.

Geoffroy was among the mourners at the sparsely attended funeral, a ceremony made all the more dreary by the knowledge that Lamarck had died without the funds to purchase even a proper burial—the Montparnasse cemetery grave he was being lowered into was a leased one, good for only five years' occupancy. "Blind, poor, forgotten, he remained alone with a glory of whose extent he himself was conscious," Geoffroy remembered of his mentor and friend, "but which only the coming ages will sanction, when shall be revealed more clearly the laws of organization."

The five-year lease on Lamarck's grave ended in 1835. His remains were exhumed and placed in a common pit. Its location has long been forgotten.

. . .

By then Cuvier himself was dead, one of nineteen thousand victims of the 1832 cholera epidemic that swept through Paris. But his ambitious graspings at fame and political favor had not gone unrewarded. He died a baron, and when the Eiffel Tower rose in 1887 it bore his name in gilded letters, a part of a gallery of "great French men of Science." To create a new position for his son Frederick Cuvier, the museum directors stripped Geoffroy of his post as head of the Jardin's zoo, a job he'd held since he created it in 1793. The humiliated Geoffroy retreated from the public eye; he died in 1844.

Geoffroy's son Isidore, who followed him into natural history, shared his father's appreciation for the Jardin's pre-revolutionary intendant. "Buffon is to the doctrine of the mutability of species what Linnaeus is to that of its fixity," Isidore wrote. "It is only since the appearance of Buffon's *Histoire Naturelle*, and in consequence thereof, that the mutability of species has taken rank among scientific questions."

Lamarck's theory of transformism, so vigorously attacked within France, evinced little contemporary attention elsewhere. One of its few appraisals had appeared in 1826, in an anonymous review published in a short-lived journal in Edinburgh, Scotland. The review was notable on two counts. First, the anonymous reviewer, seeking an English substitute for *transformisme*, decided on the term *evolution*—a coinage that gave the word its modern meaning. Second, the journal's publisher (and likely the anonymous reviewer) was Robert Jameson, a professor of natural history at the University of Edinburgh. One of his students at the time was an eighteen-year-old named Charles Darwin.

*James Edward Smith and the (imaginary) Swedish frigate,
chasing the brig carrying Linnaeus's collection to England*

TWENTY-FIVE

Platypus

WHILE FRANCE WAS EMBRACING ITS ERA OF *NÉOLINNÉISME*, A
parallel yet fundamentally different posthumous revival of Linnaeus
was taking place on the other side of the English Channel. The French
had recast Linnaeus as an egalitarian revolutionary, but the British
were beginning to celebrate him as a champion of intellectual conquest, embracing his systematics as a comfortable bulwark of empire.
After an indifferent initial reception, Great Britain had spent four

decades gradually warming to Linnaeus's ideas. But the shift from grudging acceptance to adulation had begun two days before Christmas in 1783, when a young dilettante accepted a breakfast invitation.

Twenty-four-year-old James Edward Smith was a gentleman of leisure, but uncomfortably so. Not content with being the son and heir of a wealthy Norwich wool merchant, he hoped to make his mark in the world of British savantry. He'd dabbled in natural history and briefly studied medicine at Edinburgh, but nothing had quite jelled as a vocation. At loose ends in London, he was pleased to be asked to breakfast at 32 Soho Square, the elegant London townhouse of Sir Joseph Banks.

At forty, Banks was already a towering figure in natural history. Like Smith, he'd been born to considerable wealth and drawn to savantry, but he'd pursued his interests with an unrelenting drive that rivaled that of Buffon. As a student at Oxford, disappointed that the university offered no instruction in botany, he'd personally hired a Cambridge professor to lecture on the subject for himself and his friends. By the time he was Smith's age, he was a member of the Royal Society, a consultant in botanical matters to King George III, and an accomplished expeditionary. His 1766 voyage to Newfoundland and Labrador onboard the frigate HMS *Niger* had yielded up unique specimens of hundreds of plants and thirty-four novel species of birds. Once it was clear that the twelfth edition of *Systema Naturae*, published the previous year, would be Linnaeus's last, Banks arranged his specimens according to Linnean classification himself.

His second expedition brought him a knighthood and lasting fame. In 1768, Banks headed up the naturalist contingent of a scientific voyage to the South Pacific, paying out of his own pocket to staff and equip a team of botanists, draftsmen, and artists. They were aboard the HMS *Endeavour* under the command of Lieutenant James Cook, and sailing into history. Arriving in Tahiti later that year (just months after the Bougainville expedition), the *Endeavour* expeditionaries sailed on to New Zealand in 1769, where they became only the second group of Europeans to make landfall. In 1770, they were the first Europeans to come ashore on the continent now known as Australia. Banks's bounty from all three landfalls was thousands of specimens previously unknown to Western science—the kangaroo, the acacia, the eucalyptus, and a myriad of other species. Thirteen years later, Banks, now president of the Royal Society, was still laboring to organize and classify his trove. On the morning of December 23, 1783, Banks paused his breakfast with Smith to sort through the mail. He found a letter from Uppsala, written by Johan Gustav Acrel on behalf of widow von Linné, offering her late husband's collection for sale.

Sir Joseph had little interest in the legacy of Carl Linnaeus. He had a vast collection of his own, and while he'd paid tribute to the man upon Linnaeus's death five years earlier (writing to Linnaeus the Younger that "I have always had the highest respect for that valuable man"), in private he was following with great interest the in-progress taxonomic reforms of Adanson and the younger Jussieu; with *Systema Naturae* now seventeen years stale, he was open to innovations. He flung the letter across the table to Smith, casually suggesting that the young man consider purchasing the collection himself. The conversation turned to other matters.

Smith was eager to ingratiate himself with Banks, but he was also a far greater admirer of Linnaeus than Banks might have imagined. He would later write of a transcendent moment in his childhood, when he first discovered a copy of *Systema Naturae*.

> Nor shall I ever forget the feelings of wonder excited by finding his whole system of animals, vegetables and minerals, comprised in three octavo volumes. I had seen a fine quarto volume of Buffon, on the Horse alone. I expected to find the systematical works of Linnaeus constituting a whole library; but they proved almost capable of being put, like the *Iliad*, into a nutshell. Hence a new world was opened to me.

Buffon had overwhelmed him. Linnaeus, with his tidy hierarchies, made the whole of nature seem capable of being grasped, even by a child. During his brief stint studying medicine at Edinburgh, Smith had paid rapt attention to lectures by Dr. John Hope, the first botany professor to teach the Linnean system there. In what he would come to consider a semi-mystical connection, he noted his first day of studies had been January 10, 1778, the date of Linnaeus's passing.

Smith immediately wrote to Uppsala, offering a thousand guineas—

about $250,000 in modern currency—but only if he was purchasing the entirety of Linnaeus's collection: rare books, valuable mineral specimens, and all. Acrel replied with a meticulously compiled inventory, and the deal was consummated. The British brig *Appearance* arrived from Stockholm in October of 1784, carrying twenty-six crates containing 14,000 specimens of plants, 158 of fish, 3,198 of insects, 1,564 of shells, and a library of 1,600 volumes.

Unpacking the crates, Smith found an unexpected bonus. Seeing no reason not to clean house, the widow Linnaeus had thrown in the whole of her husband's papers as well—some three thousand original letters and dozens of original manuscripts. They'd served as packing material, helping to cushion the specimens in transit.

The inclusion of the manuscripts and letters changed the nature of Smith's acquisition. He'd expected to defray the cost by selling off some of the more valuable books—he had, in fact, already promised one to Sir Joseph Banks for one hundred guineas, or a tenth of what he'd paid for the entire collection. But the addition of the archives made him feel less like a collector than the recipient of a legacy. What was he going to do with the contents of these twenty-six crates? He wasn't sure. He rented rooms in Chelsea to display them, and invited his learned friends over. The informal exhibition earned him election to the Royal Society, but the honor only intensified his quandary. Well aware of his dilettante status and wanting to be known for something more than an acquisition, he placed the collection in storage in 1785, and took off on an extended tour of the Continent.

In his three years abroad, Smith visited Italy and Holland, where he followed in Linnaeus's footsteps by picking up a fast-tracked medical degree. He also passed several times through pre-revolutionary Paris, where he met and recorded his impressions of several Jardinistes. He liked Antoine-Laurent de Jussieu, even though he "takes the lead among those who with respect to system may be called Anti-Linnaeans."

> He is a true philosopher, profound in science, ardent in the pursuit of truth, open to conviction himself. . . . His manners are gentle and pleasing, his conversation easy, cheerful, and polite. Although we differed on many points, as on the laws of nomenclature, and

the merits of the Linnean system, yet as truth was our common object, repeated and free discussions increased our esteem for each other.

He was less of a fan of Lamarck: "Equally devoted to botany, but his character is less pleasing. . . . I freely acknowledge that I shrunk from the society of a man who always had occasion to attack with violence what he knew to be my most favourite sentiments." Smith was even more repelled by Adanson, "whose knowledge of botany would procure him great reputation, were he less a slave to paradox and pedantry. He generally accosted me with some attack on Linnaeus, sometimes calling him grossly ignorant and illiterate."

Smith emerged from the tour convinced of his life's purpose: He was destined to be the keeper of Linnaeus's flame. "In spite of all opposition, the system of Linnaeus is even now become universal, every part of the world abounding with his disciples," he wrote (although this was far from the case), while

> the Families des Plantes of Monsieur Adanson, professedly written to supersede it, is only occasionally read by those who are disposed to amuse themselves with whimsical paradoxes, presenting themselves in a preposterous orthography, which renders them still more ridiculous, and unintelligible.

• • •

The initial reception of Linnean systematics in the British Isles had been a skeptical one. In 1736, the naturalist John Ammon dismissed "some systematical tables concerning Natural History, composed by Dr. Linnaeus," disbelieving that "very much if any Botanist will follow his lewd method." In 1759, Benjamin Stillingfleet considered Linnaeus already irrelevant, remarking, "His works, I imagine, are little known, except to a few virtuosi who have a more than ordinary curiosity." In 1761, Sir John Hill (who had been one of the first botanists outside Sweden to adopt the sexual system) ultimately concluded that "Linnaeus' method has pleased by its novelty; but it is false in the principles, and erroneous in his conduct of it. His discoveries have scarce done more service, than his method hurt to the science." The

English physician Richard Brookes began his 1763 work *A New and Accurate System of Natural History* by taking Linnaeus and his followers to task. "Their attempts to reduce the names of plants into a system, has rendered the study more difficult and prone to error," he wrote, "than it would have been if the Student had only used his sight for the distinguishing of plants, and his memory for registering them."

Yet 1763 was also the year pro-Linneans gained a strategic foothold. That was when the British Museum hired a visiting foreigner to organize their botanical collection, recently much expanded by bequest of specimens upon the death of Sir Hans Sloane, a wealthy and influential naturalist. Sloane, who in life had been one of Linnaeus's more notable detractors, had thoroughly documented his collection along the lines of his own preference. But the British Museum's new hire was Daniel Solander, the apostle consigned to England after intemperately falling in love with Linnaeus's daughter Elisabeth Christina.

Solander (who would later sail with Banks on board the *Endeavour*) was a spurned son-in-law, but still a faithful acolyte of Linnaeus; he wasted no time in stripping away the old designations and substituting those he'd learned in Uppsala. In time his efforts extended to every plant and animal on display. At his prompting the museum built new cabinets for storing dried plants—cabinets constructed exactly to a plan described by Linnaeus in his *Philosophia Botanica*. In botanical circles, this had the effect of decisively moving Linnean taxonomy toward the status of a *lingua franca* for British savantry. Pro-Linneans grew in influence and confidence, to the point that in 1785 the translator of an English edition of *Histoire Naturelle* refused to include Buffon's extensive commentary on the dangers of classification, because

> the chief intention of these discourses is to ridicule the authors of systematic arrangements, and particularly the late ingenious and indefatigable Sir Charles Linnaeus, whose zeal and labours in promoting the investigation of natural objects merit the highest applause.

Edward Smith was eager to stoke such sentiment. On February 26, 1788, before assembled friends in London's Marlborough Coffee

House, he announced the formation of a new learned society. Its first meeting, on April 8, enrolled twenty full-fledged members, thirty-nine foreign members, and eleven associate members. These members, in turn, elected Smith president—a position he held until the end of his life—and adopted a name: the Linnean Society of London.

This spelling was a result of Smith's other meetings in Paris, with members of the then-clandestine Société de Linné de Paris. An accurate English adaptation of the name Linnaeus would be *Linnaean*, but Smith wished to support the Parisians' more populist adoption of the name *Linné*. One of the Society's first orders of business was extending honorary memberships to the founders of the SLP. *Linnaean* it would be.

Missing from the founders' meeting was Sir Joseph Banks, who accepted only an honorary membership. He was skeptical of an organization centered around Linnaeus but welcomed it at arm's length. "I incline to think it will flourish," he wrote, "as great care is taken in the institution to keep out improper people."

Flourish it did. Membership grew to 209 by 1800, and to 463 by 1820. Smith, however, did not donate his collection to the society; instead he retained ownership, even taking it with him to his native Norwich when he opted to spend most of his time there. Still, he did his best to burnish Linnaeus's posthumous reputation. He quietly sold off the collection's rock specimens, eliminating reminders of *Systema Naturae*'s misbegotten classification of the mineral kingdom. He also indulged in outright mythologizing, encouraging a rumor that the king of Sweden, belatedly recognizing the treasure trove of the collection, had unsuccessfully sent a frigate to chase down the *Appearance* on its way to England. It is unknown who started the rumor, but despite knowing the voyage had been without incident, Smith certainly did nothing to dispel it. In an 1807 portrait included in a massive, full-color tome entitled *The Temple of Flora*, a rendition of the imaginary pursuit looms larger than the image of Smith himself. Another page of the *Temple* features a painting with clear connotations to the victory of the Société de Linné in the Jardin du Roi. A bust of Linnaeus is allegorically wreathed in laurels and floral garlands, delivered by the gods themselves.

Smith was knighted in 1814. The following year brought Waterloo

Aesculapius, Flora, Ceres, and Cupid honoring the bust of Linnaeus,
from The Temple of Flora *(1807)*

and the end of the Napoleonic Wars. Britain now held undisputed control of the seas, and new colonies ceded by the brokered peace—colonies that had previously belonged not only to France but to Spain and the Netherlands. Britain took control of Ceylon and the African Cape Colony, and expanded its sphere of influence to China via the opium trade. In 1818, the British East India Company quelled the last rebellion against its de facto control over most of India. In 1819, the United States and the United Kingdom decided to share the Oregon Country, which gave them both North American territory extending from the Atlantic to the Pacific. The Americans began building a continent-spanning nation, and the British set about building an empire.

In this new age of expansion, classification became another form of conquest. What better way to "civilize" a region than to inventory its flora and fauna, scouring them of native names and naming them anew? British naval ships, less burdened in peacetime, now had ample room for naturalists to accompany them on voyages. British citizens abroad, a rapidly growing colonial class, shipped thousands of specimens homeward. Seeing the value of aligning political interests with developments in natural history, the British government threw its support behind the Linnean Society. Prince Regent George Augustus (the effective ruler and future George IV) became the group's official patron in 1814, and Smith was granted a knighthood soon thereafter. By 1817, the man who had impulsively tossed him a letter thirty-four years earlier was diplomatically praising him. "I admire your defense of Linnaeus's natural classes. It is ingenious and entertaining, and evinces a deep skill in the mysteries of classification," Sir Joseph Banks wrote to Sir James Smith. "I fear you will differ with me in opinion when I fancy Jussieu's Natural Orders to be superior to those of Linnaeus. I do not, however, mean to allege that he has even an equal degree of merit in having compiled them."

Banks's apologetic tone was a tacit acknowledgment of a now-dominant worldview. Smith felt no need to debate the matter. In 1821, the society moved into 32 Soho Square, Banks's former townhouse and the site of their fateful breakfast.

By the time of Smith's death in 1828, amateur natural history was well established as a popular hobby in Britain, promoted by Linneans touting the pleasures of classification itself. "A person who is in the pursuit of the Class and Order of any unknown flower may be said to be upon a botanical journey," wrote John Thornton in *The Temple of Flora*. "If he can read the botanical characters impressed on it by the pen of Nature, he will certainly, following system, very soon arrive at his journey's end."

In earlier generations, the cabinet of curiosities was an elite possession, emphasizing rare (and therefore valuable) possessions. The botanical collections of Regency and Victorian Britain connoted membership not in the upper class but in the prosperous middle. Displays of pressed flowers and stick-pinned butterflies became commonplace in British households, advertising the possession of leisure hours, the ability to travel, and a level of education sufficient to affix impressive-sounding Latin names. "Thousands became attached to the pursuit," the botanist James Drummond wrote in 1849.

> It became a favorite occupation with the learned, and indeed of the relatively unlearned, to examine flowers, and refer them to their classes and orders, and finally, through this easy introduction, to ascertain their names and history; and thus a love of nature was increased, or generated, to an extent before unknown. . . . To every ramble was thus super-added a source of pleasure before unknown.

By then it was accepted fact: Nature could be contained, linguistically as well as physically. In the words of one nineteenth-century biographer, Linnaeus's achievement was no less than decoding "that language in which the history of plants is written: The botanical terms correspond to letters, the names of plants to words, and the system to the grammar." The world had the genius of Uppsala to thank for ending the benighted era of misguided complexists.

> False impressions prevailed . . . that nature could only be understood from hair-splitting interpretation. Then appeared Linné, and his activity can be likened to a fresh wind, driving away mists and showing a free prospect over a sunny landscape. Natural science, formerly a neglected child, who seldom came into view, soon became a cherished possession of high and low, old and young.

The naturalist Adolph Koelsch put it more succinctly: Linnaeus was nothing less than "God's Registrar."

· · ·

Back in Sweden, the other remnants of Linnaeus's legacy began to fade. Carl the Younger's successor as professor at Uppsala was Carl Peter Thunberg, who'd had the distinction of being the senior Linnaeus's last apostle, as well as the one who had traveled the farthest. Linnaeus had assigned him to the farthest reaches of the Far East, and it had proven the most challenging assignment of all.

In December of 1771, Thunberg joined the Dutch East India Company as a ship's surgeon. The posting took him as far as the Dutch-controlled colony of Cape Town in South Africa, where he resided for three years pursuing his secret agenda. He wished to learn the Dutch

language thoroughly enough to pass for a Netherlander, as his ultimate destination was Japan, a nation where the Dutch were the only
Westerners allowed. The final leg of his passage was underwritten by
Dutch floral enthusiasts, who expected him to bring back exotic flowers from the closed land. It was a risky proposition. "The voyage to
Japan is reckoned to be the most dangerous in all the Indies," Thunberg wrote, "and the Dutch East India Company always considers one
out of five of their ships send thither, as lost."

Thunberg arrived in the hermit nation in 1775, or rather he was
allowed to approach it as closely as possible. Japan, shut off from
all foreigners and especially from Western barbarians, maintained a
foothold for Europeans on Deshima, a tiny artificial island in the Bay
of Nagasaki, where Dutch traders were quarantined and allowed a limited trade. Thunberg's masquerade was successful, as he soon learned
that the Japanese assigned to deal with the traders spoke only snippets
of antiquated Dutch. But he was effectively imprisoned in the compound, unable to set foot on Japan itself.

To acquire botanical specimens, Thunberg had taken to covertly
examining the fodder brought in to feed the animals. Surreptitiously
trading Western medical knowledge for briefings on Japanese animals
and plants, he'd found himself on the track of the *totsu kaso*, an organism that, if real, confounded the foundations of Linnean taxonomy: "a
very singular plant, which in the summer was a crawling insect but in
winter a plant." Thunberg moved heaven and earth to obtain a specimen. When he did, "I plainly saw that it was nothing else than a cater-
pillar, which against its approaching change to a chrysalis had crept
down into the ground." Since only one Dutch ship was allowed to
depart each year, Thunberg whiled away the months until he could
leave.

He returned to Uppsala in March of 1779, little more than a year
after Linnaeus's death. When Carl the Younger passed away four
years later, the professorship was his to claim. He opposed the sale of
the Linnean collection, but was powerless to stop it.

Having watched the collection disappear, Thunberg was more than
willing to let the past continue to recede. He made no efforts to update *Systema Naturae*, or any of the work of Linnaeus Senior or Junior.
Considering Linnaeus's garden cramped, and unhappy with its loca-

tion (it was subject to flooding by the river Fyris), in 1787 Thunberg convinced King Gustav III to donate land for a new Uppsala Botanical Garden on the grounds of Uppsala Castle. After closing the original garden in 1802, Thunberg selected only some of the plants for transplantation. The process went slowly, but the new garden was opened to the public in 1807. Thunberg honored his predecessor by naming a structure on the grounds the Linneanum, holding a small ceremony to dedicate it on Linnaeus's birthday. But the garden was laid out according to his plans, not Linnaeus's. He did not choose to bring over the animals, which at the time included marmosets, cranes, peacocks, parrots, an ostrich, and a cassowary. These were left behind for others to claim.

Linnaeus's rise from that overgrown garden, a half-century earlier, had come full circle. As it had before the advent of Linnaeus, the old Uppsala garden fell once more into neglect. In coming decades the site would be reused as parkland and potato fields. Its structures became a meeting hall for student clubs, then a museum of Nordic antiquities, then a storage space for athletic equipment.

· · ·

Who, then, became Linnaeus's taxonomic successor?

Not Smith, who lovingly curated the original manuscripts, nor even the Linnean Society itself. *Systema Naturae* would not see new editions but instead became the foundation for a growing pool of crowd-sourced taxonomy, fueled by a rarely noted aspect of the Industrial Revolution. The technology of printing was undergoing rapid advancements, beginning with the invention in 1797 of lithography. The new reproduction technique not only allowed for more detail than dry-point metal engraving, but its stone-based plates could be used longer without wearing out. Printing itself became a significantly cheaper process, with more efficient press designs adding ink to both sides of paper simultaneously. Once the power of steam engines was added to the equation, presses went from turning out a few dozen sheets per hour to more than a thousand.

Suddenly, hundreds of academic institutions, learned societies, and even informal naturalist groups could afford to print specialized journals, and publish they did. At the beginning of Linnaeus's and Buffon's careers, only a handful of well-funded journals could publish with any regularity. By 1799, England was home to 35 scientific journals, France 54, and Germany a startling 304. According to the rules Linnaeus established in *Philosophia Botanica*, the right to name a species fell to its "discoverer," by which he meant the first to describe it in a publication. The boom in scientific journals ushered in something of a free-for-all, with individual naturalists coining Linnean taxonomies and rushing news of their discoveries into print as quickly as possible, the better to establish precedent. This led to occasionally unruly results and dubious compromises.

One example of this came in 1799, when a small keg arrived at the quayside docks of Newcastle-upon-Tyne in the northeast of England, addressed to that city's Literary and Philosophical Society. A servant woman was sent to fetch it. Walking back from the dock, she balanced the keg upon her head and was horrified when the keg's butt gave way, soaking her in alcohol and depositing at her feet the soggy remains of "a strange creature, half bird, half beast."

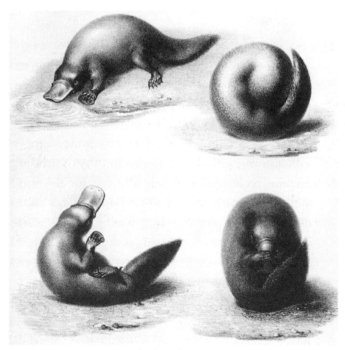

Ornithorhynchus anatinus

How did such a creature come to exist? The governor of New South Wales, who had sent the specimen from Australia, believed it was the result of frenzied copulation between multiple species in the Outback. It was called *dulaiwarrung* by indigenous Australians and is now more commonly known as the platypus. *Platypus* means "flat-footed," a surprising choice considering the feet are one of the creature's least extraordinary aspects.

Not since the Hydra of Hamburg had a specimen so clearly straddled multiple categories in Linnean taxonomy. And unlike the Hydra, it was real, albeit so bizarre that it would take several more specimens to fully convince European naturalists. Here at last was a true instance of Linnaeus's old order of *Paradoxa*—a living vindication of Buffon's assertion that systematics could not capture a Nature which "advances by imperceptible nuances so that it is impossible to describe it with full accuracy by strict classes, genera, and species."

But dismantling the Linnean hierarchy was now out of the question. Anomalous as the platypus might be, it needed to be tacked into now-standardized systematics. As soon as additional specimens made it clear the creature was not a hoax, it was Lamarck, of all people, who took up the task of fitting it into the existing hierarchy. He designated an entirely new order, *Prototheria* ("first beast"), and created the unique genus of *Monotremata*, where, after two hundred years, it has since been joined by only one other animal, the echidna. But the order and genus belonged to the class *Mammalia*, a compromise against which Geoffroy strongly protested. The platypus lays eggs, he pointed out. While it nurtures its young with a form of milk, it has no nipples; the milk is sweated into existence. How could such a creature possibly be a mammal?

More squabbles lay in wait. A British doctor named George Shaw had published a description in a 1799 edition of the *Naturalist's Miscellany* after examining a dried specimen. Still uncertain as to whether or not it constituted a hoax, he cautioned that "this paradoxical quadruped must be left to further investigation," but nonetheless named it *Platypus anatinus*, a somewhat clunky amalgam of Greek (*platypus* meaning "flat-footed") and Latin (*anatinus* meaning "like a duck"). This publication went unnoticed by the German professor Johann Blumenbach—the same Blumenbach who invented the term

Caucasian—who one year later reported his own examination of a preserved set of skin, bill, and bones in the pages of his natural history journal *Abbildungen*. Considering himself "under the necessity of christening it," he coined the name *Ornithorhynchus paradoxus*, a title equally cumbersome (it means "bird-snouted, paradoxical") but at least consistently Greek. Shaw's name clearly had precedence, but he'd not been fully confident that the animal on his examination table did in fact exist.

Given that uncertainty, was the right of species designation truly his? With no central authority to resolve the question, the scientific community continued to use both names for several years, during which the term *platypus* entered common parlance among Europeans (the aboriginal Australian name being roundly ignored). Finally, someone noticed that *Platypus* had been used in 1793 to name a genus of beetles, which invalidated Shaw's terminology: Under Linnean rules a genus name could not be used more than once, no matter how disparate the organisms in question. The platypus could not be *Platypus*.

The *Ornithorhynchus paradoxus* and the *Platypus anatinus* camps reached a compromise. The species would be called neither "flat-footed like a duck" nor "bird-snouted, paradoxical" but a combination of the two: *Ornithorhynchus anatinus*. The name was descriptively useless (it meant "bird-snouted like a duck"), but both parties were appeased.

The bizarre nature of *Ornithorhynchus anatinus* only grew clearer upon further study. It has no stomach. The male of the species is equipped with poisonous spurs. It stores nearly half of its body fat in its tail, as a mobile reserve of energy. It generates an electrical field, by which it perceives its environment. Yet instead of rendering Linnean taxonomy ridiculous, the platypus itself was seen as ridiculous. As the poet Oliver Herford wrote:

> My child, the Duck-billed Platypus
> A sad example sets for us:
> From him we learn how indecision
> Of character provokes Derision.
> This vacillating Thing, you see,
> Could not decide which he would be.
> Fish, Flesh, or Fowl, and chose all three.

With a profusion of scientific journals, improved preservation techniques, and a continued flood of new specimens, similar taxonomical clashes only increased. By 1850, the genus name *Argus* had been improperly applied to ten different genera, including two types of mollusks, two butterflies, and a spider. Botanists moved toward standardization in 1867, when they drafted and adopted the Paris Code, an international agreement on the nomenclature of plants that later became the International Code of Botanical Nomenclature. Because its lexicon needed a point of origin, the ICBN formally fixed Linnaeus's 1753 edition of *Species Plantarum*—the first botanical spinoff volume of *Systema Naturae*—as the official starting point for plant names. In 1895, a second delegation founded the International Commission on Zoological Nomenclature (ICZN), becoming the central authority for fixing species names of animals. They selected as their starting point the 1758 tenth edition of *Systema Naturae*, in which Linnaeus fully applied binomial names. Unless adopted by Linnaeus himself, species names published prior to these specific volumes were deemed to have no scientific validity. As legislated by the ICBN and the International Commission on Zoological Nomenclature, both botany and zoology officially began with Linnaeus.

While separate entities, both governing bodies agreed to mutual rules. One determination was that no genus name should be used more than once within the same kingdom, a rule that swept away nine of those *Argus* genera (leaving only the earliest usage for gastropods, dating back to 1761). Names could, however, be applied to both animals and plants: Hence *Iris* is both a flower and a kind of mantis. *Darwiniella* is both an orchid and a barnacle. *Cannabis* is both a hemp and a bird. Yet other rules seem to reinforce not clarity but rigidity. One is a prohibition against changes to binomial names, no matter what the circumstances. *Wisteria*, for instance, should have been called *Wistaria*, as its first describer named it after one Caspar Wistar, a colonial physician and author. Though the name was misspelled, it can never be corrected: *Wisteria* it is. The scientific name for the bald eagle is *Haliaeetus savigny*. It will forever contain a typo, in the form of the extraneous and accidental second *e*. Not even egregious tributes can be erased. A blind, cave-dwelling beetle discovered in Slovenia in 1937 bears the name *Anophtalmus hitleri* because its discoverer was a

Nazi entomologist. Calls to change the name have been ignored. "It was not offensive when it was proposed," the president of the ICZN said in 2022, "and it may not be offensive one hundred years from now."

. . .

Another lasting consequence of the now-dominant systematist worldview arose from a self-imposed blind spot. Because Linnaeus divided life neatly into two kingdoms, *Animalia* and *Plantae*, that demarcation hardened into a wall over time, discouraging communication between zoologists and botanists and excluding the possibility that some life belonged to neither kingdom.

Such rigid thinking was particularly acute in the case of microbial life, which by the 1830s was blooming into fantastic variety and detail under the eyes of biologists armed with increasingly powerful microscopes. As naturalists tried to populate the singular species *Chaos chaos* with discernible variations, they found themselves wondering if Buffon had perhaps been correct: that some might be animals, some might be plants, and some might be denizens of neither kingdom. Georges Cuvier decided to settle the matter for good by renaming the catch-all species *Chaos animalia*. This, of course, settled nothing, and only served to reinforce the growing divide between botany and zoology. Lamarck, a botanist compelled to become a zoologist, had coined the term *biology* to encompass a new science, the unitary study of life in all its forms. In practice the disciplines remained largely separate, with neither believing it had much to share with, or learn from, the other.

One stark example of this arose in 1838, during a chance conversation between two dinner guests in Germany. One was Matthias Schleiden, a thirty-four-year-old botanist at the University of Jena. The other was Theodor Schwann, a twenty-eight-year-old assistant at the University of Berlin's Museum of Anatomy and Zoology. The two strangers chatted while coffee was served, and Schwann mentioned an interesting development in his work: His newer, more powerful microscope was allowing him to look not only *at* individual cells but *within* them. He'd discerned that despite their different configurations, cells from different parts of a plant appeared to have a common interior anatomy, with a dense region or *nucleus* (from the Latin for "seed") more or less at its center.

Schleiden was struck by the coincidence: He was also working with tissue samples under an improved-model microscope and had also noticed this commonality. The two men immediately put down their coffee cups and strode over to Schwann's laboratory, where they examined his slides and compared notes.

Later that year, the botanist Schleiden published a theory that plants were made up *entirely* of cells or their by-products. While mature individual cells might look quite different from each other, their underlying anatomy pointed to a common origin—a theory that went a long way toward explaining how a large tree could grow from a tiny seed, or how even a single cell could be alive. The following year, the zoologist Schwann published *his* theory that "there is one universal principle of development for the elementary parts of organisms, however different, and that this principle is the formation of cells."

Neither, however, cited the other in his publication. Although Schleiden and Schwann remained cordial to each other for the rest of their careers, each stuck to his specialty. The idea of truly collaborating, of searching for further commonalities across the barrier between animal and vegetable, appears to have occurred neither to them nor to their colleagues. As zoologists and botanists explored the "cellular theory" along parallel but separate tracks, their focus remained on complex organisms. Single-celled ones, despite increasing proof of their abundance, were largely ignored.

Buffon's proposal of living organic particles ("an assemblage of these particles constitutes an animal or a plant") was, by this period, forgotten. His postulation of "in-between objects," lifeforms neither animal nor vegetable, was briefly revived in 1866, when the German naturalist Ernest Haeckel suggested a taxonomical reform. By now, even amateur microscopists were identifying many different kinds of unicellular life: molds, algae, paramecium, and bacteria, among others. Haeckel, already convinced that mushrooms and other fungi did not belong in kingdom *Animalia* because they contained no chlorophyll for photosynthesis, proposed a third kingdom: *Protista*, containing fungi and single-celled life.

The idea did not catch on. Nor, for that matter, did Haeckel's other suggestion: that naturalists develop a discipline he dubbed *ecology*, which would study environments as a whole rather than focusing on individual species. Zoologists and botanists informally divided the increasing count of microbial life between their kingdoms, arriving at a consensus that protozoa were animals and algae, bacteria, and fungi were plants. Classification and specialization remained the order of the day.

Thomas Henry Huxley

TWENTY-SIX

Laughably Like Mine

IT WAS NO SURPRISE THAT THOMAS HENRY HUXLEY AND Charles Robert Darwin became friends. In many ways, their lives ran in parallel. Both had begun their scientific careers as resident naturalists on board British warships—Darwin on the HMS *Beagle*, Huxley on the HMS *Rattlesnake*—and both initially gained recognition for their papers on marine invertebrates. Huxley was sixteen years younger, but Darwin worked so slowly that their professional arcs began to intersect. When Darwin won a medal from the Royal Society for his work on barnacles in 1853, he was congratulated by Huxley, who had not only won the same medal two years earlier but was already serving on the society's leadership council.

Darwin worked slowly due to delicate health—heart palpitations and stomach pain, often brought on by the slightest stress—but also because he could afford to do so. His father was a wealthy financier, his mother an heiress of the Wedgwood pottery empire. Huxley's more accelerated ambition arose from necessity. He was born the second of eight children in West London, to a family so crimped by poverty they could not afford to educate him past the age of ten. Apprenticed to a series of semi-quack doctors from the age of thirteen, Huxley scraped his way into a barely reputable anatomy school, then Charing Cross Hospital, then the University of London, where he won academic medals in anatomy and physiology. Clearly brilliant but still lacking funds, he dropped out to join the Royal Navy as a surgeon's mate at the age of twenty. Unlike Darwin, who spent his five years aboard the *Beagle* leisurely compiling notes, Huxley published numerous scientific papers while still aboard the *Rattlesnake*. Promptly elected to the Royal Society after landing back in England in 1850, he was appointed professor of natural history at London's Royal School of Mines in 1854, a position he would hold for the next thirty-one years. Firmly established as a central figure in British science, he regularly encouraged his slow-paced, deliberative colleague. "Darwin might be anything if he had good health," Huxley wrote, maintaining a keen, if slightly impatient, interest in what he knew of Darwin's work in progress.

Darwin vividly recalled the exact moment of his revelation, in January of 1844. "I can remember the very spot in the road, whilst in my carriage, when to my joy the solution occurred to me." He'd been a firm believer in the fixity of species, but as he excitedly wrote soon after,

> I am almost convinced (quite contrary to opinion I started with) that species are not (it is like confessing a murder) immutable. . . . I think I have found out (here's presumption!) the simple way by which species become exquisitely adapted to various ends.

Darwin's concept, which he later labeled *natural selection*, differed from Lamarck's in that he posited a different engine of species change. Rather than the direct result of adaptation, of adjusting to a new environment, Darwin believed evolutionary changes could be randomly introduced—glitches in reproduction's internal matrix. While some changes might be detrimental, others could serve adaptive purposes, making an individual more likely to thrive (and reproduce) in its environment.

This struck him as more likely than Lamarckian evolution, which he described as "absurd though clever." Yet in the fifteen years he would take to further refine the concept, he never thought of it as a refutation of Lamarck—it was entirely possible for both processes to coexist. Those fifteen years of development might have extended to a lifetime, had he not learned of the naturalist Alfred Russel Wallace's work along parallel lines. "Wallace's impetus seems to have set Darwin going in earnest, and I am rejoiced to hear we shall learn his views in full, at last," Huxley wrote in 1858. "I look forward to a great revolution being effected."

Rushed into print, *On the Origin of Species by Means of Natural Selection, or the Preservation of Favoured Races in the Struggle for Life* was published on November 24, 1859. Darwin, nervously awaiting its reception, said, "If I can convert Huxley I shall be content." The work's first chapter is practically an homage to Lamarck:

> Habit also has a decided influence. . . . The great and inherited development of the udders in cows and goats in countries where they are habitually milked, in comparison with the state of these organs in other countries, is another instance of the effect of use. Not a single domestic animal can be named which has not in some country drooping ears; and the view suggested by some authors, that the drooping is due to the disuse of the muscles of the ear, from the animals not being much alarmed by danger, seems probable.

The concept of "great and inherited development" was Lamarckian at its core. In *Origin*, Darwin made no attempts to dispute the validity of Lamarckian transformism, only to provide a parallel mechanism. In only one passage did Darwin postulate that natural selection could accomplish what transformism could not: in the case of "neuter insects" such as worker ants, which cannot mate. If they couldn't reproduce, how could their traits be passed on?

"For no amount of exercise, or habit, or volition, in the utterly sterile members of a community could possibly have affected the structure or instincts of the fertile members, which alone leave descendants," he wrote. "I am surprised that no one has advanced this demonstra-

tive case of neuter insects, against the well-known doctrine of Lamarck."

This did not disprove Lamarck's work but augmented it. Indeed,
Darwin's theories had so little quarrel with Lamarck that in 1863, the
Scottish geologist Charles Lyell described them as "a modification of
Lamarck's doctrine of development and progression."

But Lamarck was dead. The only living target for the growing outrage over theories of evolution was Darwin, and he was disinclined to
defend himself. As the controversy mounted he withdrew even further into his work, leaving Huxley (who would earn the nickname
"Darwin's bulldog") to defend his ideas in public. In a famous debate
at Oxford that took place seven months after the publication, Bishop
Samuel Wilberforce asked Huxley: Did he trace his descent from a
monkey through his grandfather, or grandmother? Huxley replied he
would be prouder of having a monkey for an ancestor than someone
who, like Wilberforce, used rhetorical skills to ridiculous ends.

While Huxley was happy to rise to a spirited defense of Darwin in
public, in private he admitted to not exactly *believing* in Darwinian
evolution, or Lamarckian transformism for that matter. He considered evolution "likely" (and Darwin's theory more likely than Lamarck's), but that was as far as he could take it. In both religious and
scientific matters Huxley was an *agnostic*, a term he'd coined himself: It
means "one who does not know" in Greek. His worldview was limited
only to facts that could be confirmed, not ideas and dogma that required a leap of faith. Evolution, if it existed, appeared to operate on
a timescale longer than a human lifespan. While future generations
might be able to confirm it, they might also be able to disprove it.
Until then, Huxley considered both theories of evolution only theories.

• • •

Darwin continued to rely upon Huxley as his research turned to the
logical question implied by *On the Origin of Species*: If evolution is the
gradual change of physical characteristics over time, by what means
were physical characteristics transmitted? By 1865 he'd written up a
working theory he called *pangenesis*, which he submitted to Huxley for
review.

Huxley responded with puzzlement. Didn't Darwin realize he was borrowing, practically to the point of plagiarism, from Buffon's theory of the *moule intérieur*, the internal shaping matrix? It turned out that while Darwin was thoroughly familiar with Lamarck, he knew nothing at all of Buffon. While vaguely aware of the name, he had never read the man.

This startled Huxley. During his years as a naturalist in the navy, his crewmates had called his specimens "Buffons," as the author's name was prominent on the copies of *Histoire Naturelle* he carried with him. "I thank you most sincerely for having so carefully considered my manuscript," Darwin replied. "It would have annoyed me extremely to have republished Buffon's views, which I did not know of but I will get the book."

Darwin found a copy of *Histoire Naturelle* and responded soon after.

"My dear Huxley," he wrote, "I have read Buffon—whole pages are laughably like mine. It is surprising how candid it makes one to see one's view in another man's words. I am rather ashamed of the whole affair. . . . What a kindness you have done me with your vulpine sharpness."

In the fourth edition of *On the Origin of Species*, published the following year, Darwin added an addendum entitled "An Historical Sketch of the Progress of Opinion on the Origin of Species." In it, he admitted that Buffon "was the first author who, in modern times, has treated [the origin of species] in a scientific spirit. But as his opinions fluctuated greatly at different periods, and as he does not enter on the causes or means of the transformation of species, I need not here enter on details."

He also acknowledged that "Lamarck was the first man whose conclusions excited much attention. . . . He upholds the doctrine that species, including man, are descended from other species. He first did the eminent service of arousing attention to the probability of all change in the organic, as well as in the inorganic world, being the result of law, and not of miraculous interposition."

Yet Darwin was now taking pains to distance himself from Lamarck. "With respect to the means of modification, he attributed something to the direct action of the physical conditions of life, something to the crossing of already existing forms, and much to use and disuse, that is, to the effects of habit. To this latter agency he seemed to attribute all the beautiful adaptations in nature, such as the long neck of the giraffe for browsing on the branches of trees."

Poor Lamarck and his giraffes. Here's a popular English translation of what Lamarck had to say about giraffes:

> This animal . . . is known to live in the interior of Africa in places where the soil is nearly always arid and barren, so that it is obliged to browse on the leaves of trees and to make constant efforts to reach them. From this habit long maintained in all its race, it has resulted that the animal's fore-legs have become longer than its hind legs, and that its neck is lengthened to such a degree that the giraffe, without standing up on its hind legs, attains a height of six metres [nearly twenty feet].

The key factor here was the phrase "From this habit long maintained," which in the original French reads *Il est résulté de cette habitude, soutenue, depuis long-temps.* While the word *habitude* can mean a simple act of repetition, it can also connote characteristic behavior. There is nothing here that conflicts with Darwin's theory of natural selection: In a species that feeds on trees, individuals with a *habitude* of consuming a broad range of foliage could very well tend to live longer than those with a more limited diet, thereby having greater opportunities to reproduce.

In fact, the giraffe is an excellent example of sexual selection. Male giraffes use their necks and heads aggressively against each other to establish dominance, not unlike the agonistic displays of bucks and rams. Those hardened tufts on the top of the head (not horns but cartilage extrusions called *ossicones*) can kill another giraffe, especially when swung with the whiplike force of a longer neck. The mating successes of victors have reinforced a tendency toward thicker skulls, denser ossicones, and stronger neck musculature, the latter of which can support necks of greater length. They are embodiments of that popular, if not entirely accurate, summation of Darwinian thought: *survival of the fittest.*

. . .

On the Origin of Species became one of the most controversial books of its century. While many of its critics rejected evolution entirely on religious grounds, others found Lamarckian inheritance at least some-what more theologically acceptable than Darwinian natural selection. A species's ability to change through striving might be read as a gift from a beneficent Creator, and at any rate it seemed preferable to Darwin's vision of cold, random chance. As the furor surrounding the book grew, the nuance that evolution might encompass *two* separate processes was lost in a sea of invective. In rushing to defend Darwin, some scientists felt compelled to attack Lamarck.

Taking their cue from Darwin's own half-hearted acknowledgment of Lamarck, Darwinists seized upon (and mocked) the image of a prehistoric short-necked giraffe, straining to reach tree branches until his neck miraculously grew. It was absurd, but equally absurd was sug-gesting Lamarck had meant anything remotely like it in the first place. The irony was that the most widespread argument against Darwin rested on a similarly absurd oversimplification, namely that he be-lieved humanity descended from monkeys, and unless he could pro-duce a "missing link"—shades of the Great Chain—between humans and apes, his theory must be wrong.

Reports of experiments that "disproved" Lamarckian evolution also reflected this misconception. The German biologist August Weismann began cutting the tails off of a population of hundreds of mice, repeating the amputation across twenty-two generations. Well into the twentieth century, biology textbooks recorded that "La-marck's hypothesis would predict that eventually mice would be born with shorter tails or no tails at all. However, Weismann's mice contin-ued to produce baby mice with normal tails. Weismann concluded that changes in the body during an individual's lifetime do not affect the reproductive cells or the offspring." Another textbook displayed a picture of a dog and a puppy, with this caption:

> Acquired characteristics are not inherited as Lamarck thought. Even though this adult Doberman has had its tail and ears cropped, you can see that its offspring still was born with long ears and tail.

Meanwhile Huxley, despite his reputation as "Darwin's bulldog," remained not a Darwinist but a complexist, holding that both theo-ries had merit and deserved to be understood in a proper historical context. "I am not likely to take a low view of Darwin's position in the history of science," he wrote in 1882, the year of Darwin's death, "but I am disposed to think that Buffon and Lamarck would run him hard in both genius and fertility. In breadth of view and in extent of knowledge these two men were giants, though we are apt to forget their services."

But Huxley was a rarity. By then, Lamarck's work was mostly seen as a regrettable misstep in the development of biology (a discipline, as noted, that Lamarck had named). Buffon was moved to the margins as well. Despite privately calling his ideas "laughably like my own," Darwin's public acknowledgment of Buffon had carried an equivocation ("but as his opinions fluctuated greatly at different periods . . .") that stopped Darwin short of claiming him as a predecessor. This, along with Buffon's known antagonism toward Linnaeus, helped give rise to what the British writer Samuel Butler called "the common misconception of Buffon, namely, that he was more or less of an elegant trifler with science, who cared rather about the language in which his ideas were clothed than the ideas themselves, and that he did not hold the same opinions for long together."

Butler was a novelist, best known for the utopian satire *Erewhon.* But in 1879 he'd paused his career in fiction to write *Evolution, Old and New,* a book-length attempt to correct that misconception. Lost to contemporary audiences, he maintained, was the fact that Buffon's seeming contradictions were rhetorical safeguards, *pro forma* verbal performances inserted to appease the Sorbonne and other censors of the day.

Buffon "sailed as near the wind as was desirable," Butler explained, citing a passage from *Histoire Naturelle* that expressed one of Buffon's more controversial ideas: that animals are not only intelligent, but in some important regards display an intelligence greater than ours. "Why do we find in the hole of the field-mouse enough acorns to keep him until the following summer?" Buffon had written. "Why do we find such an abundant store of honey and wax within the beehive? Why do ants store food? Why should birds make nests if they do not know that they will have need of them?"

> Granting . . . that animals have a presentiment, a forethought, and even a certainty concerning coming events, does it therefore follow that this should spring from intelligence? If so, theirs is assur-

edly much greater than our own. For our foreknowledge amounts
to conjecture only; the vaunted light of our reason doth but suffice
to show us a little probability; whereas the forethought of animals
is unerring, and must spring from some principle far higher than
any we know through our own experience.

Asserting intelligence in animals was, Buffon knew, highly controversial: He'd already been censured for implying they had souls. To
buffer the shock value of the paragraph, he followed immediately with

Does not such a consequence, I ask, prove repugnant alike to religion and common sense?

"This is Buffon's way," Butler pointed out. "Whenever he has
shown us clearly what we ought to think, he stops short suddenly on
religious grounds." Buffon had, of course, done the same thing with
species change, stating it boldly (*to draw through time all other organized
forms from one primordial type*), then disavowing it in the very next paragraph (*But no! It is certain from revelation*). *Histoire Naturelle* is rife with such
safeguards, necessities of their time but extraneous, if not confusing,
to later readers.

Coming as it did from the pen of a novelist specializing in satire,
Evolution, Old and New did little to rehabilitate Buffon's reputation. But
Histoire Naturelle's author was not fading into obscurity. Far from it:
While no longer taken seriously as a natural historian, Buffon remained a figure in the public imagination. He still served as a convenient foil for those burnishing Linnaeus's life history, depicted as an
unenlightened antagonist whose opposition to God's Registrar arose
not from disagreement but base motives. Why had he disputed Linnaeus? Opinions varied. The same Linnaeus biographer who attributed it to sheer ignorance ("did not understand the Linnaean system,
nor chose to give himself any trouble to understand it") added that
"this great man in the violence of his attacks and criticisms, was chiefly
hurried away by jealousy." Another, more generous-minded writer explained Buffon's "inaccuracy and thoughtlessness in his manner of
judging Linné" by attributing it to a grandiose character: "It would
seem that, cut out by nature on a large scale, it was difficult for him to

stoop to the contemplation of little things." Nevertheless, he concluded, the man had widely missed the mark. Buffon's notions of nature "cannot be that of our days."

Other scholarship was less than stellar. In his book *Anthropoid Apes*, the German naturalist Robert Hartmann acknowledged Buffon chiefly as an animal trainer, publishing as fact a fanciful account of an ape in the Montbard menagerie. "Buffon's chimpanzee . . . sat down to table like a man, opened his napkin and wiped his lips with it, made use of his spoon and fork, poured out wine and clinked glasses, fetched a cup and saucer and put in sugar, poured out tea, let it get cold before drinking it," he wrote. "He took such a fancy to one lady, that when other people approached her he seized a stick and began to flourish it about, until Buffon intimated his displeasure at such behaviour."

It is telling that Hartmann, a distinguished anatomy professor at the University of Berlin, did not bother consulting the source. Buffon

Buffon and simian servant, advertising tapioca

had only briefly mentioned (in volume fourteen of *Histoire Naturelle*) that he had once seen an orangutan on display in Paris, a sickly specimen with a heavy cough and a fondness for sweets. This heavily embroidered version caught on in the public imagination, dovetailing as it did with the predominant image of Buffon as a quaint-at-best eccentric. A French confection manufacturer began featuring Buffon in its advertisements, depicting him as pleased to be waited upon by a comically domesticated companion.

Yet Buffon remained a bestselling author, after a fashion. He was deprived of a direct heir after Buffonet's death by guillotine in 1794, and the Buffon fortune had largely drained into revolutionary coffers. The remnants of the estate were only belatedly claimed, by distant relations. Buffon's great nephew, Henri Naudault de Buffon, honored his memory by publishing selected correspondence (from the few papers that had survived Buffon's habit of burning), but neither he nor other descendants made any claims on the copyright of *Histoire Naturelle* itself. This made the massive text and its painstakingly compiled illustrations fair game for reprinting and repurposing. It became common to use Buffon's original work as the basis for a jumble of verbiage on nature, appropriating his name for marketing purposes. Take, for instance, the "edition" of *Buffon's Natural History of Man, the Globe, and of Quadrupeds*, published in 1857 in New York by Leavitt & Allen. Its preface promised "facts calculated to excite astonishment, and perhaps a higher feeling, and to yield innocent entertainment, or valuable information."

> The materials for these volumes have been drawn from celebrated naturalists; from Buffon, the most eloquent of them all, from Cuvier, Lacépède, and many others, whom it would be superfluous to enumerate. Various interesting particulars have also been gleaned from the narratives of modern voyagers and travellers.

While Buffon had passed his life's work on to Lacépède, he would have been surprised by who was also sharing the pages of *Histoire Naturelle*. The 1857 version included his surveys of human diversity and cautions against racial categories ("the dissimilarities are merely external, the alterations of nature but superficial"), but the editors tacked on a disclaimer: *Such is Buffon's arrangement of the varieties of the human race. Other naturalists, however, differ widely from him, and the systems which they have formed are numerous. According to Linnaeus . . .*

What follows is a direct quotation of *Systema Naturae*'s four racial categories, then a more recent definition of "the white race" as "characterized by regularity of form, according to the ideas which we have of beauty." Rounding out the chapter is an account of conjoined twins, "now exhibiting in the British metropolis."

For several decades, such versions sold briskly. But the conversion of *Histoire Naturelle* into a hodgepodge of sources was an attempt to maintain its relevance, and that could not be indefinitely sustained. Eventually publishers turned from adulteration to simplification, producing increasingly abridged editions that emphasized the illustrations and streamlined the language. Because of this, in the realm of public imagination Buffon increasingly became a de facto popularizer, if not a writer of texts for the young. One unauthorized edition, entitled *Buffon for Little Children, with Pictures,* boiled a few dozen entries into a primer for beginning readers. ("The squirrel has gentle manners, although it sometimes grasps small birds nearby.") As George Bernard Shaw recalled,

> One day early in the eighteen hundred and sixties, I, being then a small boy, was with my nurse, buying something in the shop of a petty newsagent, bookseller, and stationer in Camden Street, Dublin, when there entered an elderly man, weighty and solemn, who advanced to the counter, and said pompously, "Have you the works of the celebrated Buffoon?"

The incident struck him as funny, as even small children knew "the celebrated Buffoon was not a humorist, but the famous naturalist Buffon. Every literate child at that time knew Buffon's *Natural History* as well as *Aesop's Fables.* And no living child had heard the name that has since obliterated Buffon's in the popular consciousness: the name of Darwin.

"Ten years elapsed," Shaw continued. "The celebrated Buffoon was forgotten."

Louis Agassiz and Gregor Mendel

TWENTY-SEVEN

The Rhymes of the Universe

IN A BANQUETING HALL IN CAMBRIDGE, MASSACHUSETTS, IN May of 1857, America's foremost poet rose to recite a poem he'd written in honor of his friend, America's foremost scientist. Henry Wadsworth Longfellow's *The Fiftieth Birthday of Agassiz* would never outshine his *Song of Hiawatha*, written two years earlier, but several of the verses are notable nonetheless.

> And he wandered away and away
> With Nature, the dear old nurse,
> Who sang to him night and day
> The rhymes of the universe.
>
> And whenever the way seemed long,
> Or his heart began to fail,
> She would sing a more wonderful song,
> Or tell a more marvelous tale.

Smiling from his seat of honor, Longfellow's subject could not agree more with the poem's sentiment. The universe *did* rhyme, in the sense of operating along clear and orderly patterns. A steadfast belief in his ability to discern those patterns had taken him far, not only to a new continent but to the pinnacle of his profession.

Born the son of a Protestant clergyman in the Swiss village of Motier on May 28, 1807, Jean Louis Rodolphe Agassiz had been a competent but unremarkable scholar, with no ambitions beyond becoming a practicing physician. That changed in 1826 when, while studying medicine in Munich, he came to the attention of his botany professor, Carl von Martius. In his younger days, von Martius had been something of an adventurer, collaborating with the zoologist Johann Baptist von Spix on an expedition through Brazil. Spix had recently died, leaving behind a great number of uncatalogued specimens of Brazilian fish. Being a botanist and busy with his own specimens, Martius had no interest in taking on his colleague's unfinished work. Instead he foisted them off on his nineteen-year-old student.

Agassiz had no particular interest in fishes, or zoology for that matter, but he was not about to shirk the assignment. After compiling and writing (in academic Latin) *Selecta Genera et Species Piscium Quos Collegit et Pingendos Curavit Dr. J.W. de Spix*, he took advantage of a minor opportunity presented by its publication. He'd read that the great Cuvier in Paris had announced his own progress on "a complete work on all the known fishes," and in doing so had asked that naturalists send him specimens. Agassiz dedicated the Spix catalogue to Cuvier, and sent a copy to the Jardin. He then resumed his medical education, all but forgetting about the work until five years later, when his studies brought him to Paris.

Agassiz had moved there because he hoped to specialize in treating cholera, a then-mysterious disease sweeping across Western Europe: At least nineteen thousand Parisians would die of it in the following year alone. For Agassiz this represented more of a learning opportunity than a risk, since no one had yet fully grasped its infectious nature and means of transmissibility. He took up lodgings at the Hotel du Jardin du Roi, a location he'd chosen because it was a ten-minute walk to the medical school. Surprised to find it was also "not two hundred steps away from the Jardin des Plantes," he decided to send a calling card to Baron Cuvier.

That very evening he received an invitation to dine with Cuvier, who had not forgotten the book's dedication. "He greeted me with great politeness," Agassiz recalled. "To my surprise, I found myself not quite a stranger, rather, as it were, among old acquaintances." Under the mistaken impression that Agassiz was still interested in fishes, Cuvier gave him a standing invitation to dinner every Saturday. Awestruck and eager to oblige, Agassiz resumed studying ichthyology.

Soon he was dividing his time between medical school and the Jardin, where Cuvier gave him a corner in one of his laboratories and often visited to provide encouragement. Three months later, during one of their regular Saturday dinners, Cuvier drew Agassiz aside and informed him that when it came to fishes (in Agassiz's words), "I had given it so much attention, and had done my work so well, he had decided to renounce his project, and to place at my disposition all the materials he had collected and all the preliminary notes he had taken."

Suddenly anointed both as a protégé of Baron Cuvier and as one of the world's foremost authorities on fishes, Agassiz found himself in no position to refuse. He was desperately poor, too poor to take on the unpaid work, and had never even seen the ocean. But he toiled away for another three months, working directly with Cuvier in his study, until five o'clock on the evening of May 12, 1832. The next day, while walking into the French parlement, Cuvier collapsed. Overtaken by a swiftly increasing paralysis, he died that evening, a victim of the cholera epidemic Agassiz had once hoped to study.

Deprived of Cuvier's patronage, Agassiz saw few prospects for himself at the Jardin. He retreated to his native Switzerland, finding a teaching post at the Neufchatel Lyceum. The position paid little, and Agassiz's publications over the next decade on fossil fish and mollusks failed to attract much attention. In 1842, he made a bold play by publishing *Nomenclator Zoologicus*, an attempt to effectively succeed Linnaeus by providing for zoology what *Philosophia Botanica* had for botany: a set of consistent rules. "Before the immortal works of Linnaeus were produced, the nomenclature of organic bodies was not bound by any law," he wrote in tribute. "Linnaeus was the first to propose a nomenclature established by certain laws, which most of his descendants regarded as valid and inviolable." The publication did not achieve the desired effect; at least two other works named *Nomenclator Zoologicus* would be published, by different authors.

In the fall of 1846, the forty-year-old Agassiz traveled to the United States to deliver a lecture series in Boston, at the invitation of John Amory Lowell, a member of the Linnean Society and a fellow of Harvard College. He was immediately enticed to stay. As it happened, Harvard was grappling with the largest gift ever given to an American educational institution, but it came with strings attached. Up until then, Harvard was strictly a liberal-arts school issuing only bachelor of arts degrees, but the industrialist Abbott Lawrence had bequeathed $50,000 specifically to educate "men of science, applying their attainments to practical purpose"—in short, to issue bachelor of science degrees as well. With Agassiz as head, Harvard opened its Lawrence Scientific School in 1847.

Agassiz seized the opportunity to dominate American science. In August of 1849, during a five-day meeting of the American Academy for the Advancement of Science, he presented no fewer than twenty-seven formal academic papers, on topics as diverse as botany, entomology, geology, marine biology, and comparative embryology. As thousands of specimens streamed into Harvard, donated by both professional and amateur naturalists from across the growing nation, Agassiz unhesitatingly assigned them places in the Linnean hierarchy and issued scientific names. He published extensively on science in general-interest publications such as *The Atlantic* and *The Christian Record*, and began socializing with the elite intelligentsia of Cambridge and Boston. By 1857, the year Longfellow wrote a poem in his honor, Agassiz was one of the nation's most influential and celebrated academics. Later that year, the Musée/Jardin in Paris offered him Cuvier's old position. He turned it down.

Yet he remained, in spirit and outlook, an acolyte of Cuvier. When the uproar over Darwinian evolution reached the United States, he was more than happy to revisit the Sacred Ibis Debates; like his late mentor, Agassiz was a firm believer in the fixity of the species. "The monuments of Egypt have fortunately yielded skeletons of animals that lived several thousand years ago," he wrote for an American audience. "It has been found that there is no difference between them—that they agree precisely in all particulars as perfectly as the different individuals of the species now living agree together. So that we have in this fact, which has been fully investigated by Cuvier in his researches upon fossils, full evidence that time does not alter organized beings."

· · ·

On August 14, 1862, Abraham Lincoln welcomed to the White House five well-dressed men. They were educators, entrepreneurs, and clergymen—community leaders from across the Northern states, recruited just a few days earlier to advise Lincoln on an undisclosed matter. Each had hurried to Washington, aware that they were also entering into history. All five were formerly enslaved or the sons of enslaved parents, and never before had Black Americans been invited to the White House to consult on policy. The hastily assembled delegation took their seats with a sense of gravity, and high hopes. Those hopes were dashed as soon as Lincoln began to speak.

"You and we are different races," he said. "We have between us a broader difference than exists between almost any other two races. Whether it is right or wrong I need not discuss, but this physical difference is a great disadvantage to us both, as I think your race suffer very greatly, many of them by living among us, while ours suffer from your presence. In a word we suffer on each side. If this is admitted, it affords a reason at least why we should be separated. You here are free men, I suppose?"

The question surprised them. Of course they were free: The Department of the Interior had billed them as "The Delegation of Free Negroes." One was a high-ranking Freemason, another a teacher who had graduated from Oberlin College. After a moment, one answered. "Yes, sir."

"On this broad continent, not a single man of your race is made the equal of a single man of ours," Lincoln continued. "I do not propose to discuss this, but to present it as a fact with which we have to deal. We look to our condition, owing to the existence of the two races on this continent."

The president, it transpired, had called them there to propose a migration scheme. Desperate to end the Civil War, he hoped to coax secessionist states back into the Union by eliminating one factor the Confederates had found objectionable—namely, the presence of free Black Americans like them, strivers and achievers with ambitions of equality. Would they not, Lincoln asked, be better off founding a country of their own? A region in Central America (comprising much of what is now present-day Panama) could be easily claimed, he assured them, at least with the United States' military backing. Given funds and further support, they could establish settlements, build infrastructure, and free themselves of the burden of dealing with a prejudice that Lincoln considered ineradicable. Surely that was preferable to the current state of affairs?

The president's proposal was met with stunned silence. The members of the delegation had expected the conference to be an acknowledgment of Black Americans, of their right to claim a role in shaping and guiding the Union. Instead, they were being asked to voluntarily relinquish their citizenship and quietly go away. The abolitionist leader Frederick Douglass (who was not part of the delegation but was informed soon after) was scathing in his condemnation. "The President of the United States seems to possess an ever increasing passion for making himself appear silly and ridiculous," he wrote in his magazine *Douglass' Monthly*, "if nothing worse."

Unable to get free Black Americans to disappear, Lincoln then turned to the issue of their continued existence, which as far as he was concerned was deeply problematic. In March of the following year—three months after he'd signed the Emancipation Proclamation—he asked Congress to create a Freedmen's Inquiry Commission, charged with determining the "condition and capacity" of both long-standing and recently freed Black citizens.

The commissioners traveled to Canada, where they interviewed the families of escaped enslaved persons, many of whom had experienced generations of assimilation. They corresponded with army officers who'd commanded Black troops, assessing their courage and willingness to serve. And they consulted with the United States' foremost natural historian, Professor Louis Agassiz of Harvard University.

Agassiz was unabashedly, profoundly racist. While he'd begun his career believing, like his mentor Cuvier, that humans comprised a single species, during his American tenure he'd become a committed polygenist, teaching that separate races should be ascribed to separate acts of Creation. As early as 1850 he'd written on "the propriety of considering Genesis as chiefly relating to the history of the white race," adding that science held "the obligation to settle the relative rank among these races." Blacks were "submissive, obsequious, imitative." Mongolians were "tricky, cunning, and cowardly." Indigenous North Americans, on the other hand, were "indomitable, courageous, proud . . . the most striking contrast with the character of the negro race, or with the character of the Mongolian." Africa itself was "a most forcible illustration of the fact that the races are essentially distinct," since "there has never been a regulated society of black men developed on that continent, so congenial to their race. . . . Does this not indicate in this race a peculiar apathy, a peculiar indifference to the advantages afforded by civilized society?"

The Freedmen's Inquiry Commission asked Agassiz to predict what would happen if the Union won the war, conferring citizenship to millions of freed Black people. This was an important question, as the commission's charge was to make policy recommendations for a post-war government. Where, when, and how should funds be allocated to help the emancipated? Should the formerly enslaved population be educated? Should assistance programs aid them in relocating from the South to the North? Should laws be passed to ensure their integration?

Agassiz assured them that no educational programs would be necessary, since "they may but be compared to children, grown in the stature of adults, while retaining a childlike mind." No relocation initiatives were called for, since he predicted a migration in the opposite direction. Black people would instinctually leave the colder climates of the Northern states for the Southern ones, as the climate there "was particularly favorable to the maintenance and multiplication of the negro race." Lasting integration was scientifically impossible, since the mingling of the races produced "weak and lighter" hybrids unfit for either society. These would eventually die off, leaving both races purified and pursuing separate destinies.

The commissioners listened. In June they published a preliminary report, recommending that the federal government should not trouble much with educating the freed slave, as "he will somewhere find, and will maintain, his own appropriate social position." Relocation would not be necessary, as "there is no disposition in these people to go North. . . . The southern climate suits him far better than ours." Segregation, not integration, was inevitable, they concluded. "The sooner they can learn to stand alone and to make their own unaided way, the better both for our race and for theirs."

. . .

At the same time that Agassiz was trafficking in prejudices from his Harvard pulpit, an obscure citizen of Austria, far from any campus, was quietly pursuing a very different vision of what constituted scientific inquiry. He was a monk, and the pursuit was a hobby.

Born in 1822, Johann Mendel was a farmer's son who showed early promise as a scholar and hoped to become a teacher. The family encouraged his ambition, sending him to boarding school despite the significant financial burden. He was sixteen when his father fell under a rolling tree trunk while trying to clear a field, crushing his chest and leaving him unable to wrest a living out of the already-struggling forty-three-acre farm. The father sold the farm, directing what funds he could to his son, but eventually the money ran out. Seeing no further options, Johann dropped out of school and prepared to enter the clergy. In October of 1843, he entered the Königskloster Augustan Monastery in Brünn as a novitiate, choosing Gregor as his ecclesiastical name.

Brother Gregor remained with the monastery for the rest of his life. It was not a fully cloistered existence; he was allowed to walk the streets of Brünn as he pleased, and he settled into a role as teacher of the lower grades in the school affiliated with Königskloster. In his spare time, he took up a hobby of recording weather observations. When that proved an insufficient pastime, in 1856 he asked the abbot for permission to tend to an oblong patch of land, just 20 feet wide and 120 feet long, behind the monastery library. There he planted *Pisum sativum*, the ordinary garden pea, and watched the plants as they germinated, grew, and died.

Why pea plants? They mature quickly, in ten to twelve weeks, and have a long growing season lasting from February through June. With care, Mendel could produce a new generation in a single season. Another reason was their manner of blooming. Since the petals of *Pisum sativum* fully open only after each plant self-fertilizes, no free-floating pollen could drift from one pea stalk to another.

What Mendel did was interfere with that self-fertilization. First he identified plants that were distinct along fourteen criteria: the color (white or violet) and position of their flowers (in the middle of stems or at the end), the color (green or yellow) and shape (smooth or bunched) of their pods, the color (green or yellow) and shape (round or wrinkled) of the peas themselves, and the length of the stems (short or long). Then he gently pried open the emerging buds, swapped the pollen manually between varieties, and compiled the results.

In other words, Mendel was attempting to observe the invisible. Under controlled circumstances and over a lifelong timescale, with no intention to do anything more than document nature itself at work, he was effectively creating in miniature a version of Buffon's private park. The extent to which Mendel knew of Buffon's *moule intérieur*, the internal matrix theory of reproduction, is undocumented, as none of his private papers survive. But if someone wished to design an experiment specifically to test the theory, Mendel's pea-patch investigations could hardly be improved upon. And while his papers are lost, volumes from his personal library have survived. In his copy of a German translation of Darwin's *The Variation of Animals and Plants Under Domestication*, he annotated a telling passage with a bold pencil stroke. *If Buffon had assumed that his organic molecules had been formed by each separate unit throughout the body, his view and mine would have been closely similar.*

Observing his plants over the course of seven years, Mendel noticed that some traits appeared and disappeared in interesting ways: for instance, when a yellow and a green pea were interbred, their offspring plant was always yellow, yet in the next generation one out of three of those plants would produce green peas. Yellow peas, he decided, were a "dominant" trait, and green peas a "recessive" one. After growing and compiling tables on approximately twenty-eight thousand individual pea plants, he deduced further rules for the inheritance of physical traits. Since the rules dictating the emergence of one trait did not necessarily apply to another, he concluded that inheritance was dictated by a combination of forces, working in unison during germination and growth. He had no idea what those forces were—he only called them "factors"—but they clearly existed, functioning below the level of the visible. The shaping internal matrix was real.

Although he would never know the full implications of his work, Mendel's painstaking observations and analysis would go on to form the foundations of the modern discipline of genetics. It is tempting to view his achievements as the work of a singularly gifted individual, transcending his limitations as a hobbyist, but it can be argued that his breakthroughs came because of his amateur status, not in spite of it. Throughout his experiments Mendel repeatedly attempted to garner the interest of the academic community, writing letters documenting his progress and asking for advice. The replies, when they came at all, were patronizing at best. Only one botanist, a professor at the University of Munich, showed some interest in replicating his results, asking Mendel for seeds from his pea patch. The professor planted some, but didn't bother to study the plants that emerged; in the interim he'd decided the matter wasn't worthy of pursuit.

The experiment was exceedingly simple. The mindset of conducting it, at least in academic circles, was exceedingly rare.

Mendel fared only slightly better closer to home. The local Brünn Society for the Study of Natural History invited him to read his paper "Experiments on Plant Hybridization," which he presented over the course of two evenings, in February and March of 1865, to an unenthusiastic audience. The society subsequently printed it in their journal, where to no one's surprise it attracted zero notice (the publication had a circulation of 120 copies). "It requires some courage to undertake a labor of such far-reaching extent," Mendel wrote in the paper of his experiment. "It appears, however, to be the only right way by which we can finally reach the solution of a question . . . the evolution of organic forms."

Mendel continued his experiments for two more years, until a mundane matter of taxation brought them to an end. The Augustans of the monastery, slapped with a hefty death tax upon the passing of their head monk in 1867, hoped to avoid paying it again any time soon. Bypassing more senior monks, they thrust the office upon Brother Gregor, who at the age of forty-five could be assumed to have a reasonable life expectancy. With the new prelate (as the head monk was called) bogged down in administrative matters, the neglected pea patch withered, and was eventually plowed over. Mendel contented himself with less time-intensive hobbies: beekeeping, observing sun-

spots, recording the water level in the monastery's well, and continuing his long-maintained chronicle of the weather. The self-designed,
self-directed career of a complexist biologist came to a quiet end.

• • •

Louis Agassiz, meanwhile, was continuing to reassure the general
public that evolution did not exist. Even as increasing numbers of his
academic colleagues began to acknowledge the reality of species
change, he remained stalwart in his defense of the fixity of species.
"The world has arisen in some way or another," he wrote in an 1873
essay for *The Atlantic Monthly*. "Darwin's theory, like all other attempts
to explain the origin of life, is thus far merely conjectural. I believe he
has not even made the best conjecture possible in the present state of
our knowledge."

These became his final public words. The essay was at the printers
when he died of a cerebral hemorrhage on December 14, 1873, at the
age of sixty-six. "The wise of old welcome and own him of their peaceful fold," the poet John Russell Lowell wrote in tribute. "And Cuvier
clasps once more his long lost son."

In Brünn, Austria, on January 4, 1884, Gregor Mendel took his last
readings of the Königskloster monastery's thermometer, barometer,
and rain gauge. He died two days later, at the age of sixty-one. The
funeral service praised him as a priest, a teacher, an administrator, and
an amateur meteorologist, but not as a natural historian—his experimental pea-patch days were mostly forgotten. The next prelate, as a
matter of routine, fed Mendel's papers to a monastery fire.

Julian Huxley, in front of a portrait of his grandfather

TWENTY-EIGHT

Most Human of Humans

EVEN AS A VERY YOUNG CHILD, IT SEEMED CLEAR THAT JULIAN Huxley was going to follow in the footsteps of his grandfather Thomas. Born in 1887, he spent his earliest years in the fond company of the man known as "Darwin's bulldog," who was happy to dispense private lessons in natural history and encourage a spirit of inquiry. "There are some people who see a great deal and some who see very little in the same things," the elder Huxley wrote to his grandson, then age five. "When you grow up I dare say you will be one of the great-deal seers." Taking that admonition to heart, the younger Huxley began studying biology at Oxford in 1906. This meant immersing himself in an emerging new discipline at the exact time it was being named: *genetics*.

It was a discipline with a scandalous birth. At the turn of the century, Hugo de Vries, a botany professor at the University of Amsterdam, had published a paper that struck many of his colleagues as brilliantly original. In it, Vries described a process he called *mutation*, in which plant characteristics were transmitted via inheritance mechanisms he dubbed *pangenes* (later shortened to *genes*). He laid out the rules that seemed to dictate these transmissions, classifying traits as either dominant or recessive and predicting their interplay with a startling degree of accuracy.

De Vries, it transpired, had grown fascinated with a stand of evening primrose bushes growing in an abandoned potato field and begun to delve into plant inheritance. Searching the scientific literature for signs of precedent, he'd come across one of the 120 copies of the *Proceedings of the Brünn Society for the Study of Natural History* containing the findings of Gregor Mendel, the late abbot of Austria's Königskloster monastery. De Vries was astonished. Mendel had not only anticipated him by a generation, but delved so deeply and effectively into the principles of biological inheritance that his own work seemed primitive by comparison. It was a wonder that such an achievement had gone unnoticed.

It was also a temptation. When he published his own paper, De Vries drew freely from Mendel, borrowing concepts and terminology but omitting any mention of Mendel himself. Gambling that no one else would take notice of the single record of a dead monk's temporary hobby, De Vries appropriated Mendel's work as his own.

The deception came to light through the most tenuous of connections. Another botanist remembered that a professor from his student days had corresponded with a scientifical friar with similar ideas, but only to dismiss his theories and discourage him from experimenting further. The distant memory led more readers to Mendel's work, and to an embarrassed De Vries belatedly giving credit where it was due. "In the study of evolution progress had well-nigh stopped," wrote the British biologist William Bateson. "Most were content supinely to rest on the great clearing Darwin had made."

> Such was our state when two years ago it was suddenly discovered
> that an unknown man, Gregor Johann Mendel, had, alone, and

unheeded, broken off from the rest—in the moment that Darwin was at work—and cut a way through. This is no mere metaphor: it is simple fact. Each of us who now looks at his own patch of work sees Mendel's clue running through it: whither that clue will lead, we dare not surmise.

Young Huxley followed with excitement the work of Bateson and others, who were rapidly extrapolating Mendel's modest paper into a full-fledged science. But he had no plans to join them on the cutting edge, or for that matter to become a significant scientific voice like his grandfather. His inclination was to specialize in marine ornithology, focusing in particular on the courtship rituals of water birds. After taking his doctorate he moved from England to a remote corner of Texas, where he plunged into what he later called "a new province of observational and experimental work, ecology—the observation of animals in their proper surroundings." Uninterested in classification for its own sake, unmoved by the prospect of being immortalized through species names, Huxley was drawn to the relatively novel practice of quiet, patient observation in the field. "Men of intelligence are taking cameras and building watching-shelters in forest and jungle and prairie," he wrote, "where formerly they took gun and trap and killing-bottle."

That career trajectory changed abruptly with the outbreak of World War I. While he survived his tour of duty in the British Army Intelligence Corps, a great many of his colleagues did not—including his friend Geoffrey Smith, a professor of zoology at Oxford, who died in the battle of the Somme. When Oxford offered the vacant post to Huxley, he could not bring himself to turn it down. In 1927, the novelist H. G. Wells approached the young professor with a proposition: Books like *War of the Worlds* and *The Time Machine* had made Wells one of the world's bestselling authors, but before achieving popular success he'd trained as a biologist, studying under Thomas Henry Huxley. He wanted to write a book about biology geared toward a popular audience, but knowing that his credentials were out of date he sought out his old professor's grandson and offered to collaborate.

Their ensuing three-volume work, *The Science of Life*, became an international bestseller, and set Huxley on the path of celebrity. The book's resounding success coincided with the emergence of the new medium of radio, and the recently founded British Broadcasting Corporation began giving Huxley frequent opportunities to explain genetics and other innovations. By 1930 he was ranked in a public survey as one of the greatest minds of the United Kingdom, his fame clearly eclipsing that of his younger brother, the novelist Aldous Huxley. He was among the first of what is now a familiar type of media personality: the science popularizer, with both academic credentials and a flair for expressing ideas in clear, accessible language.

But Huxley's celebrity grew increasingly controversial. Boundlessly confident that our transformed understanding of biology could make the world a better place, he considered it his duty to pit science against prejudice. He did not endear himself to conservatives in the American South, when he and Wells observed that "in some of the more backward regions of the United States . . . there is a formidable campaign for the penalization of any biological teaching that may seem to run counter to the Bible." He drew broader outrage with his observation that birth control had the advantage of separating sexuality from reproduction, allowing people to "consummate the sexual function with those they love, but to fulfill the reproductive function with those whom on perhaps quite different grounds they admire." But he courted genuine furor when he attempted to dismantle the pseudoscience of race.

Huxley watched in dismay as Adolf Hitler, riding a wave of virulent antisemitism and racial mythos, rose to absolute power in Germany in 1934. The following year, as Nazi Germany passed laws granting special privileges to "Aryans" and began sending minority groups to concentration camps, Huxley joined with the anthropologist A. C. Haddon to write *We Europeans*, an excoriation of supremacist thought. "In a scientific age, prejudice and passion seek to clothe themselves in a garb of scientific responsibility," they wrote. "A vast pseudo-science of 'racial biology' has been erected which serves to justify political ambitions, economic ends, social grudges, class prejudices." Race, Huxley insisted, had no biological existence—at best, one might talk about ethnic groups, but even those groups had no inherent qualities beyond appearances.

To Huxley these were simple facts, but in the Neville Chamberlain era of British appeasement it was seen as needlessly antagonistic to point out (as Huxley did) that a German "as blond as Hitler . . . as tall as Goebbels, as slender as Goering" would hardly resemble the Nazis' Aryan ideal. When war broke out, Huxley found his perspective coldly received for a different reason: Recruitment and rationing were driven by emotional appeals, not science. The Allied countries had bigots in abundance, and they would not be motivated to fight against their own prejudices. Propaganda posters extolled the fight against "Hitlerism," not racism.

Huxley was not the only one stung by this calculation. In 1943, the American United Services Organization, or USO—best known for staging entertainment for soldiers in the field, featuring the likes of Bob Hope and Marlene Dietrich—decided to pair their performances with the distribution of a pamphlet designed to remind soldiers of what they were fighting for. They approached two Columbia University professors, Ruth Benedict and Gene Weltfish, to write the text. The resulting work, entitled *The Races of Mankind*, presented in plain language what they considered to be indisputable truths. The fight against Hitler, they concluded, was a fight against the folly of racial prejudice.

> The movements of people over the face of the earth inevitably produce race mixture and have produced it since before history began. No one has been able to show that this is necessarily bad. . . . Race prejudice is, after all, a determination to keep a people down, and it misuses the label "inferior" to justify unfairness and injustice.

The pamphlet made its way to soldiers fighting in the trenches, but also into the hands of Congressman Andrew Jackson May of Kentucky, the powerful head of the House Military Affairs Committee. Outraged by a little book that, among other things, depicted Adam and Eve as dark-skinned, May led a group of politicians in demanding that the pamphlet not only be withdrawn but declared subversive— a demand that succeeded, ruining the authors' careers in the process.

Sidelined by hepatitis during the war, Huxley recovered enough to take an active role in the 1945 formation of the United Nations, and

THE PEOPLES OF THE EARTH ARE ONE FAMILY.

From The Races of Mankind (1943)

was subsequently asked to serve as the first director-general of a key sub-organization: the United Nations Educational, Scientific and Cultural Organization, or UNESCO. Its goal was to promote world peace through international education and intellectual cooperation, a mission that Huxley took seriously. One of his priorities was, once again, setting the record straight: From a biological perspective, race did not exist.

In 1950 he threw the full weight of his organization behind the subject, co-authoring *The Race Question*, an official UNESCO document that sought to remove the concept from serious scientific discourse. "The biological fact of race and the myth of 'race' should be distinguished," it read in part.

> For all practical social purposes "race" is not so much a biological phenomenon as a social myth. The myth "race" has created an

enormous amount of human and social damage. In recent years it has taken a heavy toll in human lives and caused untold suffering. It still prevents the normal development of millions of human beings and deprives civilization of the effective co-operation of productive minds. The biological differences between ethnic groups should be disregarded from the standpoint of social acceptance and social action.

In the ensuing uproar, the prevailing sentiment was that Huxley had gone too far. Once again, American politicians led the attack, denouncing the statement and demanding that Huxley be fired from UNESCO. The statement was not retracted, but his term of office was suddenly shortened; what had been a six-year tenure was cut to two years. Huxley left the post soon thereafter.

. . .

In a 1959 paper published in the journal *Systematic Zoology*, a botanist named William Stearn proposed the awarding of a new and singular honor to the memory of Carl Linnaeus. As he observed, Linnaeus's *Systema Naturae*

> looks a rather dull and formal reference book, to be taken off the shelf merely to check the spelling of a name and its typification. It is a little hard to see it in its full significance as a work wherein a single mind sought to grasp and to record succinctly the distinguishing features of all the genera and all the species of animals and plants upon the face of the earth and in its waters. The mere thought of doing this today, when an army of systematists cannot accomplish it, is staggering.

Still, Stearn noted, this towering work had one glaring omission.

As previously noted, the concept of type specimen is central to Linnean taxonomy. Since a species is defined by a physical description, that description necessarily requires looking at a physical object (either a preserved specimen or a detailed illustration). This object becomes the "type," the fixed standard by which all subsequent specimens are identified as "typical" of the species.

The methodology had relaxed somewhat in the first part of the twentieth century, with some scientists substituting instead a syntype, a listing of several examples of a species, none of which had priority over the other. But Stearn was writing in light of a recent crackdown. The keepers of the International Code of Botanical Nomenclature had announced a return to Linnean orthodoxy: All new species would henceforth again be defined by a single instance only, called either a holotype (if chosen by the first describer of the species) or a lectotype (if chosen at a later date). Linnaeus's specimens became holotypes as soon as he used them in compiling *Systema Naturae*, which was why James Edward Smith had been so keen to acquire the collection in 1783: Its specimens were *the* definitive examples—the embodiments, so to speak—of their respective species.

Yet Linnaeus had neglected to collect the type specimen of one important species. He had never supplied a holotype for *Homo sapiens*, and for that matter had defined the species only with the terse phrase *nosce te ipsum*, Latin for "know yourself." Stearn proposed a novel means of correcting this omission. Since "the specimen most carefully studied and recorded by the author is to be accepted as the type," he wrote, the appropriate lectotype was obvious. Linnaeus had presumably examined himself for decades, even if only by glancing in the mirror while shaving.

"Clearly," Stearn concluded, "Linnaeus himself . . . must stand as the type of his *Homo sapiens!*"

The suggestion was playfully phrased—papers in *Systematic Zoology* rarely employ exclamation marks—but taken seriously nonetheless. At a convocation seven years later, the International Committee on Zoological Nomenclature declared the body of Linnaeus (still entombed in Uppsala Cathedral) the lectotype of *Homo sapiens*. Officially, Carl Linnaeus is the most human of humans, the baseline measure of us all.

The gesture was intended to honor the founder of Linnean taxonomy, by placing him prominently within his own creation. But it brings into stark relief the limitations and distortions of the taxonomy it seeks to celebrate. The concept of "type specimen" itself inherently extracts idealized, universal qualities from one example—the very thing Buffon had railed against, insisting that "in Nature, only the individual exists."

Equally problematic is the elevation of any single sample above

The "official" type specimen of humanity, Linnaeus himself

others of its kind, an inevitable by-product of declaring it definitive. It is impossible to excuse away the fact that in 1959, with World War II only fourteen years in the past, the International Committee on Zoological Nomenclature failed to address the implications of naming a Nordic European as the quintessential human being. Compounding the matter is the fact that the committee neglected to designate an allotype, a counterpart specimen of a different gender. The official archetype of humanity is white, Swedish, and male.

· · ·

Julian Huxley saw a dead end in this manner of thinking. "The older descriptive work in biology," he and Wells had written in *The Science of Life*, "based only on collection, classification, and anatomy, seems to be unlikely to achieve much more without cooperation with more modern branches of research." Like his grandfather, Huxley had worked to reconcile Darwinian evolution with the Linnean hierarchy, and like his grandfather he'd found himself struggling to make everything fit.

The problem was the past, or more specifically the growing fossil record. Categorizing by external physical differences was simple enough with living species: One simply sorted by similarities. But placing extinct species within hierarchical groupings meant comparing skeletal structures alone, and there the demarcations were not so apparent. Which resemblances constituted an evolutionary relationship, and which did not? Some ancient specimens might resemble contemporary ones on the species level but share other features common to a different genera, order, or class entirely.

For instance, the skull of *Indohyus*, a land-based genus that lived in Asia about fifty million years ago, resembles nothing so much as the skull of a raccoon. It was only when a paleontologist accidentally broke an *Indohyus* skull that the distinctive structure of its inner ears was noticed, pointing toward a very different lineage. A living *Indohyus* likely looked very much like a raccoon, but today we understand it to be the oldest known member of the infraorder *Cetacea*, which includes the dolphin and the great blue whale. It is the small, land-dwelling ancestor of those aquatic species. *Fallaces sunt rerum species*, to re-quote the philosopher Seneca. *Appearances are deceptive.*

The burgeoning fossil record also posed another question: When, exactly, does one taxonomic category end and another begin? This problem came to be known as "the boundary paradox," and it underscored a fundamental incompatibility between evolution and conventional taxonomy. The nested boxes of Linnaeus seem logical when viewed through the lens of fixity, but from the perspective of deep time and continual change their walls become tenuous. For instance, the genus *Homo* is believed to have descended from the genus *Australopithecus*. But this implies that at some point in evolution, there must have been at least one generation that straddled both groups, simultaneously belonging to both *Australopithecus* and *Homo*. Unlike other entities of reason, such as the fifty United States, it's exceedingly difficult to declare clear boundaries.

In 1957—just one year before the proposed elevation of Linnaeus to type specimen—Huxley proposed new taxonomy aimed at eliminating such quandaries. Why not scrap the nested boxes entirely and build an organizational structure dictated solely by evolution? Organisms descended from a single ancestor could form groups he dubbed *clades*. The point where evolution divided such groups he called *cladogenesis*: the birth of a new clade. The discipline would be called cladistics.

It was an intriguing idea, but a far-fetched one at the time. For cladistics to work, it would require a new science—one that Huxley may have optimistically considered on the horizon but that did not yet exist. Equipped with increasingly powerful microscopes, geneticists following through on Gregor Mendel's insights had discovered intriguing, threadlike cellular components they named *chromosomes* ("colorful parts") because they were best perceived when dyes were applied to the cells. Were these the mechanisms by which genetic information was inherited? Science could peer no further until 1953, when researchers discovered that chromosomes themselves were composed of long strings of deoxyribonucleic acid, or DNA, arranged in tight strands of double helixes.

This insight is celebrated as one of the twentieth century's best-known scientific breakthroughs—the three men credited with its discovery, James Watson, Francis Crick, and Maurice Wilson, shared the Nobel Prize in Medicine. (A fourth collaborator, Rosalind Franklin, was not so honored.) But knowing the structure was not the same as understanding how the structure worked—how it communicated genetic information. For a taxonomy such as cladistics to become viable, we would need to understand a chemical language, the language that deoxyribonucleic acid spoke to each succeeding generation.

· · ·

The fusion of biology and chemistry began in earnest not long after the rediscovery of Mendel's work, when early "bio-chemists" examined known chemistry in light of its possible genetic implications. But the discipline could trace its roots to 1869, when a Swiss doctor named Friedrich Miescher placed the bandages of hospital patients under a microscope and discovered a substance in the nucleus of white blood cells, which he named *nuclein*. By 1878, researchers had isolated the substance's unique chemical component and named it nucleic acid. By 1929, two kinds of nucleic acid had been found, one in the yeast of plants and one in the thymus glands of animals. Given the division between botany and zoology, these were studied separately until 1933, when a marine biologist discovered the presence of both in sea urchin eggs. He also noticed that they occupied distinct locations within the cell. The one associated with animals was centrally located in the nucleus itself, while the other remained in the cytoplasm, the outside region bounded by the membrane.

Further tests confirmed that the two substances seemed present in both animal and vegetable life. The acids were renamed to reflect the sugars present in their chemical building blocks. Yeast nucleic acid, containing ribose sugar, became ribonucleic acid, or RNA. Thymus nucleic acid, deploying a deoxygenated version of the same, became deoxyribonucleic acid, or DNA.

The parallel tracks of genetics and microbiology began to converge in 1961, when Francis Crick (who'd shared the Nobel Prize for discovering the double helix) joined three other scientists, Sydney Brenner, Leslie Barnett, and R. J. Watts-Tobin, to conduct an experiment on a bacteriophage, a strain of virus chosen for its relatively simple structure of less than two hundred genes. By precisely applying a mutation-causing chemical, they were able to both erase and create connections between base pairs, the bonded molecules stretching across the twisted helix to connect both strands. These targeted changes revealed a crucial fact: If only one or two of the base pairs were changed via mutation, the genes they corresponded to remained functional. But if a group of three base pairs was changed, the gene stopped working.

Three base pairs. This was the unit of change on the genetic level. Crick and his colleagues named it the *codon*, correctly believing it formed the basis of the genetic code. To the few who retained a detailed understanding of *Histoire Naturelle*, the double helix and its codons bore a familiar resemblance to the internal matrix and encoded traits of Buffon's *moule intérieur* theory. It would be an overreach to claim that Buffon had foreseen DNA and RNA—his microscope could barely perceive single-celled organisms, and the replicative mechanisms of biology were more complex, by orders of magnitude, than his roughly sketched principles of the process he'd labeled reproduction. But it had been a singular accomplishment to produce such a rough sketch two hundred years earlier, armed only with patient observation and attention to the invisible.

Early codon-reading machines, also known as DNA sequencers, were up and running by 1970. They were slow and inefficient at first, taking weeks to analyze even a small section of chromosomes, but grew faster as the technology progressed. Julian Huxley died in 1975, knowing that his vision of a genetically based taxonomy was beginning to move into the realm of possibility. The first genome—the mapping of an organism's complete collection of codons—was completed in 1977. Our way of viewing the world, and the world itself, began to rapidly change.

Buffon's Gloriette, Jardin des Plantes

TWENTY-NINE

A Large Web or Rather a Network

WHEN HE CAN SPARE THE TIME, DR. JAN MEES IS AN EXORCIST of scientific ghosts. A marine biologist, Mees's full-time job is directing the Flanders Marine Institute in Ostend, Belgium, but his side project is serving as co-leader of an international group of volunteer scientists. Together, they're working to build a World Register of Marine Species, a comprehensive database vetting the accuracy of scientific names for aquatic life. It was a daunting task from the beginning, but Mees and his colleagues have been surprised by how much of their work involves not adding to the register but subtracting from it.

After analyzing 418,850 species, the project has eliminated 190,400 of them—more than 45 percent—as redundant. The species *felís* and *Triakis semifasciata*, for instance, are in fact both the leopard shark. *Octopus rooseveltii*, named in 1941 as a tribute to then-president Franklin Delano Roosevelt, already existed as *Octopus oculifer*, named in 1909. No one knows how many previous generations of marine biologists devoted their careers to studying a notional species, not realizing that their peers were studying it under different names.

One of the most prolific over-classifiers, they've discovered, was Louis Agassiz, who eagerly declared and named North American species based on little evidence, and even less scientific rigor. Viewing a single sample of fossilized teeth, Agassiz determined they defined not three different species but three completely new genera of fish. They've since been matched to teeth belonging to the fossil of a single individual fish, collapsing an entire branch of his imagined taxonomy. Hundreds of Agassiz's species identifications have been abandoned on similar grounds.

Presently, the greatest single instance of systematist ephemerality is the case of *Litteronia saxatilis*, a shore-dwelling sea snail commonly known in English as the rough periwinkle. Easily found on beaches on both sides of the North Atlantic, the rough periwinkle long ago developed the ability to blend into its immediate marine environment by varying the size, the coloration, and even the shape of its shell. The mollusk inside the shell is unchanged, but two centuries of classifiers, not only focusing on external morphology but accepting empty shells as definitive type specimens, made the rough periwinkle the "champion of taxonomic redundancy," according to Dr. Mees. Since its initial naming in 1792, *Litteronia saxatilis* has been identified, named, and catalogued as a separate species or subspecies no fewer than 112 times. The World Registry of Marine Species has quietly erased those identifications, retroactively updated the scientific literature to point to the original designation, and moved on.

Why so many duplications? Most are products of the Linnean emphasis on appearance, which allowed minor physical variations to be interpreted as novel species. Some constitute genuinely separate discoveries, obscured by the difficulty of reviewing printed scientific literature in the days before digitization and database searches. Still more are the results of overzealous imaginations, seeking a measure of immortality by declaring a "discovery" where none existed. To estimate how many officially designated species are mistakes or duplicates, on land as well as underwater, John Alroy, a paleobiologist at Maquarrie University in Sydney, Australia, has devised a "flux ratio," a predictive metric extrapolated from the number of corrections already made to taxonomic data. Professor Alroy's flux ratio predicts that ultimately, 24 to 31 percent of all current species names will eventually be discarded as redundant.

We've also come to realize that while some species exist under multiple names, other named species don't exist at all. In the tenth edition of *Systema Naturae*, in the order *Vermes*, Linnaeus included the Shot, or *Furia infernalis*, the creature he was convinced had nearly killed him as a student in 1728. He described it as a flying worm as thin as a human hair, which

> quickly buries itself under the skin, leaving a black point where it had entered; which is soon succeeded by the most excruciating pain, inflammation and gangrene of the part, swooning, and death. This all happens in the course of a day or two, frequently within a few hours, unless the animal be immediately extracted, which is effected with great caution and difficulty, by applying a poultice of curds or cheese; or carefully dissecting between the muscles where it has entered.

Linnaeus had given it scientific existence, and that had been good enough for other reports of *Furia infernalis* attacks to start trickling in. A priest in Lapland attested he'd narrowly avoided the bite of one, which landed on his plate. A young servant girl had not been so lucky: A Shot had bitten her finger, and she would surely have died had an onlooker not immediately amputated the infected digit. After reports of an 1823 *Furia infernalis* attack that killed some five thousand reindeer in Lapland, Finland imposed a ban of fur from the region in hopes of preventing its spread. No one, however, could produce a specimen. The contemporary consensus is that Linnaeus was likely bitten by a horsefly.

We've also been subject to a different kind of illusion: seeing one species where in fact there are several. One of the more surprising results of genomic analysis is the discovery in 2021 that the giraffe, believed to be a single species since Linnaeus named it *Giraffa camelopardelis* in 1758, is in fact four distinct species, genetically distinct for at least a million years. This startled zoologists, who are now rushing to reevaluate everything thought to be known about what seemed a familiar animal. "To put our results into perspective, the genetic differences between the distinct giraffe species are similar to those between polar and brown bears," says Dr. Axel Janke, a geneticist at Goethe University. "We've clearly completely forgotten what a giraffe is."

The finding has also galvanized conservationists. As a single spe-cies, giraffes were already classified as "vulnerable" on the Interna-tional Union for Conservation of Nature's list of threatened species. Now understood as separate species and therefore separate popula-tions, at least three of the four meet the criteria for reclassification as "endangered" or "critically endangered." "Giraffes are assumed to have similar ecological requirements across their range, but no one really knows," says Dr. Julian Fennessey, director of the Giraffe Conserva-tion Foundation. "It would be ignorant to ignore these new findings."

· · ·

With its imposing Palladian architecture and central courtyard, Lon-don's Burlington House elegantly plays its role as a temple of learning. The massive three-century-old building, formerly the residence of the Earl of Burlington, now holds the Royal Academy of Arts, the Royal Astronomical Society, the Royal Society of Chemistry, and the Geological Society of London, as well as a half-dozen other learned societies under the benevolence of the Crown. In one corner, tucked away in an unobtrusive arcade of columns, is the Linnean Society.

While it retains its original name, the society has long since grown beyond its founder James Edward Smith's vision of enshrining the Linnean worldview. It is now one of the world's foremost centers for publishing contemporary biological research, with membership a coveted honor for biologists around the world. It deals frankly with its past, acknowledging and disavowing malign aspects of its legacy. "Linnaeus' work on the classification of man forms one of the eighteenth-century roots of modern scientific racism," it notes on its website. "The Linnean Society intends to confront the consequences of scientific racism."

The society does, however, maintain a physical link to its name-sake. The surviving specimens of Linnaeus's personal collection—the ones that Smith did not auction off, give away, or simply discard—are carefully preserved at Burlington House. Access to this collection (as well as to the trove of Linnaeus's original manuscripts that Smith re-ceived in the bargain), is limited and closely monitored. Permitted visitors are ushered into an elevator that descends into an under-ground complex, then led into the display area of the Strong Room, both a security vault and a fail-safe controlled environment. If the streets of London were being firebombed above (as they were in the Blitz raids of World War II), the Strong Room is so thickly reinforced that its temperature and humidity would vary by less than one degree. Here, in glass-topped drawers that slide neatly from custom-built cabinets, reside the remnants of Linnaeus's original collection. "All the natural objects preserved in spirits of wine exist no more," lamented one visitor in 1888. "The mammals have disappeared, the birds have flown, and the fishes slipped away, probably caused by the several removals." But the cocoa plant (*Theobroma cacao*) remains, as does the slender-leaved *Buffonia* and the twinflower *Linnea borealis*. It is a singular experience, to look at a carefully curated plant or insect and know that Linnaeus once held it in his hand. And to know that (by the rules of conventional taxonomy), this particular plant or insect represents the quintessence of its species, just as Linnaeus purportedly represents *Homo sapiens*.

While the Linnean Society preserves these as historical artifacts, it's ready to move beyond the limitations they've come to represent. A myriad of lifeforms cannot be pressed under glass or pinned to display cases. They are constantly adapting and diverging, defying our attempts to attach fixed labels. A retrospective of the society, published on the three-hundredth anniversary of Linnaeus's birth, concluded with the statement, "Species names are not static. . . . The species, called by its Linnean two-word name, is a practical category for entities we wish to talk about."

> Remember that Linnaeus was a great innovator, but that not all his ideas worked or have been carried forward to today. In our attempts to come to grips with the challenges facing us and the other species with which we share the planet, it is worth reflecting on the importance of such innovation and on the necessity of looking forward and adapting our outlook as new challenges arise.

As we've come to realize, "the other species with which we share the planet" exist in mind-boggling numbers. Even with the elimination of duplicated or imaginary species, we're forced to confront the fact that we've barely begun the tally of life's diversity. As advances in genetics now make clear, the number of extant species—as defined by genetically distinct reproductive populations—is vastly greater than previously imagined.

How much greater? That number keeps revising upward. In 2011, the most exhaustive survey of biodiversity yet undertaken came up with an estimate of 8.7 million species, of which only 1.2 million had been catalogued thus far. That estimate meant that 86 percent of all land-dwelling species, and 91 percent of all aquatic species, remained undiscovered. Yet those numbers were dwarfed by a preliminary report five years later from the National Science Foundation. An NSF project called Dimensions of Biodiversity is using gene sequencing to identify species down to the microbial scale, and while it will take years for full results to emerge, project members already estimate the total number on Earth is more than twenty orders of magnitude greater than previously understood. "Until now, we haven't known whether aspects of biodiversity scale with something as simple as the abundance of organisms," reports Dr. Kenneth J. Locey, a postdoctorate fellow at Indiana University and a Dimensions of Biodiversity researcher. "As it turns out, the relationships are not only simple but powerful, resulting in the estimate of upwards of one trillion species."

One trillion species. That would mean we've discovered and recorded only one one-thousandth of one percent of all possible entries in a catalogue of life.

. . .

As hard as that number is to imagine, it is harder still to imagine that many species fitting into the Linnean taxonomical hierarchy, which is already tottering to accommodate the 1.2 million species thus far identified. Extensive modifications to the system have already been adopted: Linnaeus's original three kingdoms (animal, vegetable, mineral) have been joined by five more: fungi, chromista, protozoa, archaea, and bacteria. The original five nesting boxes (kingdom, class, order, genus, species), which Linnaeus considered perfect and complete, are usually presented as seven, with the addition of phylum (inserted between class and kingdom) and family (between order and genus). But this is the simplified version. In reality, biologists have been compelled to balloon the hierarchy to no fewer than twenty-one distinct categories.

Subspecies

Species

Subgenus

Genus

Subtribe

Tribe

Subfamily

Family

Superfamily

Infraorder

Suborder

Order

Superorder

Cohort

Infraclass

Subclass

Class

Superclass

Subphylum

Phylum

Kingdom

*The original five levels of Linnean taxonomy, with current additional
levels interspersed*

Some taxonomists have deployed a twenty-second category, beneath the level of kingdom, although its name has not been standardized. It is variously referred to as dominion, superkingdom, realm,
empire, and domain. Under any name, it is intended to accommodate
our increased understanding of the diversity of unicellular life. A
group of microbes called hemimastigotes, for instance, differs from
other microbes as much as a chanterelle mushroom differs from a
chimpanzee. To account for such evolutionary distance, the hierarchy
requires augmentation beneath what has, from the beginning, been
understood as its very root.

Extended as it is, Linnean-based taxonomy does not include cultivars, a term coined in 1923 to describe distinctive plant varieties that
nevertheless retain the genome of the original species. Pumpkins,
zucchinis, and yellow squash are all cultivars of the species *Curcurbita
pepo*. Six other vegetables commonly found in the produce section—

broccoli, cabbage, Brussels sprouts, cauliflower, kohlrabi, and kale—
are taxonomically the same plant, cultivars of the wild mustard *Brassica
oleracea*. Since 1953 these have been documented separately under the
International Code of Nomenclature for Cultivated Plants, which also
catalogues *grexes*, a separate term for hybridized orchids.

And then there are viruses, classified separately by the International Committee on the Taxonomy of Viruses. There's no consensus
on where they fit in the current taxonomical system, if at all. Some
biologists argue that they are not technically alive, as they have no
metabolism and cannot replicate without a host. Others, pointing to
recent discoveries of "megaviruses" that rival bacteria in size and genetic complexity, believe they occupy at least a border zone between
life and non-life.

It is becoming pointless to add further boxes to the nesting hierarchy, further branches of kingdoms to the common trunk of life. As the
philosopher Marc Ereshefsky wrote in *The Poverty of the Linnaean Hierarchy*, "Linnaeus's motivation for assigning species binomial names has
become obsolete. Even his sexual system and the method of logical
division has been abandoned. . . . All that remains of Linnaeus's original systems is a hierarchy of categorical ranks and the use of binomial
names."

It is, in fact, something of a misnomer to continue to call it the Linnean hierarchy. The existing infrastructure (including the category
family, borrowed from Michel Adanson) reflects the gradual acceptance of *Genera Plantarum*, Antoine-Laurent de Jussieu's rival system of
plant classification. While little noted in the eighteenth century and
largely ignored in the nineteenth, Jussieu's work gained gradual acceptance throughout the twentieth century. By 1981, seventy-six of
Jussieu's classifications of botanical genera had been incorporated
into standard taxonomy, while only eleven of Linnaeus's remained. In
2005, the International Botanical Congress made the shift away from
the Linnean sexual system official, moving the recognized starting
point of genera names from Linnaeus's 1753 edition of *Species Plantarum*
(the first botanical spinoff volume of *Systema Naturae*) to August 4,
1789, the publication date of Jussieu's *Genera Plantarum*. The present
taxonomy of plants is a hybrid, one that could rightly be called the
Linnaeus-Jussieu-Adanson system.

But no one is worrying about placing labels on our current system of labels. Instead, biologists are proposing and debating the contours of an entirely new system to replace it.

. . .

Can we build a system without objective bias?

Julian Huxley's concept of cladistics, a taxonomy based on genetic diversion, seems to fit the bill, but there are complications. The boundary of cladogenesis—when one species population becomes incapable of breeding with its source species—is difficult to pinpoint, particularly in the fossil record: There is, as yet, no consistent means of identifying which aspects of genetic drift decisively affect reproduction. And reproduction itself is proving an increasingly fluid boundary. As we've discovered, many species have evolved means of thriving without resorting to conventional fertilization. Several kinds of salamanders and frogs are both unisexual and parthenogenic, their eggs developing without male intercession. Others survive through kleptogenesis, the "stealing" of sperm from the males of another species—a process that fertilizes the egg without incorporating the male's genes. One spectacular instance of resourcefulness occurs in a cluster of five species of salamanders in the genus *Ambystomia*. Each species is unisexual, but together they form a reproductive complex, borrowing genes from one another to trigger reproduction while remaining distinct species.

Yet life is more resourceful still. We've also come to recognize that some species are not individual organisms but the interaction of several. Lichens, for instance, straddle two kingdoms. Linnaeus labeled lichens the *rustici pauperini* (poor trash) of the vegetable world, but placed them firmly in kingdom *Plantae*. In the late twentieth century, biologists discovered that most lichens are not a single species but a symbiotic colony of the fungus ascoycota (kingdom *Fungi*) and algae (kingdom *Plantae*). A more recent study established that at least some lichens incorporate a third organism, basidiomycete, which is a yeast. Far from being rustici pauperini, lichens display some of the most sophisticated interactions of lifeforms yet discovered.

Other boundaries in biological perception have begun to be dismantled. The blue whale, long considered Earth's most massive lifeform, has been dwarfed by the discovery of a thirteen-million-pound organism dubbed Pando. A resident of central Utah since at least the last Ice Age, Pando is so enormous it occupies 108 acres. Yet it went unnoticed until 1976, when researchers from the University of Colorado found what appeared to be a forest of aspen trees was really forty thousand clones of the same tree, interconnected at the roots.

Pando is truly a single organism, as the clones do not propagate by seeding. Instead, when one tree begins to die it replenishes by sending signals through the root structure, and a new clone emerges. Biologists are both elated at the identification of Pando and frustrated that we failed to recognize it sooner; human incursion into its domain may have stopped its replenishment, leading to a slow decline and eventual death. While possibly the oldest organism, Pando is not the largest, although it was considered such until 2015, when a single instance of the honey mushroom fungus (*Armillaria ostoyoe*) was found to extend underground across 2,385 acres in Oregon's Malheur National Forest. The "humongous fungus," as it's affectionately called, has since yielded the title to an even more recently perceived organism. In 2022, biologists determined that an underwater stand of Australian seagrass (*Posidonia australis*) constituted one specimen, sprouted some 4,500 years ago from a single seed. It's now grown to slightly over 49,000 acres, or 64 percent larger than the city of San Francisco. Unless humans hinder it as they did Pando, it should continue to thrive and grow.

Humans have also been compelled to perceive ourselves as more than merely *Homo sapiens*. When the genome of *Homo neanderthalensis* was sequenced in 2010, researchers were surprised to find Neanderthal genes present in a large amount of the current population. Recent samples estimate that the average person of primarily European ancestry is, genetically speaking, 1.7 percent Neanderthal; those of primarily Eastern Asian ancestry, 1.83 percent. Even more surprisingly, those of primarily African descent have been found to have a genome averaging 0.5 percent Neanderthal—a paradigm-upsetting discovery, as Neanderthals are believed to have emerged outside that continent. It indicates that instead of a diaspora of *Homo sapiens* spreading outward from Africa, a significant number of our ancestors returned there after intermingling with *Homo neanderthalensis*.

Another species, *Homo denisova*, has also contributed to our gene pool. Despite the fact that our only evidence of the species was discovered in Siberia and Tibet, aboriginal Australians carry 3 to 5 percent Denisovian DNA in their genome. Natives of Papua New Guinea carry 7 to 8 percent. We are, on the whole, not only a hybrid of species but the result of several vectors of migration. Our ancestry appears "almost as a spider web of interactions," concludes Omer Gokumen, a geneticist at the University of Buffalo, "rather than a tree with distinct branches."

Further complicating matters is the evolution of evolution itself. While Darwin's theory of natural selection—biological change through random mutations—has been abundantly confirmed through genomic analysis, the same analysis has led to the surprising return of Lamarck's theory of directed variation. While it appears to remain true that an organism does not change in direct response to its environment—a giraffe cannot will itself to have a longer neck—researchers in 2003 determined that although environmental factors do not change genes, they can change the *expression* of those genes, activating some and deactivating others. Furthermore, at least some of these expression patterns—now called *epigenetics*—appear capable of being passed on to subsequent generations, allowing Lamarck and Darwin to coexist after all.

In sum, life appears to exult in blurring the boundaries we place upon it. Buffon's observation from two and a half centuries ago seems more relevant than ever.

> This chain is not a simple thread which is only extended in length, it is a large web or rather a network, which, from interval to interval, casts branches to the side in order to unite with the networks of another order.

Not even Julian Huxley's cladistics can meaningfully encapsulate all of life. While genome-sequencing technology gets faster and cheaper each year, it also produces cladistic connections that strain our ability to grasp at the larger whole. In classical taxonomy, birds presently remain in Linnaeus's original class of *Aves*. In the separate Linnean class of *Reptilia*, alligators and crocodiles belong to the order *Crocodylia*, while lizards and snakes currently occupy the order *Squamata*. To a mindset based on morphology, this makes perfect sense: Crocodiles, lizards, and snakes resemble each other far more closely than any of them resemble birds. But cladistics traces monophyletic evolution—that is, descent from a common ancestor. Despite their appearance, *Crocodylia* are the closest living relatives of *Aves*, both having emerged from the cladistic group *Pseudosuchia* around a quarter of a billion years ago. An alligator is more closely related to a peacock than it is to a Komodo dragon. That Komodo dragon is more closely related to you.

Other surprises of cladistics: The lotus is less related to the water lily than it is to the sycamore tree. Roses and figs are closely related. The papaya's nearest relative is that multiple-cultivar *Brassica oleracea*, known simultaneously as cabbage, broccoli, Brussels sprouts, cauliflower, and kale.

Some of the most disorienting results of cladistic taxonomy are underwater. The genus of *Cancer*, or crab, is now understood to be a broad collection of genetically distant species—in other words, not a genus at all. The advantages of developing a crablike anatomy are such that multiple monophyletic lines (organisms descending from a common ancestor) evolved into a convergent body shape, creating close resemblances in spite of vastly dissimilar origins. And while Linnaeus ultimately recast the class *Pisces*, or fish, to exclude whales and other cetaceans, so many evolutionary paths have adapted to free-swimming undersea life that *Pisces* has now been retired entirely. What we informally group as *fish* represent more than a dozen different monophyletic lines of descent, so genetically diverse that, as some biologists have enjoyed pointing out, "fish" do not really exist. If one drew a cladistic circle broad enough to encompass all fish, the circle would include humans as well.

• • •

Yet humans need "fish," or something very like the concept of "fish," to make sense of the world. Words are not just units of speech; they are units of thought. The decisions we make in organizing the world tend to disappear once we've made them, but they're inevitably encoded in language.

Take colors, for instance. Among English-speaking people, we distinguish pink as a separate color from red. In the Malaysian language (Bahasa Malay), however, there is no pink. There is only *merah*, red. You can try to approximate a concept of pink by describing it as "light red," but even then you'd be imprecise: The closest term in everyday use is *merah muda*, which means literally "young red," and which can designate a bright red as well as a light one. You can convey the sense of pink by evoking a pink object, such as *merah jambu*, or "red like a guava." The only problem is that you're now describing not a range of shades but one shade in particular: the pink of a guava skin. Speaking Malaysian, of course, does not confer color-blindness, but to an Anglophone such vagueness can seem like an awkward and inaccurate approach to color. Why not just coin a word for "pink" and have done with it?

Yet English does exactly the same thing. We have no equivalent of pink for the blue portion of the spectrum. If you are not content with the vagueness of "light blue," you have no choice but to get hyper-specific—"robin's-egg blue," for instance. In other words, there's a chromatic gap in our language just as big as the one in Malaysian. But if you're like most native English speakers, you've probably never noticed it. In contrast, Russian speakers learning English notice it right away: Their language divides up our "blue" into two colors, the paler *goluboy* and the darker *siniy*. What's fascinating is that these distinctions are more than technicalities. They become hardwired into our brains. Neuroscientists have found that native Russian speakers are measurably faster than native English speakers at distinguishing dark-blue shades from lighter ones, presumably because, in their minds, the difference between *goluboy* and *siniy* is clear-cut.

As language-bound human beings, struggling to understand life's complexities, we still require the semantic construct of agreed-upon labels. A team of biologists, Francine Pleijel and George Rouse, have proposed the LITU, or "least-inclusive taxonomic unit," to replace the concept of species entirely. These would be provisional identities, which they describe as "statements about the current state of knowledge (or lack thereof)"—snapshots rather than static points, documented under the assumption that they might change as more genetic information becomes available. In other words, the LITU is remarkably similar to Buffon's original concept of species—as an entity of reason rather than a physical fact. (It's worth noting that Dr. Pleijel is a researcher at the contemporary Jardin des Plantes.)

This would move us away from lectotypes, allotypes, and other type specimens, a shift Pleijel and Rouse strongly advocate. According to them, "making taxonomists decide that a few dead specimens represent a species is an extravagant extrapolation that has no place in science." Scientists are "forced by the existing codes of nomenclature to describe organisms as species when in fact they generally have no idea of what is going on in nature."

· · ·

At a conference in Mexico in the year 2000, the chemist Paul Crutzen proposed the declaration of a new geological epoch. Current standardized geology places us within the Holocene epoch, which succeeded the Pleistocene era approximately twelve thousand years ago. But Crutzen, awarded the Nobel Prize for his work analyzing Earth's atmosphere, argued that humanity was wreaking profound change upon the planet, to the extent that it had ushered in an entirely new geological period. Owing to the "major and still growing impacts of human activities on earth and atmosphere," he wrote in a joint statement delivered afterward, "it seems to us more than appropriate to emphasize the central role of mankind in geology and ecology by proposing to use the term 'Anthropocene' for the current geological epoch."

Anthropocene means "*the epoch of humans.*" In 2009, Crutzen's speech prompted the formation of the Anthropocene Working Group, an interdisciplinary task force charged with making recommendations for formal recognition of the epoch. In the course of their investigations, one member, the epistemologist and historian Jacques Grinevald, brought something to their attention: An "epoch of humans" had already been declared.

In *The Epochs of Nature*, the 1774 essay both included in *Histoire Naturelle* and published as a stand-alone volume, Buffon had argued that human-driven environmental change had proceeded to the point that it represented the "seventh and last epoch, when the power of man has assisted that of Nature." This power, he concluded, was not universally positive. "The most despicable condition of the human species is not that of the savage," he wrote,

> but that of those nations that are a quarter policed, which have always been the real curse of human nature, and which civilized peoples still have trouble to contain today. They have, as we have

said, ravaged the first happy land, they tore out the seeds of con-
tentment. . . . Cast your eyes on the annals of all the peoples, you
will count there twenty centuries of desolation for a few years of
peace and repose.

Buffon did not anticipate modern global warming: Pollution on
such a scale as to cause it was unimaginable in 1774. But he did believe
that human habitation had already permanently changed the climate,
and that the planet was not an inexhaustible resource. The rapid colo-
nization and exploitation of other continents by European nations
struck him as particularly troubling. "The English . . . did they not
make a great mistake in extending too far the limits of their colonies?"
he wrote. "The ancients seem to me to have had saner ideas about
these matters; they planned emigrations only when their population
became too great, and when their lands and their commerce no longer
sufficed for their needs."

Struck by the parallels, the Anthropocene Working Group commissioned an English translation of *The Epochs of Nature* and republished it anew. Why did a team of scientists, tasked with defining the
present day, treat with urgency the words of a man who lived a quarter
of a millennium ago? "It really is an extraordinary book," they wrote in
their introduction. "In some ways, the sciences have come full circle."

> It has become increasingly clear that one has to understand not
> just the parts (in minute detail) of the whole, but also the "whole"
> itself. . . . Among the most important incarnations of the whole is
> that of the Earth, the planet that we still entirely rely upon for
> our existence. Buffon's vision of the Earth and—perhaps more
> particularly—the way he developed it, may have lessons for us yet.

This was just one of the more recent episodes in the ongoing rediscovery of Buffon, itself part of a burgeoning renaissance in complexist
biology. In 1959, the anthropologist Loren Eisley commemorated the
centennial of *On the Origin of Species* by acknowledging the debt Darwin
owed to Buffon. Describing Buffon's theory of species change as
"nothing more than a rough sketch of evolution," Eisley added that
"Buffon managed, albeit in a somewhat scattered fashion, at least to

mention *every significant ingredient which was to be incorporated into Darwin's great synthesis of 1859.* [Italics in the original.] It is a great pity that his ideas were scattered and diffused throughout the vast body of his *Natural History* with its accounts of individual animals." He continued,

> Not only did this concealment make his interpretation difficult, but it lessened the impact of his evolutionary ideas. . . . However, almost everything necessary to originate a theory of natural selection existed in Buffon. It needed only to be brought together and removed from the protective ecclesiastical coloration which the exigencies of his time demanded.

Other reassessments followed. "Buffon asked almost all of the questions that science has since been striving to answer," the historian Otis Fellows wrote in 1970.

> Those who have looked with care, agree . . . that his glory lies in what he prepared for his successors: bold and seminal views on the common characters of life's origin, laws of geographical distribution, a geological record of the earth's evolution, extinction of old species, the successive appearance of new species, the unity of the human race.

In 1971, the botanist Frans Stafleu—one of Linnaeus's most respected biographers—acknowledged the ways in which Buffon's vision exceeded that of his rival. "Linnaeus mentions the law of generation which accounts for the production of identical unchanging forms, but that is as far as time goes," Stafleu wrote.

> In Buffon we encounter, however, a biologist of exceptional insight, almost free from traditional thought and religious scruples, a highly intelligent though somewhat speculative mind, exquisitely original and in many ways far ahead of his time. . . . The introduction of the historical element by Buffon was of the greatest importance for the evolution of biology as an independent science.

∙ ∙ ∙

The Jardin des Plantes still thrives in Paris. It remains the mother campus of the National Museum of Natural History, whose fourteen locations are spread throughout France. The institution that Buffon finessed through the controversies of the Enlightenment, that Lamarck saved and Geoffroy defended during the French Revolution, is today both a respected research center and a public attraction, welcoming nearly two million visitors each year. The maps of Paris have been redrawn several times over the centuries, but the Jardin's current address provides a fitting tribute to its history. It is now bounded on one edge by Rue Buffon, and Rue Cuvier on another. Its western boundary is a street that changes names. Up until Rue Cuvier, it is called Rue Linné. As it touches the Jardin, it becomes Rue Geoffroy Saint-Hilaire. This, appropriately, prevents the intersection of Rue Linné and Rue Buffon.

The heroic statue of Buffon remains. It now occupies a new place of honor, presiding over the Hall of Evolution, an exhibition displaying the growth and change of species through time—a parade through the shards of the broken lens of fixity, one of Buffon's keystone achievements. Few of the visitors passing by know the statue is now a reliquary as well as a monument. The crystal urn containing his cerebellum, removed and measured after his death, was placed in the pedestal in 1870 (in Montbard his other remains, scattered during the revolution, were reassembled and reinterred in 1971).

The Gloriette, Buffon's most personal monument, is intact and recently restored. If you climb up the winding path to its location on the summit of the Jardin's only hill, you are rewarded with a view of well-tended greenery, the river Seine, and Paris beyond. Visible to your right are the museum's research halls—the direct descendant of Buffon's Cabinet du Roi—which contain only a small portion of the museum's collection of more than sixty-two million specimens. The museum's herbarium still contains specimens collected by Tournefort, Commerson, and Lamarck. It publishes a botanical journal named *Adansonia*, in honor of Michel Adanson.

In the middle distance, slightly to the left, lies the zoological garden that Geoffroy founded in 1793. His original grounds constitute just a small percentage of the four zoos the institution now operates. One of them is the Paris Zoological Park, four miles away in the Bois de Vincennes, where a strange and wonderful creature is on display. It is *Physarum polycephalum*, a bright-yellow but otherwise unremarkable-looking tree slime.

Physarum polycephalum

Physarum polycephalum defies the foundations of Linnean thought. It is an ordinary tree slime, common in European and North American forests, classified and labeled in 1822 but otherwise ignored until 1970, when a teaching assistant at Iowa State University discovered a surprising characteristic: The slime not only had an immune system but one that functioned externally rather than internally. A sample of *polycephalum*, taken from a rotting elm log, kept infections at bay by secreting an antiviral substance so potent that when sprayed upon crops it was 100 percent effective in eliminating tobacco mosaic, a potentially devastating virus that blighted not only tobacco plants but tomatoes, peppers, and cucumbers as well.

Over the course of the past five decades, more of the organism's extraordinary aspects have come to light. As we've now discovered, it is neither an animal, a plant, nor a fungus. It can hibernate for years at a time. It has no musculature, yet it moves itself at a brisk 1.6 inches an hour. It is, somehow, a single-celled organism. (The *Guinness Book of World Records* proclaims it the largest cell on the planet.) If separated, segments are fully capable of operating independently, then reintegrating into the whole. They can even merge seamlessly into different specimens, gathered from different locations. Individual existence, it appears, is optional.

Its seeming simplicity belies an extraordinarily complicated sex life. Instead of two genders, male and female, *Physarum polycephalum* has 720 distinct forms of mating pairs, methods of inducing genetic variety that are the functional equivalent of genders. Sexuality and reproduction, as we've come to understand, accommodate a multitude of themes and variations.

Physarum polycephalum is also capable of learning. In order to most efficiently find food sources, it spreads itself out in a pattern both expanding and self-correcting, until it has covered the maximum amount of territory with the least amount of resources. By this measure, it can be seen as intelligent: The highly efficient networks it creates can find the quickest path out of a labyrinth or the shortest routes to connect multiple locations. In one experiment, it was presented with multiple food sources (in this case, oat flakes), placed in a pattern that replicated the geographic locations of Tokyo and thirty-six towns in the surrounding region. The slime mold reached out to all food sources with pathways that nearly replicated the Japanese rail system connecting those locations—a system carefully designed by humans to operate as efficiently as possible.

Despite its lack of a central nervous system, much less a brain, it's also capable of remembering. Somehow, it manages to retain what it's learned. If placed in the same labyrinth weeks apart, it will recognize the maze and re-create its previous escape route. Even a small piece of the original will do the same.

We don't fully understand *polycephalum*'s intelligence, but that's not keeping us from collaborating with it. In fact, we've recently recruited it to help us explore the cosmos. Current astrophysical theory holds that following the Big Bang, all matter in the universe dispersed in a pattern creating filaments between adjacent galaxies. Physical evidence of this dispersal is difficult to discover, since the filaments consist of thin, diffused streams of hydrogen gas. Astronomical instruments can detect these filaments, but only if pointed directly at them.

How to point the instruments in the right direction? By predicting in advance where these streams will be. To do that, astrophysicists have turned to *polycephalum*, harnessing the same efficiency it uses to solve mazes and re-create the Tokyo metropolitan train system. Using an artificial intelligence program designed to emulate the spore as closely as possible, they've been feeding it galactic maps and asking it to make connections. "A slime mold creates an optimized transport network, finding the most efficient pathways to connect food sources," observes Dr. Joseph Burchett, the project's chief researcher. "In the cosmic web, the growth of structure produces networks that are also, in a sense, optimal. The underlying processes are different, but they produce mathematical structures that are analogous."

So far, the project has traced the connections between more than thirty-seven thousand galaxies. It's just getting started, but it's already demonstrated the power of shifting our perception of the living. Of approaching nature not as static objects to be inventoried, but as dynamic, interdependent manifestations of a greater whole.

To exist is to coexist. To be is to be in conversation.

In Buffon's own words, "Nature is not a thing, for this thing would be everything."

> Nature is not a being, for that being would be God; but one can consider her an immense living power, which embraces everything, which animates everything. . . . Nature is herself a perpetually living finished product, a worker ceaselessly active, who knows how to employ everything, who in working by herself always on the same resources, far from exhausting them, renders them inexhaustible.

Acknowledgments

ENCOUNTERING AND EXPLORING THE STORIES IN THIS BOOK comprised a task of several years. The necessary patience was far from mine alone. I'm deeply grateful to Michael Carlisle, my agent, and Andy Ward, my editor, who championed and shepherded the work through a myriad of drafts and permutations. My gratitude further extends to Michael Mungiello, Lyndsey Blessing, Greg Mollica, Ella Laytham, Carrie Neill, Azraf Khan, Kaeli Subberwal, Stuart Calderwood, Mark Birkey, Simon Sullivan, Mark Maguire, Armando Veve, Bill Hamilton, Kristin Cochrane, Tim Rostron, Moritz Volk, Sebastian Ritscher, and Jon Riley.

I'm thankful beyond measure for the indomitable support of my partner, Julia Scott, who, in addition to her loving forbearance, contributed her French translation and editing skills to this book. I want to equally thank our extended family: Patricia Roberts, Eden Roberts, and Jesse Roberts, as well as Gloria Roberts, Moana Roberts, Phil Schacht, Michael Scott, and Laura Scott. I benefited greatly from the counsels and encouragement of Tom Barbash, Peter Orner, Cathryn Ramin, Oscar Villalon, Julia Flynn Siler, Elizabeth Stix, Geoff Shandler, Louis B. Jones, Brett Hall Jones, Lisa Alvarez, and Andrew Tonkovich. My heartfelt appreciation also goes to David Clark, Drew Pearce, Bruce Grossan, Christopher Michel, Bari Levinson, David Rosane, Austin Hughes, Nicole Laby, Revi Airborne-Williams, and Mark Fassett. I reserve special thanks for Jeremy Cline, Michael Levinson, Scott Lucas, Erik Peterson, Gary Rudman, and Aaron Stern, long-standing members of a men's support group masquerading as a poker game. Your absent-minded Professor cherishes his seat at the table.

For Your Exploration

France

The Jardin des Plantes in Paris remains free and open to the public daily. Tickets to the Ménagerie, the zoo within the garden, are available online, as are tickets to exhibits and special events of the Muséum National d'Histoire Naturelle. To visit the restored Gloriette, Buffon's forged-iron gazebo, follow the spiral path above the Grand Amphithéâtre du Muséum.

www.jardindesplantesdeparis.fr/en
www.mnhn.fr/en
57 Rue Cuvier, 75005 Paris
+33 1 40 79 56 01

The Musée & Parc Buffon de Montbard occupies the site of Buffon's former estate in Montbard, Bourgogne. Admission to both the grounds and the museum are free. From April through September, tickets may be purchased for guided tours that include access to Buffon's writing studio and tower workspace.

www.musee-parc-buffon.fr
museeparcbuffon@montbard.com
Rue du Parc 21500
Montbard, France
+03 80 92 50 42 / 50 57

Sweden

A restored version of Linnaeus's botanical garden and home, on the original site in central Uppsala's Svartbäcksgatan district, is open for tours and events from the beginning of May through the end of September. During the rest of the year, the home (now a museum) is open only to groups with advance booking. Both guided and self-guided tours are available.

> www.botan.uu.se/our-gardens/the-linnaeus-garden/visit-the-garden
> bokning@botan.uu.se
> Villavägen 8
> S-752 36 Uppsala, Sweden
> +46 18 471 28 38

The estate of Hammarby, Linnaeus's residence 15 kilometers (9.5 miles) southeast of Uppsala, has been maintained in a remarkable state of preservation, with Linnaeus's original furniture and decorations still in place. Administered by the same organization, it is also open from May through September. The surrounding grounds, now a Cultural Heritage Reserve, are open year-round.

> www.botan.uu.se/our-gardens/linnaeus-hammarby
> bokning@botan.uu.se
> +46 18 471 28 38

England

The library and archives of the Linnean Society of London may be visited by appointment Tuesdays through Fridays, from 10 A.M. to 5 P.M. Viewings of Linnaeus's original collections are limited, and must be requested a minimum of two weeks in advance.

> www.linnean.org/research-collections
> library@linnean.org
> Burlington House, Piccadilly, London, UK W1J 0BF
> 020 7434 4479

Notes and Sources

INTRODUCTION | *Savants*

xi *"Objects are distinguished and known"* · Carl Linnaeus, *Systema Naturae (First Edition)* (Leyden: 1735).

xi *"The true and only science"* · Jacques Roger, *Buffon: A Life in Natural History* (Ithaca, N.Y.: Cornell University Press, 1997), 83.

PRELUDE | *The Mask and the Veil*

3 *"a luster rarely accorded"* · Jacques Roger, *Buffon: A Life in Natural History* (Ithaca, N.Y.: Cornell University Press, 1997), 433.

3 *First came a crier and six bailiffs* · Charles Coulston Gillispie, *Science and Polity in France at the End of the Old Regime* (Princeton, N.J.: Princeton University Press, 1980), 144.

4 *"You remember, gentlemen"* · Henri Nadault de Buffon, *Correspondance Inédite de Buffon* (Paris: Hachette, 1860), 615.

4 *"four bright lamps"* · Otis E. Fellows and Stephen F. Milliken, *Buffon* (New York: Twayne Publishers, 1972), 16.

4 *"the last to vanish"* · Charles Augustin Saint-Beuve, *Causeries du Lundi* (London: George Routledge & Sons, 1909), 32.

4 *"The history of science presents"* · John Herbert Eddy, "Buffon, Organic Change, and the Races of Man," diss., The University of Oklahoma, 1977, 20.

5 *"Monsieur Buffon has never spoken"* · Fellows and Milliken, *Buffon*, 65.

5 *"the whole extent of Nature"* · John Lyon and Phillip R. Sloan, *From Natural History to the History of Nature: Readings from Buffon and His Critics* (Notre Dame, Ind.: University of Notre Dame Press, 1981), 307.

6 *"Good writing is good thinking"* · John Galsworthy, "The New Spirit in the Drama," *The Living Age*, 1913, 264.

6 *the most popular nonfiction author in French history* · Roger (*Buffon*, 184) calls *Histoire Naturelle* "the most widespread work of the eighteenth century, beating . . . even the better-known works of Voltaire and Rousseau." Fellows, tallies 52 complete editions of the work published in France alone, and more than 325 full or par-

tial editions published in translation. In a survey of French private libraries conducted in the nineteenth century, *Histoire Naturelle* was the second-most common item found on shelves, exceeded only by a French-language encyclopedia. See Otis Fellows, *From Voltaire to la Nouvelle Critique: Problems and Personalities* (Geneva: Librairie Droz, 1970), 16.

7 *"God himself has guided him"* · Carl Linnaeus, *Vita Carolia Linnaei* (1760), trans. Ann-Mari Jönsson, 146.

7 *"The greatest obstacles to the advancement"* · Samuel Butler, *Evolution, Old and New* (London: Hardwicke and Bogue, 1879), 138.

8 *"Nature, displayed in its full extent"* · Georges-Louis Leclerc le Comte de Buffon, *Histoire Naturelle des Oiseaux* (Paris, 1770), vol. 1, trans. Phillip R. Sloan, 394.

ONE | *Of the Linden Tree*

13 *"Flowers became Carl's first and choicest"* · Theodor Magnus Fries and Benjamin Daydon Jackson, *Linnaeus: The Story of His Life* (London: H.F. & G. Witherby, 1923), 7.

14 *"The Bible is the Book"* · "Sketch of the Life of Carl von Linne," Edward L. Morris, *The Bicentenary of the Birth of Carolus Linnaeus,* Annals of the New York Academy of Sciences, 47.

14 *"The guests seated themselves"* · Richard Pulteney, *A General View of the Writings of Linnaeus* (London: J. Mawman, 1805), 512.

15 *"preferring stripes and punishments"* · Ibid., 513.

16 *"Flora seems to have lavished all her beauties"* · Ibid., 511.

16 *"hoping to hear from the preceptors"* · Ibid., 514.

TWO | *A Course in Starvation*

21 *"What are you examining?"* · Linnaeus recounted each exchange of this conversation, although he did not cast it as dialogue. Ibid., 517.

22 *"His eyes were full of plants"* · John Muir, "Linnaeus," in *The World's Best Literature*, ed. John W. Cunliffe and Ashley Thorndike (New York: The Knickerbocker Press, 1917), 9078.

23 *"He loved me not as a student"* · Heinz Goerke, *Linnaeus* (New York: Scribner, 1973), 17.

23 *"the thickness of a human hair"* · John Wright, *The Naming of the Shrew: A Curious History of Latin Names* (London: Bloomsbury, 2014), 29.

25 *"obliged to trust to chance"* · Pulteney, *A General View of the Writings of Linnaeus*, 517.

THREE | *The Salt-Keeper's Son*

27 *"One can cite of his childhood"* · Fellows and Milliken, *Buffon*, 40.

33 *"As for me, I shall do whatever lies"* · Ibid., 46.

33 *"showed from the beginning a great disposition"* · Saint-Beuve, *Causeries du Lundi*, 32.

34 *"any gentleman who struck another"* · John Gideon Millingen, *The History of Duelling* (London: Samuel Bentley, 1841), 189.

FOUR | *Vegetable Lambs and Barnacle Trees*

36 *"We immediately started talking"* · Wilfrid Blunt, *Linnaeus: The Compleat Naturalist* (Princeton, N.J.: Princeton University Press, 2001), 31.
36 *"tall, slow and serious"* · Gunnar Broberg, *"Brown-Eyed, Nimble, Hasty, Did Everything Promptly": Carl Linnaeus 1708–1778* (Uppsala: Uppsala University, 1990), 9.
37 *"the legs of the Birde hanging out"* · John Gerard, *The Herbal, or Generall Historie of Plantes* (London: Adam Islip, 1636), 1587.
38 *"fruit whereof is a wool exceeding in beauty"* · Herodotus and George Rawllinson, *The History: A New English Version* (New York: D. Appleton, 1889), 410.
39 *"This is the highest pitch of humane reason"* · Arthur O. Lovejoy, *The Great Chain of Being: A Study of the History of an Idea* (Cambridge, Mass.: Harvard University Press, 1976), 232.
42 *"To know a flighty Latin word"* · Lisbet Koerner, *Linnaeus: Nature and Nation* (Cambridge, Mass.: Harvard University Press, 2009), 34.
42 *could all discuss the same tree* · Unfortunately, Theophrastus himself could not have joined in the conversation. Being Greek, he would likely have used the word *asvéstis.*
44 *"I am no poet"* · Blunt, *Linnaeus: The Compleat Naturalist,* 32.

FIVE | *Several Bridegrooms, Several Brides*

45 *"See how every bird"* · Ibid.
50 *he still planned to kill Rosén* · Dietrich Heinrich Stoever, *The Life of Sir Charles Linnaeus* (London: E. Hobson, 1794), 41.

SIX | *The Greater Gift of Patience*

52 *"The Duke of Kingston has hitherto"* · Roger, *Buffon: A Life in Natural History,* 9.
52 *"According to the law of custom"* · Edward Gibbon, *The Miscellaneous Works of Edward Gibbon, Esq.* (London: B. Blake, 1837), 71.
54 *"He is an extremely likable man whose intelligence"* · Catherine Ostler, *The Duchess Countess* (New York: Atria Books, 2022), 132.
57 *"Rome is at this hour in all its glory"* · Fellows and Milliken, *Buffon,* 43.
58 *approximately thirty million dollars today* · Translating past amounts into contemporary values is notoriously subjective, as prices of different items change at different rates. I have pegged this estimate to the cost of hiring laborers at the time, as that quickly became Buffon's chief expenditure.
60 *"If you were to cover my gardens"* · Roger, *Buffon: A Life in Natural History,* 24.

60 *"Through these keen, reasoned and sustained experiments"* · Lee Alan Dugatkin, *Mr. Jefferson and the Giant Moose: Natural History in Early America* (Chicago: University of Chicago Press, 2019), 15.

60 *"Buffon's originality here is considerable"* · Roger, *Buffon: A Life in Natural History*, 40.

61 *"Genius," he quipped* · Ibid., 4.

61 *"I sigh for the tranquility"* · Ibid., 30.

64 *"I have long been predicting"* · Stephen F. Milliken, *Buffon and the British* (New York: Columbia University Press, 1965), 56.

64 *"neither passion nor weakness"* · Voltaire, *A Philosophical Dictionary* (London: John and Henry L. Hunt, 1824), 109.

64 *"Give me some linen"* · Henri Nadault de Buffon, *Correspondance Inédite de Buffon* (Paris: Hachette, 1860), 7.

65 *"There, in a bare room"* · Saint-Beuve, *Causeries du Lundi*, 37.

65 *"He walked around thinking"* · Nadault de Buffon, *Correspondance Inédite de Buffon*, 628.

66 *"I burn everything"* · Lyon and Sloan, *From Natural History to the History of Nature: Readings from Buffon and His Critics*, 362.

66 *"was my whole pleasure"* · Ibid., 371.

SEVEN | Now in Blame, Now in Honor

67 *"Never have I known a worse road"* · Blunt, *Linnaeus: The Compleat Naturalist*, 45.

68 *"the whole land laughs and sings"* · Ibid., 41.

68 *"All the elements were against me"* · Daniel C. Carr, *Linnæus and Jussieu* (West Strand, U.K.: John W. Parker, 1844), 87.

68 *"the gnats kept inflicting their stings"* · Carl Linnaeus, *Lachesis Lapponica, or, a Tour in Lapland* (London: White and Cochrane, 1811), 142.

69 *"like all people addicted to fishing"* · Carl Linnaeus, *Travels* (New York: Charles Scribner's Sons, 1979), 23.

69 *"lowly, insignificant, disregarded"* · Londa Schiebinger, *Plants and Empire: Colonial Bioprospecting in the Atlantic World* (Cambridge, Mass.: Harvard University Press, 2004b), 202.

72 *"abundantly sufficient for all the animals"* · Society of Gentlemen in Scotland, *Encyclopaedia Britannica* (1771), 425.

75 *"Close to the road"* · Linnaeus, *Lachesis Lapponica*, 191.

76 *"Oh, how many weary steps we took"* · Blunt, *Linnaeus: The Compleat Naturalist*, 67.

76 *"I said good-bye to Uppsala Academy"* · Fries and Jackson, *Linnaeus: The Story of His Life*, 129.

EIGHT | The Seven-Headed Hydra of Hamburg

78 *"All that this skillful man thinks"* · Carl Linnaeus, *Musa Cliffortiana* (Vienna: A.R.G. Ganter Verlag K. G., 2007), 16.

79 *"Many people said it was the only one"* · Blunt, *Linnaeus: The Compleat Naturalist*, 89.

80 *"a natural truth"* · Peter Dance, *Animal Fakes and Frauds* (Berkshire, U.K.: Sampson Low, 1976), 35.

80 *"in no way a work of art"* · Paula Findlen, "Inventing Nature: Commerce, Art, and Science in the Early Modern Cabinet of Curiosities," in *Merchants and Marvels: Commerce, Science and Art in Early Modern Europe*, ed. Pamela H. Smith and Paul Findlen (New York: Routledge, 2002), 319.

80 *"O Great God"* · Norah Gourlie, *The Prince of Botanists: Carl Linnaeus* (London: H.F. & G. Witherby, 1953), 119.

81 *"Linnaeus must hasten his departure"* · Ibid.

81 *"bloaters, bilberries, and degrees"* · Blunt, *Linnaeus: The Compleat Naturalist*, 94.

81 *"in order to frequent the famous"* · Ibid., 96.

82 *"Our tears showed what joy we felt"* · Ibid., 98.

83 *the equivalent of $8,000* · The less extravagant among them rented pineapples by the evening, to display as prestige objects during dinner parties.

83 *"such a wealth of plants"* · Goerke, *Linnaeus*, 29.

84 *"I happen to have two copies"* · Blunt, *Linnaeus: The Compleat Naturalist*, 101.

85 *"When I saw the lifeless"* · Donald Culross Peattie, *Green Laurels: The Lives and Achievements of the Great Naturalists* (New York: Simon & Schuster, 1936), 109.

85 *"The first step in wisdom" (and subsequent quotes)* · Linnaeus, *Systema Naturae (First Edition)*.

92 *"showed the first signs of flowering"* · Linnaeus, *Musa Cliffortiana*, 70.

92 *"Nature has never granted"* · Ibid., 89.

92 *no small amount of praise for himself* · The work begins with a verse. ("O Banana, more beautiful than your beautiful name . . . the credit, which is due to LINNAEUS alone, must be remembered by posterity.")

94 *"that goddess of the ancients"* · Ibid., 109.

NINE | *An Abridgment of the World Entire*

98 *"one has found the means"* · E. C. Spaary, *Utopia's Garden: French Natural History from Old Regime to Revolution* (Chicago: University of Chicago Press, 2000), 24.

99 *"Facies Americana."* · Roger, *Buffon: A Life in Natural History*, 51.

99 *"Behold Charles Linnaeus, the prince of botany"* · Stoever, *The Life of Sir Charles Linnaeus*, 132.

100 *"the man who has thrown all botany into confusion"* · Blunt, *Linnaeus: The Compleat Naturalist*, 113.

100 *"I have never pretended"* · Stoever, *The Life of Sir Charles Linnaeus*, 111.

100 *"become a Frenchman"* · Pulteney, *A General View of the Writings of Linnaeus*, 534.

101 *"I am the lost child"* · Roger, *Buffon: A Life in Natural History*, 32.

TEN | *Loathsome Harlotry*

102 *"received me as a stranger"* · Blunt, *Linnaeus: The Compleat Naturalist*, 130.

103 *"loathsome harlotry"* · Ibid., 121.

104 *"What a fool have I been"* · James Edward Smith, *A Selection of the Correspondence of Linnaeus and Other Naturalists, from the Original Manuscripts* (London: Longman, Hurst, Rees, Orme and Brown, 1821), 320.

104 *"Aha! said I, Esculapius"* · Ibid., 335.

105 *"A night with Venus"* · Blunt, *Linnaeus: The Compleat Naturalist*, 131.

105 *"My adverse fate"* · Smith, *A Selection of the Correspondence of Linnaeus and Other Naturalists, from the Original Manuscripts*, 335.

105 *"alas, almost all the young men"* · Goerke, *Linnaeus*, 36.

106 *"I have succeeded in obtaining quickly"* · Frans A. Stafleu, *Linnaeus and Linnaeans: The Spreading of Their Ideas in Systematic Botany, 1735–1789* (Utrecht, The Netherlands: A. Oosthoek's Uitgeversmaataschappij, N.V. First edition, 1971), 17.

106 *"All the medical world"* · Roger, *Buffon: A Life in Natural History*, 46.

107 *"The Intendancy of the Jardin du Roi"* · Ibid., 45.

108 *"I now grew fond again of plants"* · Stoever, *The Life of Sir Charles Linnaeus*, 145.

108 *"Botany is very difficult"* · Linnaeus, *Musa Cliffortiana*, 27.

108 *"I was obliged to publish"* (and subsequent quotes) · Carl Linnaeus, *Systema Naturae (Second Edition)* (Stockholm: 1740).

110 *"beautiful in body"* · Michael Keevak, *Becoming Yellow: A Short History of Racial Thinking* (Princeton, N.J.: Princeton University Press, 2011), 24.

110 *"white like us"* Ibid., 27.

110 *"the color of Africans"* · Ibid., 29.

110 *"they of the most inward provinces"* · Ibid., 33.

110 *"more yealow . . . like unto the Almans"* · Ibid., 32.

111 *"Rosén, who cannot even recognize a nettle"* · Goerke, *Linnaeus*, 39.

112 *"By God's Grace I am now released"* · Mary Gribbin and John Gribbin, *Flower Hunters* (Oxford: Oxford University Press, 2008), 49.

112 *"You may now devote yourself entirely to the service of Flora"* · James Edward Smith, *Memoir and Correspondence* (London: Longman, Rees, Orme, Brown, Green and Longman, 1832), 425.

113 *"more like an owl's nest"* · Goerke, *Linnaeus*, 41.

113 *"gives honey without bees"* · Eva Crane, *The World History of Beekeeping and Honey Hunting* (New York: Routledge, 1999), 493.

ELEVEN | *The Quarrel of the Universals*

115 *"He carries himself marvelously"* · Roger, *Buffon: A Life in Natural History*, 32.

115 *"His manner to public men"* · Alpheus Spring Packard, *Lamarck, the Founder of Evolution* (New York: Longmans, Green, 1901), 22.

116 he hired a woman · By some accounts Basseporte was already painting in the Jardin (in an unofficial capacity), but Buffon made the position official.

116 *"Nature gives plants their existence"* · Uncredited, "Necrologe des Artistes et des Curieux," *Revue universelle des arts* 13 (1861), 142.

117 *"The king's cabinet is not rich"* · Mary Terall, *Catching Nature in the Act: Réaumur and the Practice of Natural History in the Eighteenth Century* (Chicago: University of Chicago Press, 2014), 193.

118 *"subsist (in the nature of things)"* · Ralph M. McInerny, *A History of Western Philosophy* (Notre Dame, Ind.: University of Notre Dame Press, 1963), 357.

119 *"The abstract does not exist"* · Roger, *Buffon: A Life in Natural History*, 133.

120 *"If there are no cleavages in nature"* · Lovejoy, *The Great Chain of Being: A Study of the History of an Idea*, 231.

120 *"Linnaeus' method is of all the least sensible"* · Eddy, "Buffon, Organic Change, and the Races of Man," 30.

120 *"The study of nature supposes two qualities"* · Lyon and Sloan, *From Natural History to the History of Nature: Readings from Buffon and His Critics*, 98.

124 *"It is not in such terms"* · Roger, *Buffon: A Life in Natural History*, 211.

124 *"all the objects presented to us"* · Fellows and Milliken, *Buffon*, 83.

125 *"The [first three] parts"* · Lyon and Sloan, *From Natural History to the History of Nature: Readings from Buffon and His Critics*, 307.

125 *"I look forward to the subsequent"* · Letter of Linnaeus to Jussieu, archived as L1352, Box 510, Linnaean correspondence, Uppsala University Library. Latin trans. by the author.

TWELVE | *Goldfish for the Queen*

129 *"He came to Uppsala quite young"* · Fries and Jackson, *Linnaeus: The Story of His Life*, 232.

130 *The ill-fated American bear* · Linnaeus wept at the loss, and for the rest of his life kept a portrait of Sjupp on prominent display in his household.

130 *"officers of Flora"* · Koerner, *Linnaeus: Nature and Nation*, 49.

131 *a trumpet was sounded* · The field troops of these herbaciones were not just volunteers. They paid Linnaeus for the privilege, considering it money well spent for specimens to add to their private collections.

131 *"I decided to take him into"* · IK Foundation & Company, *The Linnaeus Apostles: Global Science and Adventure* (London: IK Foundation & Company, 2008), 1013.

132 *"He lived with me"* · Ibid.

132 *"I have botanically described the most beautiful plant"* · Julien d'Offray de la Mettrie, *L'Homme Plante* (Potsdam, Germany: Chretien Frederic Voss, 1748), 16.

133 *"a public malice without cause"* · Letter of Linnaeus to Count Sten Carl Bielke, dated April 24, 1745.

133 *"You gratify your enemies"* · Smith, *A Selection of the Correspondence of Linnaeus and Other Naturalists, from the Original Manuscripts*, 381.

133 *"I have ever loved you"* · Ibid., 393.

134 *"I had rather been without his apology"* · Stoever, *The Life of Sir Charles Linnaeus*, xiii.

134 *"We that admire you are much concerned"* · Ann-Mari Jonsson, "Linnaeus' International Correspondence. The Spread of a Revolution," in *Languages of Science in the*

Eighteenth Century, ed. Britt-Louise Gunnarsson (Berlin: De Gruyter Mouton, 2011), 183.

135 *"To acquire a tea bush"* · IK Foundation & Company, *The Linnaeus Apostles: Global Science and Adventure* (London: IK Foundation & Company, 2010), 164.

137 *"wrapped up in leaves or paper"* · Ibid., 361.

138 *"You would scarcely believe how many"* · Smith, *A Selection of the Correspondence of Linnaeus and Other Naturalists, from the Original Manuscripts*, 315.

138 *"Now is the time"* · Koerner, *Linnaeus: Nature and Nation*, 115.

138 *"a sort of hexagonal mirror"* · John Scott, "On the Burning Mirrors of Archimedes, and on the Concentration of Light Produced by Reflectors," *Proceedings of the Royal Society of Edinburgh* 6 (1869), 233.

139 *"extremely large, or more likely mythical"* · Roger, *Buffon: A Life in Natural History*, 52.

139 *Frederick the Great of Prussia* · Fellows and Milliken, *Buffon*, 57.

139 *"Buffon! There is nothing"* · Roger, *Buffon: A Life in Natural History*, 53.

139 *"Without any previous knowledge"* · Gibbon, *The Miscellaneous Works of Edward Gibbon, Esq.*, 243.

139 *Resurrecting Archimedes's mirror* · In the process of tinkering with focused light, Buffon designed a new kind of concave lens with stepped degrees of thickness, the better to concentrate a beam's power. He never built the lens—the mirror did not need it—but he did publish his notes. In 1819, when Augustin-Jean Fresnel debuted the improved lens still in use in lighthouses today, he was shocked to find that Buffon had anticipated him by seven decades. He apologized for having "broken through an open door."

140 *"do not displease the Divine teacher"* · Phillip R. Sloan, "The Buffon-Linnaeus Controversy," *Isis* 67, no. 3 (1976).

140 *Buffonia* · Additionally, Linnaeus could take comfort in the fact that *Bufo* was already in use as a genus name for "toad."

140 *"very slender pretensions to botanical honor"* · *The Naturalist* (London: Whittaker and Co., 1838), 394.

THIRTEEN | *Covering Myself in Dust and Ashes*

141 *an opening salvo* · This was an expansion of his essay *On the Manner of Studying and Considering Natural History*, delivered to the Académie des Sciences six years earlier.

141 *"established on undoubted testimony"* · Georges-Louis Leclerc le Comte de Buffon, *The Natural History of Animals, Vegetables, and Minerals (Histoire Naturelle)*, trans. W. Kenrick and J. Murdoch (London: T. Bell, 1775).

142 *"celebrated for their beauty"* · Georges-Louis Leclerc le Comte de Buffon, *Buffon's Natural History (Barr's Buffon)*, trans. J. S. Barr (London: J. S. Barr, 1792), 256.

142 *"their complexions beautiful"* · Ibid., 220.

142 *"of all mankind, perhaps the most miserable"* · Ibid., 232.

142 *"Black or brown hair begins"* · Georges-Louis Leclerc le Comte de Buffon, *Natural*

History, General and Particular (Histoire Naturelle), trans. William Smellie (London: W. Strahan and T. Cadell, 1785), 127.

142 *"If blackness was the effect"* · Georges-Louis Leclerc le Comte de Buffon, *Buffon's Natural History (Barr's Buffon)*, trans. J. S. Barr (London: J. S. Barr, 1792), 307.

143 *"I am inclined to believe, therefore"* · Ibid., 338.

143 *"ought to be considered as accidental"* · Ibid., 306.

144 *"This attack is directed straight"* · Lyon and Sloan, *From Natural History to the History of Nature: Readings from Buffon and His Critics*, 217.

144 *"Everyone knows that spirit"* · Ibid., 220.

145 *"which heat necessarily underwent"* · Georges-Louis Leclerc le Comte de Buffon, *Buffon's Natural History (Barr's Buffon)*, trans. J. S. Barr (London: J. S. Barr, 1792), 94.

145 *"a book whose venom"* Lyon and Sloan, *From Natural History to the History of Nature: Readings from Buffon and His Critics*, 237.

145 *"I think of acting differently"* · Saint-Beuve, *Causeries du Lundi*, 36.

145 *"because it contained principles and maxims"* · Roger, *Buffon: A Life in Natural History*, 187.

146 *"I abandon whatever in my book"* · Fellows and Milliken, *Buffon*, 82.

146 *"I have extricated myself"* · Ibid.

146 *"It is better to be humble"* · Roger, *Buffon: A Life in Natural History*, 188.

147 *"charming, gentle, pretty rather than beautiful"* · Ibid., 205.

147 *"I will worry even less about criticisms"* · Ibid., 206.

FOURTEEN | *The Only Prize Available*

148 *"The Ariadne's thread of botany is system"* · Carl Linnaeus, *Philosophia Botanica* (Oxford, U.K.: Oxford University Press, 2005), 113.

148 *"instruction of a uniform and a new"* · Koerner, *Linnaeus: Nature and Nation*, 43.

149 *"incapacitated my mind and spirit"* · Linnaeus, *Philosophia Botanica*, 6.

149 *"the digest of the Science of Botany"* · Ibid.

149 *"The calyx is the bedroom"* · Ibid., 105.

149 *"We reckon the number of species"* · Ibid., 113.

149 *"the species are very constant"* · Ibid., 115.

149 *"have established all the classes"* · Ibid., 23.

149 *"have determined the truly proper names"* · Ibid., 26.

149 *"If you do not know the names of things"* · Ibid., 169.

150 *"words one and a half feet long"* · Ibid., 214.

150 *"Without the concept of a genus"* · Ibid., 44.

151 *"are to be banished from the Commonwealth of Botany"* · Ibid., 172.

151 *"the specific name ought to be derived"* · Ibid., 221.

151 *"mostly variable and rarely constant"* · Ibid.

151 *"zeal for subtleties"* · Ibid.

151 *"generic names should not be misused"* · Ibid., 182.

151 *"private individuals have applied absurd names"* · Ibid., 169.

152 *"No sane person introduces primitive generic names"* · Ibid., 172.

152 *"As I am now occupied"* · Ibid., 6.

152 *"The starting-point must be to marvel"* · Ibid., 332.

FIFTEEN | *Durable and Even Eternal*

155 *"The case will be the same"* · Lyon and Sloan, *From Natural History to the History of Nature: Readings from Buffon and His Critics*, 112.

155 *"We believe we have had sufficient reason"* · Ibid., 114.

156 *"whose natural qualities have been matured"* · Buffon, *Natural History, General and Particular (Histoire Naturelle)*, 307.

156 *"Buffon had first dazzled as a poet"* · Geoffroy Saint-Hilaire, 1838, *Biographical Fragments (Fragments Biographiques): Previous Studies on the Life, Works and Doctrines of Buffon* (Paris: F. D. Pillot, 1838), 8.

156 *"noble, dignified, with a magnificent appositeness"* · Saint-Beuve, *Causeries du Lundi*, 42.

156 *"I will say nothing against a method"* · Terall, *Catching Nature in the Act*, 194.

157 *"Well-written works are the only ones"* · François Pierre Guillaume Guizot, *The Nations of the World: France* (New York: Peter Fenelon Collier, 1898), 223.

157 *"I am every day learning to write"* · Ibid.

158 *"skull of a monstrous calf"* · Georges-Louis Leclerc le Comte de Buffon, *Histoire Naturelle* (Paris: L'Imprimerie Royale, 1753), 543.

158 *"all the labour of our country depends upon him"* · Ibid., 336.

SIXTEEN | *Baobab-zu-zu*

160 *"Young man, you have studied enough"* · Uncredited, 1844, *Linnæus and Jussieu*, 59.

160 *"Senegal is of all white settlements"* · Adanson I (Pittsburgh: The Hunt Botanical Library, 1963), 14.

160 *"He was . . . robust and healthy."* · Ibid., 36.

161 *"the blood oftentimes opened itself"* · John Pinkerton, *A General Collection of the Best and Most Interesting Voyages and Travels in All Parts of the World* (Longman, 1814), 663.

161 *"Nay, they went so far"* · Ibid., 659.

162 *"I leave you the freedom"* · Letter of February 20, 1752. Xavier Carteret, "Michel Adanson in Senegal (1749–1754): A Great Naturalistic and Anthropological Journey of the Enlightenment," *Revue d'Histoire des Sciences* 65 (January–June 2012), xiii.

162 *"As soon as we leave our temperate countries"* · Letter of February 20, 1752. Ibid.

162 *"an instructive and natural method"* · Eddy, "Buffon, Organic Change, and the Races of Man," 35.

163 *"one sees clearly"* · Lyon and Sloan, *From Natural History to the History of Nature: Readings from Buffon and His Critics*, 102.

163 *"I have found a manner of describing"* · Carteret, "Michel Adanson in Senegal (1749–1754): A Great Naturalistic and Anthropological Journey of the Enlightenment," v.

163 *"precious assemblage . . . worthy to be acquired"* · Adanson I, 50.

164 *"Let us multiply observations"* · Ibid., 39.

164 *"Some modern botanists call barbarians"* · Michel Adanson, *Familles des Plantes* (New York: J. Kramer-Lehre, 1966), 142.

164 *"This is what made Monsieur de Buffon"* · Ibid.

165 *"I saw the natural method"* · Adanson I, 50.

166 *"He spoils and destroys"* · Ibid., 51.

SEVENTEEN | So Many New and Unknown Parts

169 *"I found that I was now"* · Pehr Kalm, *Travels into North America* (Warrington, U.K.: William Eyres, 1770), 31.

169 *"I invited him to lodge at my house"* · Letter from Benjamin Franklin to James Logan, October 20, 1748.

169 *"It came from New England"* · Peter Kalm (sic), *Peter Kalm's Travels in North America* (New York: Wilson-Erickson, 1937), 643.

169 *"It is the most rapid water"* · Paula Robbins, *The Travels of Peter Kalm: Finnish-Swedish Naturalist Through Colonial North America, 1748–1751* (Fleischmanns, N.Y.: Purple Mountain Press, 2007), 119.

170 *"five times worse than the Lapland"* · Ibid., 107.

170 *"are false and treacherous"* · IK Foundation & Company, *The Linnaeus Apostles: Global Science and Adventure* (London: IK Foundation & Company, 2008), 780.

170 *"Our friend Mr. Kalm, goes home"* · Robbins, *The Travels of Peter Kalm: Finnish-Swedish Naturalist Through Colonial North America, 1748–1751*, 152.

170 *"Take burning firebrands and throw them"* · Koerner, *Linnaeus: Nature and Nation*, 152.

170 *"beautiful and ripe fruit"* · Nancy Pick, "Linnaeus Canadensis," *The Walrus*, November 12, 2007, https://thewalrus.ca/2007-11-science.

171 *"peculiar friendship and kindness"* · Ibid.

171 *"like a lamp whose oil is consumed"* · Fries and Jackson, *Linnaeus: The Story of His Life*, 230.

171 *"a double death"* · Koerner, *Linnaeus: Nature and Nation*, 147.

172 *"It was a matter of complete indifference"* · Blunt, *Linnaeus: The Compleat Naturalist*, 191.

172 *"colic and pains throughout my body"* · Jorge M. Gonzalez, "Pehr Löfling: Un Apóstol de Linné en Tierras de Venezuela," *Ciencia y Tecnología* (2018), https://www.meer .com/es/36201-pehr-lofling.

172 *"The great Vulture is dead"* · Blunt, *Linnaeus: The Compleat Naturalist*, 192.

174 *"I went to visit the forest"* · IK Foundation & Company, *The Linnaeus Apostles: Global Science and Adventure*, 1518.

175 *in a fragile mental state* · Stephanie Pain, "The Forgotten Apostle," *New Scientist* 195, no. 2615 (2007).

175 *"might prolong human life"* · Johan Beckmann, *Schwedische Reise: 1765–1766* (Uppsala: Almqvist & Wiksells, 1911), 110.

176 *"The death of many whom I have induced"* · Fries and Jackson, *Linnaeus: The Story of His Life*, 285.

EIGHTEEN | *Governed by Laws, Governed by Whim*

177 *"a work which for a knowledge of nature"* · Gourlie, *The Prince of Botanists: Carl Linnaeus*, 238.
177 *"My hand is too weary"* · Gribbin and Gribbin, *Flower Hunters*, 61.
178 *"Monsieur Linnaeus made it first a badger"* · Roger, *Buffon: A Life in Natural History*, 324.
178 *"A general character"* · Londa Schiebinger, *Nature's Body: Gender in the Making of Modern Science* (New Brunswick, N.J.: Rutgers University Press, 2004a), 46.
179 *"some small relationship between the number"* · Ibid.
179 *"Man is a reasonable being"* · Roger, *Buffon: A Life in Natural History*, 240.
180 Homo sapiens americanus, *etc.* · Carl Linnaeus, *Systema Naturae (Tenth Edition)* (Stockholm: 1758).
181 *epilepsy was caused by washing* · Koerner, *Cultures of Natural History*, 158.
181 *aquavit to a puppy's fur* · Sten Lindroth, "The Two Faces of Linnaeus," in *Linnaeus: The Man and His Work*, ed. Tore Frangsmyr (Sagamore Beach, Mass.: Science History Publications/USA, 1994), 39.
181 *swallows slept through winter* · Ibid., 38.
181 *"care must be taken"* · Lyon and Sloan, *From Natural History to the History of Nature: Readings from Buffon and His Critics*, 111.
182 *"It is neither the number"* · Adanson, *Familles des Plantes*, 132.
183 *"a female wolf I kept"* · Fellows and Milliken, *Buffon*, 160.
183 *nine hybrid offspring* · Buffon, 1776, III, *Supplément à l'Histoire Naturelle*, 3.
183 *"The dissimilarities are merely external"* · Georges-Louis Leclerc le Comte de Buffon, *Natural History, General and Particular (Histoire Naturelle)*, trans. William Smellie (London: W. Strahan and T. Cadell, 1785), 394.
184 *"blanc appears to be the primitive color"* · Buffon, 1749, III, *Histoire Naturelle*, 502.
184 *the earliest humans were dark-skinned Africans* · Georges-Louis Leclerc le Comte de Buffon, *Supplément à l'Histoire Naturelle* (Paris: L'Imprimerie Royale, 1775), 564.
184 *"the most temperate climate"* · Stephen Jay Gould, *The Mismeasure of Man* (New York: W. W. Norton, 1996), 71.
184 *"the ancient continent"* · Buffon, *Buffon's Natural History (Barr's Buffon)*, 213.
185 *"Their sufferings demand a tear"* · Ibid., 292.

NINETEEN | *A General Prototype*

187 *"to betoken nature which is continued"* · Gourlie, *The Prince of Botanists: Carl Linnaeus*, 246.
188 *"If we observe God's works"* · Linnaeus, *Systema Naturae (First Edition)*.
188 *mammalian eggs* · The Dutch savant Regnier de Graaf, examining the reproductive tracts of rabbits in 1672, described ovarian follicles (which are roundish), but actual mammalian eggs would not be observed until 1827.
189 *"it is exclusively the male semen that forms"* · Matthew Cobb, *Generation* (New York: Bloomsbury, 2008), 205.
189 *"that which suppose the thing already done"* · Lyon and Sloan, *From Natural History to the History of Nature: Readings from Buffon and His Critics*, 227.

189 *"If we do not succeed in explaining"* · Fellows and Milliken, *Buffon*, 91.

189 *not female sperm* · The most likely explanation is that they were witnessing cells detached from the follicular epithelium.

190 *"There is in Nature"* · Roger, *Buffon: A Life in Natural History*, 297.

190 internal matrix · Georges-Louis Leclerc le Comte de Buffon, *The Epochs of Nature* (Chicago: University of Chicago Press, 2018), ix.

191 *"If a physician were to attempt today"* · Fellows and Milliken, *Buffon*, 98.

191 *"One day he escaped and descended"* · Ibid., 160.

TWENTY | *Breaking the Lens*

192 *Linnaeus semi-retired* · He continued to lecture privately, as a separate source of income.

192 *"never helped me in botany"* · Fries and Jackson, *Linnaeus: The Story of His Life*, 344.

193 *"inquiring less after Flora"* · Blunt, *Linnaeus: The Compleat Naturalist*, 177.

193 *"Under those disadvantages"* · Stoever, *The Life of Sir Charles Linnaeus*, 279.

194 *"your eldest daughter"* · Blunt, *Linnaeus: The Compleat Naturalist*, 193.

195 *"at fifty he began to shuffle"* · Fries and Jackson, *Linnaeus: The Story of His Life*, 301.

195 *"Even when I was with him"* · Blunt, *Linnaeus: The Compleat Naturalist*, 172.

195 *"I owe Joseph three or four volumes"* · Nadault de Buffon, *Correspondance Inédite de Buffon*, 7.

195 *"How miserably Monsieur Buffon falls foul of him"* · James Edward Smith, *A Selection of the Correspondence of Linnaeus and Other Naturalists, from the Original Manuscripts* (London: Longman, Hurst, Rees, Orme and Brown, 1821), 546.

196 *"the whole sum of the species"* · Koerner, *Linnaeus: Nature and Nation*, 45.

196 *"who has created everything"* · Ibid.

197 *"we shall find, through all of Nature"* · Georges-Louis Leclerc le Comte de Buffon, *Buffon's Natural History (Barr's Buffon)*, trans. J. S. Barr (London: J. S. Barr, 1792), 220.

198 *"The prodigious mammoth no longer exists anywhere"* · Antonello Gerbi, *The Dispute of the New World: The History of a Polemic, 1750–1900* (Pittsburgh: University of Pittsburgh Press, 2010), 16.

198 *"One can slip ideas into"* · Ibid., 53.

198 *"hidden resemblance"* · Butler, *Evolution, Old and New*, 88.

199 *"If we consider each species"* · Ibid., 103.

200 *later English translations* · Ibid., 91.

200 *"not intending the modern sense"* · Gould, *The Mismeasure of Man*, 407.

201 *"These changes are only made slowly"* · Packard, *Lamarck, the Founder of Evolution*, 205.

TWENTY-ONE | *My Cold Years*

203 *"there are bred certain minute creatures"* · George Rosen, *A History of Public Health* (Baltimore: Johns Hopkins University Press, 1993), 19.

204 *"We will find a great number of indeterminate species"* · Buffon, 1749, I, *Histoire Naturelle*, 13.

204 *"We are led to conclude that there exists"* · Georges-Louis Leclerc le Comte de Buffon, *Natural History, General and Particular (Histoire Naturelle)*, trans. William Smellie (London: W. Strahan and T. Cadell, 1785), 18.

206 *variations on Edenic originals* · Koerner, *Linnaeus: Nature and Nation*, 44.

206 *"First earth little"* · James Lee Larson, "Reason and Experience: An Inquiry into Systematic Description in the Work of Carl Von Linné," diss., University of California, 1965, 132.

206 *"if Holy Scripture would allow"* · Blunt, *Linnaeus: The Compleat Naturalist*, 182.

206 *"But this cold season"* · Knut Hagberg, *Carl Linnaeus* (New York: Cape, 1952), 244.

206 *"Many things have happened"* · Blunt, *Linnaeus: The Compleat Naturalist*, 23.

207 *"the treachery of the Chinese"* · Koerner, *Linnaeus: Nature and Nation*, 137.

207 *"Is it possible?"* · Ibid., 138.

207 *"tea was first seen away from China"* · Ibid.

207 *"notwithstanding the great extent"* · Stoever, *The Life of Sir Charles Linnaeus*, 283.

207 *"Linnaeus himself was very sensible"* · Ibid., 284.

208 *"Diseases are CURED by diseases"* · Carl Linnaeus, *Clavis Medicinae Duplex* (Stockholm: Lars Salvius, 1766), 5.

208 *"our learned and industrious author"* · Carl Linnaeus, *Clavis Medicinae Duplex: The Two Keys of Medicine* (London: IK Foundation & Company, 2012), 15.

209 *"This life? This world?"* · Carl Linnaeus, *Nemesis Divina*, trans. Eric Miller (Lanham, Md.: University Press of America, 2002), 37.

209 *"Daily we die"* · Ibid.

210 *"I must kneel for his majesty"* · Koerner, *Linnaeus: Nature and Nation*, 167.

210 *"somewhat aged, not large man"* · Ibid., 16.

211 *"Buffon, the antagonist of Linnaeus"* · Ibid., 29.

211 *"without pretty figures"* · Ibid., 28.

211 *"Buffon did not extend the boundaries of science"* · Charles Augustin Saint-Beuve, *Portraits of the Eighteenth Century, Historic and Literary* (New York: G. P. Putnam's Sons, 1905), 264.

211 *"Aha! Is it you sitting there, Carl?"* · Gunnar Broberg, *Linnaeus: Progress and Prospects in Linnaean Research* (Uppsala: Almqvist & Wiksell International, 1980), 110.

212 *"Linné limps, can hardly walk"* · Fries and Jackson, *Linnaeus: The Story of His Life*, 333.

212 *"He ventured to go a few steps"* · Ibid., 335.

212 *"He had forgotten his own name"* · Ibid., 336.

213 *"Put me in the coffin unshaved"* · Ibid., 339.

213 *"It was a dark and still evening"* · Ibid., 340.

213 *"It makes no difference to me"* · Linnaeus, *Nemesis Divina*, 36.

TWENTY-TWO | *The Price of Time*

216 *Buffon steered the appointment* · While the appointment was formally facilitated by Jussieu, no such action was taken on behalf of the Jardin without Buffon's tacit approval.

217 *"carrying, even on those laborious excursions"* · Maurice Thiéry, *Bougainville: Soldier and Sailor* (London: Grayson & Grayson, 1932), 214.

217 *"What presumption to lay down"* · Michael Ross, *Bougainville* (London: Gordon & Cremonesi, 1978), 118.

218 *"With tears in her eyes"* · John Dunmore, *Monsieur Baret: First Woman Around the World, 1766–68* (Auckland: Heritage Press, 2002), 101.

219 *"not being wholly Linnaean"* · Stafleu, *Linnaeus and Linnaeans: The Spreading of Their Ideas in Systematic Botany, 1735–1789*, 323.

220 *"How could you reconcile"* · Buffon, *The Epochs of Nature*, 15.

221 *"for a long time the seas"* · Ibid., 51.

221 *"when the Elephants"* · Ibid., 87.

221 *"proceed with circumspection"* · Fellows and Milliken, *Buffon*, 83.

221 *"When the Sorbonne picked petty quarrels"* · Roger, *Buffon: A Life in Natural History*, 423.

222 *"No man has known the price of time"* · Nadault de Buffon, *Correspondance Inédite de Buffon*, 629.

222 *"on that day"* · Stoever, *The Life of Sir Charles Linnaeus*, 128.

222 *"His delivery was fluent, but mixed"* · Ibid., 406.

222 *"wretched boy"* · Fries and Jackson, *Linnaeus: The Story of His Life*, 339.

223 *"It was singular that the lady"* · Stoever, *The Life of Sir Charles Linnaeus*, 292.

223 *"Poh! My father's successor"* · Ibid., 293.

224 *newly ennobled* · By the end of his life, Buffon would also acquire the titles Marquis de Rougemont, Vicompte de Quincy, and Seigneur of La Mairie, Les Harans, and Les Berges.

224 *"the strange irony"* · Fellows and Milliken, *Buffon*, 62.

225 *"It would have given me greater pleasure"* · James David Draper, *Augustin Pajou: Royal Sculptor, 1730–1809* (New York: Metropolitan Museum of Art, 1998), 283.

225 *"Glory . . . becomes no more than an object"* · Roger, *Buffon: A Life in Natural History*, 201.

225 *"absolutely naked, enveloped only by a drapery"* · Draper, *Augustin Pajou: Royal Sculptor, 1730–1809*, 286.

226 *"unfortunate marriage of actual and symbolic"* · Ibid.

226 *"He carries himself straight and tall"* · Ibid., 287.

227 *"Let no naturalist steal a single plant"* · Fries and Jackson, *Linnaeus: The Story of His Life*, 343.

227 *"looking at age seventy-eight"* · Miriam Claude Meijer, "The Collaboration Manqué: Petrus Camper's Son at Montbard, 1785–1787," accessed March 6, 2023, https://petruscamper.com/buffon/montbard.htm.

229 *"He was remarkably tall"* · Nadault de Buffon, *Correspondance Inédite de Buffon*, 594.

229 *"a Natural Method comprising"* · Jacques Picard, "Encyclopédistes Méconnus (3ème Partie): Michel Adanson et Son Projet d'Encyclopédie du Vivant," Dicopathe,

accessed March 3, 2023, https://www.dicopathe.com/encyclopedistes-meconnus
-3eme-partie-michel-adanson-et-son-projet-dencyclopedie-du-vivant/.

231 *"Style is the only passport"* · Butler, *Evolution, Old and New*, 76.

231 *"Far from becoming discouraged"* · Georges-Louis Leclerc le Comte de Buffon, *Supplément à l'Histoire Naturelle* (Paris: L'Imprimerie Royale, 1776), 33.

232 *"to carry it to France"* · Thomas Jefferson, *The Writings of Thomas Jefferson: 1816–1826* (New York: G. P. Putnam's Sons, 1899), 332.

232 *"Monsieur De Buffon offers his thanks"* · Letter on file in the Library of Congress, dated December 21, 1785.

233 *"It was Buffon's practice to remain"* · Jefferson, *The Writings of Thomas Jefferson: 1816–1826*, 331.

233 *"the best informed of any naturalist"* · Thomas Jefferson, *Notes on the State of Virginia* (Boston: Wells and Lilly, 1829), 56.

233 *"the animals common both"* · Ibid., 47.

234 *"Instead of entering into an argument"* · Jefferson, *The Writings of Thomas Jefferson: 1816–1826*, 331.

234 *"This communication of elephants"* · Buffon, *The Epochs of Nature*, 103.

235 *"I attempted . . . to convince him"* · Jefferson, *The Writings of Thomas Jefferson: 1816–1826*, 331.

235 *"It would be an acquisition here"* · Dugatkin, *Mr. Jefferson and the Giant Moose: Natural History in Early America*, 96.

235 *"a very troublesome affair"* · Ibid., 98.

236 *"a proper catastrophe"* · Ibid., 99.

236 *"an infinitude of trouble"* · Ibid.

236 *"stuffed and placed on his legs"* · Ibid.

236 *"This operation of Nature"* · Georges-Louis Leclerc le Comte de Buffon, *Histoire Naturelle des Minéraux* (Paris: L'Imprimerie Royale, 1786), 156.

236 *"Nothing has more retarded the progress"* · Fellows and Milliken, *Buffon*, 166.

237 *"all that is best in love"* · Ibid., 34.

238 *"A virtue existing solely"* · Buffon, *Natural History, General and Particular (Histoire Naturelle)*, 414.

239 *"Sleep had abandoned him"* · Nadault de Buffon, *Correspondance Inédite de Buffon*, 636.

240 *"What kindness!"* · Fellows and Milliken, *Buffon*, 65.

240 *"I love you and I will love you"* · Vicompte D'Haussonville, *The Salon of Madame Necker*, trans. Henry M. Trollope (London: Chapman and Hall, 1882), 285.

241 *"submit to her judgement"* · Ibid., 297.

241 *"Ah, God!"* · D'Haussonville, *The Salon of Madame Necker*, 285.

241 *"But for the union of souls"* · Ibid.

241 *"When I see him"* · Mark Gambier-Parry, *Madame Necker: Her Family and Her Friends* (Edinburgh: William Blackwood and Sons, 1913), 83.

242 *"I feel myself dying"* · Roger, *Buffon: A Life in Natural History*, 425.

242 *"You are still a charmer to me"* · D'Haussonville, *The Salon of Madame Necker*, 303.

242 *"at a time when a warmer sun was gilding"* · Nadault de Buffon, *Correspondance Inédite de Buffon*, 612.

242 *a gift from Catherine the Great* · Roger, *Buffon: A Life in Natural History*, 425.

244 *"a general and immense natural work"* · Bernard-Germain-Étienne Lacépède, *Histoire Naturelle des Quadrupedes Ovipares et des Serpens* (1788).

244 *"when I become dangerously ill"* · Stephen Jay Gould, *The Lying Stones of Marrakesh* (Cambridge, Mass.: Harvard University Press, 2011), 84.

244 *"Dear Ignatius"* · Nadault de Buffon, *Correspondance Inédite de Buffon*, 613.

244 *three teaspoons of Alicante wine* · Ibid., 614.

244 *forty minutes past midnight* · Gillispie, *Science and Polity in France at the End of the Old Regime*, 143.

244 *"of a slightly larger size"* · Roger, *Buffon: A Life in Natural History*, 433.

245 *engraved label reading* Cerebellum of Buffon · Draper, *Augustin Pajou: Royal Sculptor, 1730–1809*, 285.

245 *"Buffon died Wednesday"* · Roger, *Buffon: A Life in Natural History*, 434.

245 *"I go every week to the Jardin"* · Suzanne Curchod Necker, *Mélanges Extraits des Manuscrits* (Paris: Charles Pougens, 1798), 327.

TWENTY-THREE | *Germinal, Floreal, Thermidor, Messidor*

250 *"gifted with a lively and penetrating mind"* · Pascal Duris, *Linné et la France (1780–1850)* (Geneva: Librairie Droz, 1993), 37.

251 *"the old errors and its prejudices"* · Ibid., 135.

252 *"plain stone monument"* · Stoever, *The Life of Sir Charles Linnaeus*, 252.

253 *"I myself witnessed the extravagant tumult"* · Étienne Geoffroy Saint-Hilaire, *Biographical Fragments (Fragments Biographiques): Previous Studies on the Life, Works and Doctrines of Buffon* (Paris: F. D. Pillot, 1838), 34.

254 *"Buffon, whose writings seduced"* · Duris, *Linné et la France (1780–1850)*, 134.

254 *"receive with health, Monsieur, the homage"* · Letter of September 21, 1771. Smith, *A Selection of the Correspondence of Linnaeus and Other Naturalists, from the Original Manuscripts*, 552.

256 *"I die innocent!"* · Alison Johnson, *Louis XVI and the French Revolution* (Jefferson, N.C.: McFarland & Company, 2013), 200.

256 *"an annex of the king's palace"* · Packard, *Lamarck, the Founder of Evolution*, 30.

258 *"giving orders to nature"* · Gillispie, *Science and Polity in France at the End of the Old Regime*, 161.

260 *"I will undertake the responsibility for your inexperience"* · Butler, *Evolution, Old and New*, 239.

261 *"of insects, of worms, and microscopic animals"* · Packard, *Lamarck, the Founder of Evolution*, 37.

261 *"The law of 1793 had prescribed"* · Ibid.

263 *"Citizens, my name is Buffon!"* · Roger, *Buffon: A Life in Natural History*, 377.

263 *"Come and fill the place"* · George Henry Lewes, *Studies in Animal Life* (New York: Harper & Brothers, 1860), 140.

264 *"the driest chronological facts"* · Sara Lee, *Memoirs of Baron Cuvier* (London: Longman, 1833), 11.

265 *"I am known, then"* · Ibid., 14.

265 *"I have just found a pearl"* · Ibid.

265 *"Geoffroy and Cuvier knew no jealousy"* · Lewes, *Studies in Animal Life*, 141.

266 *"filled with gross mistakes"* · Packard, *Lamarck, the Founder of Evolution*, 48.

266 *"Obtain, O Linné, this immortality!"* · Pascal Duris, "Linné et le Jardin du Roi," Hypotheses, April 16, 2012, https://objethistoire.hypotheses.org/88, 14.

267 *"It is well known that Buffon"* · Stoever, *The Life of Sir Charles Linnaeus*, 251.

269 *"The revolution is over"* · Andrew Roberts, *Napoleon and Wellington* (New York: Simon & Schuster, 2002), 3.

TWENTY-FOUR | *Transformism and Catastrophism*

271 *"He was then at his maturity"* · Lewes, *Studies in Animal Life*, 144.

272 *"the conditions of life"* · Butler, *Evolution, Old and New*, 210.

273 *"correlation of parts"* · Richard Hertwig, *General Principles of Zoology* (New York: Henry Holt and Company, 1896), 16.

274 *"these animals are perfectly similar"* · Toby A. Appel, *The Cuvier-Geoffroy Debate: French Biology in the Decades Before Darwin* (Oxford, U.K.: Oxford University Press, 1987), 82.

275 *"we certainly do not observe"* · Caitlin Curtis, Craig D. Millar, and David M. Lambert, "The Sacred Ibis Debate: The First Test of Evolution," *PLOS Biology* (2018), 5.

275 *"It is undeniable that his position of hostility"* · William A. Locy, *Biology and Its Makers* (New York: Henry Holt and Company, 1908), 415.

276 *"Linnaeus allotted four classes of inhabitants"* · Johan Friedrich Blumenbach, *De Generis Humani Varietate Nativa Liber* (Gottingen: 1775), 41.

276 *"five principal varieties of mankind"* · Johan Friedrich Blumenbach, *The Anthropological Treatises of Johann Friedrich Blumenbach* (London: Longman, Green, 1865), 264.

277 *"I have allotted the first place"* · Ibid.

277 *"Never has a single head done more harm"* · Robert Gordon Latham, *The Natural History of the Varieties of Man* (London: John Van Voorst, 1850), 108.

277 *"the most degraded of human races"* · Gould, *The Mismeasure of Man*, 69.

277 *"the Caucasian [race], to which we belong"* · George Cuvier, *The Animal Kingdom, Arranged in Conformity with Its Organization* (London: Wm. S. Orr and Co., 1854), 49.

278 *"There are but two ways of accounting"* · Charles White, *An Account of the Regular Gradation in Man* (London: C. Dilly, 1799), 52.

279 *"Our wise men have said"* · Voltaire, *Les Lettres d'Amabed* (London, 1769), 53.

279 *"Doth M. Buffon think it sufficient"* · Lord Kames, *Sketches of the History of Man*, vol. 1 (Edinburgh: Creech, 1774), 73.

279 *"which Napoleon considered unworthy"* · Richard W. Burkhardt, *The Spirit of System: Lamarck and Evolutionary Biology* (Cambridge, Mass.: Harvard University Press, 1977), 10.

280 *"Reptiles . . . build a branching sequence"* · Gould, *The Lying Stones of Marrakesh*, 140.

280 *"Why not proclaim an important truth?"* · Burkhardt, *The Spirit of System: Lamarck and Evolutionary Biology*, 202.

281 *"I shall see Buffon again"* · Gillispie, *Science and Polity in France at the End of the Old Regime*, 160.

281 *"Monsieur Adanson devoted himself to his great work"* · Cuvier, *The Edinburgh New Philosophical Journal*, April–July 182, 177.

283 *"too great indulgence"* · Gould, *The Lying Stones of Marrakesh*, 117.

283 *"Blind, poor, forgotten, he remained alone"* · Packard, *Lamarck, the Founder of Evolution*, 68.

284 *"Buffon is to the doctrine of the mutability"* · Isidore Geoffroy Saint-Hilaire, *Histoire Naturelle Générale des Règnes Organiques* (Paris: Libraire de Victor Masson, 1854), 383.

TWENTY-FIVE | *Platypus*

287 *"I have always had the highest"* · Smith, *A Selection of the Correspondence of Linnaeus and Other Naturalists, from the Original Manuscripts*, 574.

287 *"Nor shall I ever forget"* · Smith, *Memoir and Correspondence*, 324.

288 *"takes the lead among those"* · Ibid., 330.

289 *"In spite of all opposition"* · James Edward Smith, *Translation of Linnæus's Dissertation on the Sexes of Plants* (Dublin: Luke White, 1786), xii.

289 *"some systematical tables concerning Natural History"* · Linnaeus, *Travels*, 98.

289 *"His works, I imagine, are little known"* · Benjamin Stillingfleet, *Literary Life and Select Works of Benjamin Stillingfleet* (London: Longman, 1811), 186.

289 *"Linnaeus' method has pleased by its novelty"* · Smith, *A Selection of the Correspondence of Linnaeus and Other Naturalists, from the Original Manuscripts*, 46.

290 *"Their attempts to reduce the names"* · Richard Brookes, *A New and Accurate System of Natural History, Volume VI* (London: J. Newberry, 1763), v.

290 *"the chief intention of these discourses"* · Georges-Louis Leclerc le Comte de Buffon, *Natural History, General and Particular (Histoire Naturelle)*, trans. William Smellie (London: W. Strahan and T. Cadell, 1785), xviii.

291 *"I incline to think"* · John Gascoigne, *Joseph Banks and the English Enlightenment: Useful Knowledge and Polite Culture* (Cambridge, U.K.: Cambridge University Press, 2003), 200.

293 *"I admire your defense"* · Smith, *A Selection of the Correspondence of Linnaeus and Other Naturalists, from the Original Manuscripts*, 578.

293 *By the time of Smith's death* · Surprisingly for one so devoted to Linnaeus, Smith left behind instructions to sell his collection to the highest bidder. Unwilling to let the impetus for their creation fall again into private hands, the Linnean Society borrowed the money to purchase it for 3,150 pounds—a sum so large the Society would not pay off the debt until 1861.

293 *"A person who is in the pursuit"* · John Thornton, *The Temple of Flora* (London: Thornton, 1807), unnumbered.

293 *"Thousands became attached to the pursuit"* · James L. Drummond, *Observations of Natural Systems of Botany* (London: Longman, Brown, Green, and Longman, 1849), 1.

294 *"that language in which the history"* · Fries and Jackson, *Linnaeus: The Story of His Life*, 370.

295 *"The voyage to Japan is reckoned"* · Carl Peter Thunberg, *Travels in Europe, Africa, and Asia* (London: F. and C. Rivington, 1796), 8.

295 *"a very singular plant"* · IK Foundation & Company, *The Linnaeus Apostles: Global Science and Adventure* (London: IK Foundation & Company, 2007), 300.

298 *"this paradoxical quadruped"* · George Shaw, *Vivarium Naturae, or the Naturalist's Miscellany* (London: Royal Society of London), unnumbered.

299 *"under the necessity of christening it"* · Jacob W. Gruber, "Does the Platypus Lay Eggs? The History of an Event in Science," *Archives of Natural History* 18, no. 1.

299 *"My child, the Duck-billed Platypus"* · Oliver Herford, *The Simple Jography, or How to Know the Earth and Why It Spins* (New York: John W. Luce and Company, 1908), 95.

301 *"It was not offensive when it was proposed"* · "On the Troubles of Naming Species," *The Economist*, September 21, 2022, www.economist.com/science-and-technology/2022/09/21/on-the-troubles-of-naming-species.

302 *"there is one universal principle of development"* · Theodor Schwann, *Microscopical Researches into the Accordance in the Structure and Growth of Animals and Plants* (London: Sydenham Society, 1847), 165.

302 *"an assemblage of these particles constitutes"* · Buffon, *Natural History, General and Particular (Histoire Naturelle)*, 371.

303 *a consensus* · This division endured well into the later twentieth century. Even in the face of ample evidence that bacteria were not plants, academic biologists interested in studying them were still required to register with their university's botanical department.

TWENTY-SIX | *Laughably Like Mine*

305 *"Darwin might be anything"* · Thomas Henry Huxley, *Life and Letters of Thomas Henry Huxley* (London: Macmillan and Co., 1913), 137.

305 *"I can remember the very spot"* · Charles Darwin, *The Life and Letters of Charles Darwin* (New York: Appleton and Company, 1898), 69.

305 *"I am almost convinced"* · Charles Darwin, *Charles Darwin's Letters: A Selection 1825–1859* (Cambridge, U.K.: Cambridge University Press, 1998), 81.

306 *"absurd though clever"* · Packard, *Lamarck, the Founder of Evolution*, 74.

306 *"Wallace's impetus seems to have set"* · Huxley, *Life and Letters of Thomas Henry Huxley*, 230.

306 *"If I can convert Huxley"* · Ibid., 179.

306 *"Habit also has a decided influence"* · Charles Darwin, *On the Origin of Species* (London: John Murray, 1869), 12.

307 *"a modification of Lamarck's doctrine"* · Charles Darwin, *The Life and Letters of Charles Darwin* (London: John Murray, 1887), 14.

308 *"I thank you most sincerely"* · Ibid., 44.

308 *"My dear Huxley"* · Ibid., 45.

308 *"was the first author who"* · Darwin, *On the Origin of Species*, xv.

309 *"This animal . . . is known to live"* · Jean Baptiste Lamarck, *Zoological Philosophy*, trans. Hugh Elliot (London: Macmillan and Co., 1914), 122.

310 *"Lamarck's hypothesis would predict"* · Michael T. Ghislen, "The Imaginary Lamarck: A Look at Bogus "History" in Schoolbooks," *The Textbook Letter*, September–October 1994.

310 *"Acquired characteristics are not inherited"* · Ibid.

310 *"I am not likely to take a low view"* · Packard, *Lamarck, the Founder of Evolution*, 74.

311 *"the common misconception of Buffon"* · Butler, *Evolution, Old and New*, 97.

311 *"sailed as near the wind"* · Ibid., 91.

312 *"this great man in the violence"* · Stoever, *The Life of Sir Charles Linnaeus*, 128.

312 *"inaccuracy and thoughtlessness in his manner"* · Saint-Beuve, *Causeries du Lundi*, 38.

313 *"Buffon's chimpanzee"* · Robert Hartmann, *Anthropoid Apes* (New York: D. Appleton and Company, 1886), 267.

314 *"facts calculated to excite astonishment"* · Georges-Louis Leclerc le Comte de Buffon, and Others, *Buffon's Natural History of Man, the Globe, and of Quadrupeds* (New York: Leavitt & Allen, 1857), iii.

315 "Such is Buffon's arrangement" · Ibid., 135.

315 *"One day early"* · George Bernard Shaw, *Back to Methuselah: A Metabiological Pentateuch* (New York: Brentano's, 1921), vii.

TWENTY-SEVEN | *The Rhymes of the Universe*

316 *"And he wandered away and away"* · Botsford Comstock, *Handbook of Nature-Study* (Ithaca, N.Y.: Comstock Publishing Co., 1935), 2.

317 *"not two hundred steps away from the Jardin"* · Elizabeth Cabot Cary Agassiz, *Louis Agassiz: His Life and Correspondence* (1887), 162.

318 *"He greeted me with great politeness"* · Ibid.

318 *"I had given it so much attention"* · Ibid., 167.

318 *"Before the immortal works of Linnaeus"* · Agassiz, 1842–6, Nomenclator Zoologicus preface, unnumbered.

319 *"men of science, applying their attainments"* · Anonymous, "Miscellaneous Intelligence: Arts and Sciences at Harvard," *American Journal of Science and the Arts* IV (1847), 295.

319 *"The monuments of Egypt have fortunately"* · Louis Agassiz, "The Diversity of Origin of Human Races," *The Christian Examiner and Religious Miscellany* XIV (1850), 116.

320 *"You and we are different races"* · Henry J. Raymond, *History of the Administration of President Lincoln* (New York: J. C. Derby & N. C. Miller), 469.

321 Lincoln considered ineradicable · As early as the Lincoln-Douglas debates of 1858, he'd spoken of "a physical difference between the white and black races which I believe will forever forbid the two races living together on terms of social and political equality."

321 *"The President of the United States seems"* · Frederick Douglass, *The Portable Frederick Douglass* (New York: Penguin Publishing Group, 2016), 479.

321 *"the propriety of considering Genesis as chiefly relating"* · Agassiz, "The Diversity of Origin of Human Races," 138.

322 *"they may but be compared to children"* · Stephen Jay Gould, *The Panda's Thumb: More Reflections in Natural History* (New York: W. W. Norton, 1980), 174.

322 *"he will somewhere find"* · American Freedmen's Inquiry Commission, *Preliminary Report Touching the Condition and Management of Emancipated Refugees* (United States Government Publication, 1863), 34.

324 "If Buffon had assumed" · Mendel's Annotations in His German Translation of Darwin's (1868) *The Variation of Animals and Plants Under Domestication*, v. 2, supplemental file to D. J. Fairbanks, "Mendel and Darwin: Untangling a Persistent Enigma," *Heredity* 124 (2020).

325 *"It requires some courage to undertake a labor"* · Theodore Dru Alison Cockerell, *Zoölogy: A Textbook for Colleges and Universities* (Yonkers-on-Hudson, N.Y.: World Book Company, 1920), 42.

326 *"The world has arisen in some way"* · Louis Agassiz, "Evolution and the Permanence of Type," *The Atlantic Monthly* (1874), 101.

326 *"The wise of old welcome"* · James Russell Lowell, *The Poetical Works of James Russell Lowell* (Boston: Houghton, Mifflin and Company, 1890), 444.

TWENTY-EIGHT │ *Most Human of Humans*

327 *"There are some people who see"* · Leonard Huxley, *Thomas Henry Huxley: A Character Sketch* (London: Watts & Company, 1920), 115.

328 *"In the study of evolution"* · William Bateson, *Mendel's Principles of Heredity: A Defence* (Cambridge, U.K.: Cambridge University Press, 1902), v.

329 *"a new province of observational and experimental work"* · Herbert George Wells, Julian Huxley, and George Phillip Wells, *The Science of Life* (London: Cassell & Company Ltd., 1934), 22.

329 The Science of Life · Wells's son George is also credited as a co-author. Although he published no subsequent work, the younger Wells did go on to work as a zoologist.

330 *"in some of the more backward regions"* · Ibid.

330 *"consummate the sexual function"* · Paul T. Phillips, "One World, One Faith: The Quest for Unity in Julian Huxley's Religion of Evolutionary Humanism," *Journal of the History of Ideas* 68, no. 4 (2007).

330 Huxley watched in dismay · It's worth noting that a younger Huxley was a vocal supporter of "eugenics," which he perceived as the benign deployment of genetic knowledge to improve humanity. This naivete did not survive the rapid escalation of Nazi "eugenics" into full-fledged genocide.

330 *"In a scientific age"* · Julian Huxley and A. C. Haddon, *We Europeans* (New York: Harper & Brothers, 1936), vii.

331 *"as blond as Hitler"* · Ibid., 13.

331 ruining the authors' careers · Benedict died soon after the war, but Weltfish, forced to testify before the House Un-American Activities Committee about her supposed Communist sympathies, was fired by Columbia.

332 *"The biological fact of race and the myth"* · Julian Huxley and Others, *The Race Question* (Paris: UNESCO, 1950), 8.

333 *"looks a rather dull and formal"* · W. T. Stearn, "The Background of Linnaeus's Contributions to the Nomenclature and Methods of Systematic Biology," *Systematic Zoology* 8, 1 (1959).

335 *white, Swedish, and male* · The zoologist and paleontologist Edward Drinker Cope (1840–1897) is occasionally misidentified as the type specimen of *Homo sapiens*. While Cope may have wished for that distinction—his will specified that his skeleton be preserved and displayed for science—it was not granted.

335 *"The older descriptive work in biology"* · Wells, Huxley, and Wells, *The Science of Life*, 22.

TWENTY-NINE │ *A Large Web or Rather a Network*

341 *"champion of taxonomic redundancy"* · Lewis Smith, "Seaside Snail Most Misidentified Creature in the World," *The Guardian*, March 12, 2015, https://www.theguardian.com/environment/2015/mar/12/seaside-snail-most-misidentified-creature-in-the-world.

342 *"quickly buries itself under the skin"* · Carl Linnaeus, *A General System of Nature, Through the Three Grand Kingdoms* (London: Lackington, Allen and Co., 1806), 57.

342 *produce a specimen* · The Swedish Academy of Sciences even offered a reward in gold for a single specimen of *Furia infernalis*. There were no claimants.

342 *"To put our results into perspective"* · Giraffe Conservation Foundation, "New Genomic Level Analysis Confirms Four Species of Giraffe and a Need to Prioritise Their Conservation," Giraffe Conservation Foundation, May 5, 2021, https://giraffeconservation.org/2021/05/05/4-giraffe-species-confirmed.

343 *"Giraffes are assumed to have"* · Ibid.

343 *"Linnaeus' work on the classification"* · Isabelle Charmantier, "Linnaeus and Race," The Linnean Society, September 3, 2020, www.linnean.org/learning/who-was-linnaeus/linnaeus-and-race.

344 *"Species names are not static"* · Sandra Knapp, "Naming Nature: The Future of the Linnaean System," *The Linnean Special Issue* 8 (2008), 175.

345 *"Until now, we haven't known whether"* · Stephanie Pappas, "There Might Be 1 Trillion Species on Earth," Live Science, May 5, 2016, https://www.livescience.com/54660-1-trillion-species-on-earth.html.

347 *"Linnaeus's motivation for assigning species"* · Marc Ereshefsky, *The Poverty of the Linnaean Hierarchy: A Philosophical Study of Biological Taxonomy* (Cambridge, U.K.: Cambridge University Press, 2008), 214.

349 *"humongous fungus"* · Vince Patton, "Oregon Humongous Fungus Sets Record as Largest Single Living Organism on Earth," OPB, February 12, 2015, https://www.opb.org/television/programs/oregon-field-guide/article/oregon-humongous-fungus/.

350 *"almost as a spider web"* · Carl Zimmer, "DNA Analysis Reveals Ancient Secrets," *New York Times*, February 11, 2020.

350 *"This chain is not a simple thread"* · Georges-Louis Leclerc le Comte de Buffon, *Histoire Naturelle des Oiseaux* (Paris: 1770), 394.

352 *"statements about the current state"* · Francine Pleijel and George Rouse, "Least-Inclusive Taxonomic Unit: A New Taxonomic Concept for Biology," *Proceedings of the Royal Society B* 267, no. 1443 (2000).

353 *"major and still growing impacts"* · Paul J. Crutzen and Eugene F. Stoermer, "The "Anthropocene," in *The Future of Nature: Documents of Global Change*, ed. Libby Robin, Paul Warde, and Sverker Sorlin (New Haven, Conn.: Yale University Press, 2013), 483.

353 *"The most despicable condition of the human species"* · Buffon, *The Epochs of Nature*, 125.

354 *"It really is an extraordinary book"* · Ibid., viii.

354 *"nothing more than a rough sketch"* · Loren Eisley, *Darwin's Century: Evolution and the Men Who Discovered It* (New York: Anchor Books, 1961), 39.

355 *"Buffon asked almost all"* · Otis Fellows, *From Voltaire to la Nouvelle Critique: Problems and Personalities* (Geneva: Librairie Droz, 1970), 26.

355 *"Those who have looked with care"* · Ibid., 25.

355 *"Linnaeus mentions the law of generation"* · Stafleu, *Linnaeus and Linnaeans: The Spreading of Their Ideas in Systematic Botany, 1735–1789*, 303.

359 *"A slime mold creates"* · Tim Stephens, "Astronomers Use Slime Mold Model to Reveal Dark Threads of the Cosmic Web," UC Santa Cruz, March 10, 2020, https://news.ucsc.edu/2020/03/cosmic-web.html.

359 *"Nature is not a thing"* · Roger, *Buffon: A Life in Natural History*, 329.

Bibliography

Adanson I. Pittsburgh: The Hunt Botanical Library, 1963.

Adanson, Michel. *Familles des Plantes.* New York: J. Kramer-Lehre, 1966.

Agassiz, Elizabeth Cabot Cary. *Louis Agassiz: His Life and Correspondence,* 1887.

Agassiz, Louis. "The Diversity of Origin of Human Races." *The Christian Examiner and Religious Miscellany* XIV (1850).

———. "Evolution and the Permanence of Type." *The Atlantic Monthly* (1874).

American Freedmen's Inquiry Commission. *Preliminary Report Touching the Condition and Management of Emancipated Refugees.* United States Government Publication, 1863.

Anonymous. "Miscellaneous Intelligence: Arts and Sciences at Harvard." *American Journal of Science and the Arts* IV (1847).

Appel, Toby A. *The Cuvier-Geoffroy Debate: French Biology in the Decades Before Darwin.* Oxford, U.K.: Oxford University Press, 1987.

Bateson, William. *Mendel's Principles of Heredity: A Defence.* Cambridge, U.K.: Cambridge University Press, 1902.

Beckmann, Johan. *Schwedische Reise: 1765–1766.* Uppsala: Almqvist & Wiksells, 1911.

Blumenbach, Johan Friedrich. *De Generis Humani Varietate Nativa Liber.* Gottingen: 1775.

———. *The Anthropological Treatises of Johann Friedrich Blumenbach.* London: Longman, Green, 1865.

Blunt, Wilfrid. *Linnaeus: The Compleat Naturalist.* Princeton, N.J.: Princeton University Press, 2001.

Broberg, Gunnar. *Linnaeus: Progress and Prospects in Linnaean Research.* Uppsala: Almqvist & Wiksell International, 1980.

———. *"Brown-Eyed, Nimble, Hasty, Did Everything Promptly": Carl Linnaeus 1708–1778.* Uppsala: Uppsala University, 1990.

Brookes, Richard. *A New and Accurate System of Natural History.* London: J. Newberry, 1763.

Buffon, Georges-Louis Leclerc le Comte de. *Histoire Naturelle.* Paris: L'Imprimerie Royale, 1753.

——— *Histoire Naturelle des Oiseaux.* Paris: 1770.

———. *Supplément à l'Histoire Naturelle.* Paris: L'Imprimerie Royale, 1775.

———. *The Natural History of Animals, Vegetables, and Minerals (Histoire Naturelle).* Translated by W. Kenrick and J. Murdoch. London: T. Bell, 1775.

———. *Supplément à l'Histoire Naturelle.* Paris: L'Imprimerie Royale, 1776.

———. *Natural History, General and Particular (Histoire Naturelle).* Translated by William Smellie. London: W. Strahan and T. Cadell, 1785.

———. *Histoire Naturelle des Minéraux*. Paris: L'Imprimerie Royale, 1786.

———. *Buffon's Natural History (Barr's Buffon)*. Translated by J. S. Barr. London: J. S. Barr, 1792.

———. *The Epochs of Nature*. University of Chicago Press, 2018.

Buffon, Georges-Louis Leclerc le Comte de, and Others. *Buffon's Natural History of Man, the Globe, and of Quadrupeds*. New York: Leavitt & Allen, 1857.

Burkhardt, Richard W. *The Spirit of System: Lamarck and Evolutionary Biology*. Cambridge, Mass.: Harvard University Press, 1977.

Butler, Samuel. *Evolution, Old and New*. London: Hardwicke and Bogue, 1879.

Carr, Daniel C. *Linnæus and Jussieu*. West Strand, U.K.: John W. Parker, 1844.

Carteret, Xavier. "Michel Adanson in Senegal (1749–1754): A Great Naturalistic and Anthropological Journey of the Enlightenment." *Revue d'Histoire des Sciences* 65 (2012): 5–25.

Cobb, Matthew. *Generation*. New York: Bloomsbury, 2008.

Cockerell, Theodore Dru Alison. *Zoölogy: A Textbook for Colleges and Universities*. Yonkers-on-Hudson, N.Y.: World Book Company, 1920.

Comstock, Anna Botsford. *Handbook of Nature-Study*. Ithaca, N.Y.: Comstock Publishing Co., 1935.

Crane, Eva. *The World History of Beekeeping and Honey Hunting*. New York: Routledge, 1999.

Crutzen, Paul J., and Eugene F. Stoermer. "The Anthropocene." In *The Future of Nature: Documents of Global Change*, edited by Libby Robin, Paul Warde, and Sverker Sorlin. New Haven, Conn.: Yale University Press, 2013.

Curtis, Caitlin, Craig D. Millar, and David M. Lambert. "The Sacred Ibis Debate: The First Test of Evolution." *PLOS Biology* (2018).

Cuvier, George. *The Animal Kingdom, Arranged in Conformity with Its Organization*. London: Wm. S. Orr and Co., 1854.

D'Haussonville, Vicomte. *The Salon of Madame Necker*. Translated by Henry M. Trollope. London: Chapman and Hall, 1882.

Dance, Peter. *Animal Fakes and Frauds*. Berkshire: Sampson Low, 1976.

Darwin, Charles. *On the Origin of Species*. London: John Murrary, 1869.

———. *The Life and Letters of Charles Darwin*. London: John Murray, 1887.

———. *The Life and Letters of Charles Darwin*. New York: Appleton and Company, 1898.

———. *Charles Darwin's Letters: A Selection 1825–1859*. Cambridge, U.K.: Cambridge University Press, 1998.

Douglass, Frederick. *The Portable Frederick Douglass*. New York: Penguin Publishing Group, 2016.

Draper, James David. *Augustin Pajou: Royal Sculptor, 1730–1809*. New York: Metropolitan Museum of Art, 1998.

Drummond, James L. *Observations of Natural Systems of Botany*. London: Longman, Brown, Green, and Longman, 1849.

Dugatkin, Lee Alan. *Mr. Jefferson and the Giant Moose: Natural History in Early America*. Chicago: University of Chicago Press, 2019.

Dunmore, John. *Monsieur Baret: First Woman Around the World, 1766–68*. Auckland: Heritage Press, 2002.

Duris, Pascal. *Linné et la France (1780–1850)*. Geneva: Librairie Droz, 1993.

———. "Linné et le Jardin du Roi." Hypotheses, April 16, 2012. https://objethistoire
.hypotheses.org/88.

Economist, The. "On the Troubles of Naming Species." September 21, 2022. www
.economist.com/science-and-technology/2022/09/21/on-the-troubles-of-naming
-species.

Eddy, John Herbert. "Buffon, Organic Change, and the Races of Man," diss., The
University of Oklahoma, 1977.

Eisley, Loren. *Darwin's Century: Evolution and the Men Who Discovered It.* New York: Anchor
Books, 1961.

Ereshefsky, Marc. *The Poverty of the Linnaean Hierarchy: A Philosophical Study of Biological Taxonomy.* Cambridge, U.K.: Cambridge University Press, 2008.

Fairbanks, D. J. "Mendel and Darwin: Untangling a Persistent Enigma." *Heredity* 124
(2020): 263–73.

Fellows, Otis. *From Voltaire to la Nouvelle Critique: Problems and Personalities.* Geneva: Librairie Droz, 1970.

Fellows, Otis E., and Stephen F. Milliken. *Buffon.* New York: Twayne Publishers, 1972.

Findlen, Paula. "Inventing Nature: Commerce, Art, and Science in the Early Modern
Cabinet of Curiosities." In *Merchants and Marvels: Commerce, Science and Art in Early Modern Europe,* edited by Pamela H. Smith and Paul Findlen. New York: Routledge,
2002.

Fries, Theodor Magnus, and Benjamin Daydon Jackson. *Linnaeus: The Story of His Life.*
London: H.F. & G. Witherby, 1923.

Galsworthy, John. "The New Spirit in the Drama." *The Living Age,* 1913.

Gambier-Parry, Mark. *Madame Necker: Her Family and Her Friends.* Edinburgh: William
Blackwood and Sons, 1913.

Gascoigne, John. *Biographical Fragments (Fragments Biographiques): Previous Studies on the Life,
Works and Doctrines of Buffon.* Paris: F. D. Pillot, 1838.

———. *Joseph Banks and the English Enlightenment: Useful Knowledge and Polite Culture.* Cambridge, U.K.: Cambridge University Press, 2003.

Geoffroy Saint-Hilaire, Isidore. *Histoire Naturelle Générale des Règnes Organiques.* Paris: Libraire de Victor Masson, 1854.

Gerard, John. *The Herbal, or Generall Historie of Plantes.* London: Adam Islip, 1636.

Gerbi, Antonello. *The Dispute of the New World: The History of a Polemic, 1750–1900.* Pittsburgh: University of Pittsburgh Press, 2010.

Ghislen, Michael T. "The Imaginary Lamarck: A Look at Bogus "History" in Schoolbooks." *The Textbook Letter,* September–October 1994.

Gibbon, Edward. *The Miscellaneous Works of Edward Gibbon, Esq.* London: B. Blake, 1837.

Gillispie, Charles Coulston. *Science and Polity in France at the End of the Old Regime.* Princeton, N.J.: Princeton University Press, 1980.

Giraffe Conservation Foundation. "New Genomic Level Analysis Confirms Four Species of Giraffe and a Need to Prioritise Their Conservation." Giraffe Conservation
Foundation, May 5, 2021. https://giraffeconservation.org/2021/05/05/4-giraffe
-species-confirmed.

Goerke, Heinz. *Linnaeus.* New York: Scribner, 1973.

Gonzalez, Jorge M. "Pehr Löfling: Un Apóstol de Linné en Tierras de Venezuela." *Ciencia y Tecnología* (2018). https://www.meer.com/es/36201-pehr-lofling.

Gould, Stephen Jay. *The Panda's Thumb: More Reflections in Natural History*. New York: W. W. Norton, 1980.

———. *The Mismeasure of Man*. New York: W. W. Norton, 1996.

———. *The Lying Stones of Marrakesh*. Cambridge, Mass.: Harvard University Press, 2011.

Gourlie, Norah. *The Prince of Botanists: Carl Linnaeus*. London: H.F. & G. Witherby, 1953.

Gribbin, Mary, and John Gribbin. *Flower Hunters*. Oxford: Oxford University Press, 2008.

Gruber, Jacob W. "Does the Platypus Lay Eggs? The History of an Event in Science." *Archives of Natural History* 18, no. 1, 51–123.

Guizot, François Pierre Guillaume. *The Nations of the World: France*. New York: Peter Fenelon Collier, 1898.

Hagberg, Knut. *Carl Linnaeus*. New York: Cape, 1952.

Hartmann, Robert. *Anthropoid Apes*. New York: D. Appleton and Company, 1886.

Herford, Oliver. *The Simple Jography, or How to Know the Earth and Why It Spins*. New York: John W. Luce and Company, 1908.

Herodotus, and George Rawlinson. *The History: A New English Version*. New York: D. Appleton, 1889.

Hertwig, Richard. *General Principles of Zoology*. New York: Henry Holt and Company, 1896.

Huxley, Julian, and A. C. Haddon. *We Europeans*. New York: Harper & Brothers, 1936.

Huxley, Julian, and Others. *The Race Question*. Paris: UNESCO, 1950.

Huxley, Leonard. *Thomas Henry Huxley: A Character Sketch*. London: Watts & Company, 1920.

Huxley, Thomas Henry. *Life and Letters of Thomas Henry Huxley*. London: Macmillan and Co., 1913.

IK Foundation & Company. *The Linnaeus Apostles: Global Science and Adventure*. London: IK Foundation & Company, 2007.

———. *The Linnaeus Apostles: Global Science and Adventure*. London: IK Foundation & Company, 2008.

———. *The Linnaeus Apostles: Global Science and Adventure*. London: IK Foundation & Company, 2010.

Jefferson, Thomas. *Notes on the State of Virginia*. Boston: Wells and Lilly, 1829.

———. *The Writings of Thomas Jefferson: 1816–1826*. New York: G. P. Putnam's Sons, 1899.

Johnson, Alison. *Louis XVI and the French Revolution*. Jefferson, N.C.: McFarland & Company, 2013.

Jonsson, Ann-Mari. "Linnaeus' International Correspondence. The Spread of a Revolution." In *Languages of Science in the Eighteenth Century*, edited by Britt-Louise Gunnarsson. Berlin: De Gruyter Mouton, 2011.

Kalm, Peter. *Travels into North America*. Warrington, U.K.: William Eyres, 1770.

———. *Peter Kalm's Travels in North America*. New York: Wilson-Erickson, 1937.

Keevak, Michael. *Becoming Yellow: A Short History of Racial Thinking*. Princeton, N.J.: Princeton University Press, 2011.

Knapp, Sandra. "Naming Nature: The Future of the Linnaean System." *The Linnean Special Issue* 8 (2008): 175.

Koerner, Lisbet. *Linnaeus: Nature and Nation*. Cambridge, Mass.: Harvard University Press, 2009.

Lacépède, Bernard-Germain-Étienne. *Histoire Naturelle des Quadrupedes Ovipares et des Serpens*. 1788.

Lamarck, Jean Baptiste. *Zoological Philosophy*. Translated by Hugh Elliot. London: Macmillan and Co., 1914.

Larson, James Lee. "Reason and Experience: An Inquiry into Systematic Description in the Work of Carl Von Linné," diss., University of California, 1965.

Latham, Robert Gordon. *The Natural History of the Varieties of Man*. London: John Van Voorst, 1850.

Lee, Sara. *Memoirs of Baron Cuvier*. London: Longman, 1833.

Lewes, George Henry. *Studies in Animal Life*. New York: Harper & Brothers, 1860.

Lindroth, Sten. "The Two Faces of Linnaeus." In *Linnaeus: The Man and His Work*, edited by Tore Frangsmyr. Sagamore Beach, Mass.: Science History Publications/USA, 1994.

Linnaeus, Carl. *Systema Naturae (First Edition)*. Leyden: 1735.

———. *Systema Naturae (Second Edition)*. Stockholm: 1740.

———. *Systema Naturae (Tenth Edition)*. Stockholm: 1758.

———. *Vita Carolia Linnaei*. 1760.

———. *Clavis Medicinae Duplex*. Stockholm: Lars Salvius, 1766.

———. *A General System of Nature, Through the Three Grand Kingdoms*. London: Lackington, Allen and Co., 1806.

———. *Lachesis Lapponica, or, a Tour in Lapland*. London: White and Cochrane, 1811.

———. *Travels*. New York: Charles Scribner's Sons, 1979.

———. *Nemesis Divina*. Translated by Eric Miller. Lanham, Md.: University Press of America, 2002.

———. *Philosophia Botanica*. Oxford: Oxford University Press, 2005.

———. *Musa Cliffortiana*. Vienna: A.R.G. Ganter Verlag K. G., 2007.

———. *Clavis Medicinae Duplex: The Two Keys of Medicine*. London: IK Foundation & Company, 2012.

Locy, William A. *Biology and Its Makers*. New York: Henry Holt and Company, 1908.

Lovejoy, Arthur O. *The Great Chain of Being: A Study of the History of an Idea*. Cambridge, Mass.: Harvard University Press, 1976.

Lowell, James Russell. *The Poetical Works of James Russell Lowell*. Boston: Houghton, Mifflin and Company, 1890.

Lyon, John, and Phillip R. Sloan. *From Natural History to the History of Nature: Readings from Buffon and His Critics*. Notre Dame, Ind.: University of Notre Dame Press, 1981.

McInerny, Ralph M. *A History of Western Philosophy*. Notre Dame, Ind.: University of Notre Dame Press, 1963.

Meijer, Miriam Claude. "The Collaboration Manqué: Petrus Camper's Son at Montbard, 1785–1787." Accessed March 6, 2023. https://petruscamper.com/buffon/montbard.htm.

Mettrie, Julien d'Offray de la. *L'Homme Plante*. Potsdam, Germany: Chretien Frederic Voss, 1748.

Milliken, Stephen F. *Buffon and the British*. New York: Columbia University Press, 1965.

Millingen, John Gideon. *The History of Duelling*. London: Samuel Bentley, 1841.

Muir, John. "Linnaeus." In *The World's Best Literature*, edited by John W. Cunliffe and Ashley Thorndike. New York: The Knickerbocker Press, 1917.

Nadault de Buffon, Henri. *Correspondance Inédite de Buffon*. Paris: Hachette, 1860.

Naturalist, The. London: Whittaker and Co., 1838.

Necker, Suzanne Curchod. *Mélanges Extraits des Manuscrits*. Paris: Charles Pougens, 1798.

Ostler, Catherine. *The Duchess Countess*. New York: Atria Books, 2022.

Packard, Alpheus Spring. *Lamarck, the Founder of Evolution*. New York: Longmans, Green, 1901.

Pain, Stephanie. "The Forgotten Apostle." *New Scientist* 195, no. 2615 (2007): 41–45.

Patton, Vince. "Oregon Humongous Fungus Sets Record as Largest Single Living Organism on Earth." OPB, February 12, 2015. https://www.opb.org/television/programs/oregon-field-guide/article/oregon-humongous-fungus/.

Peattie, Donald Culross. *Green Laurels: The Lives and Achievements of the Great Naturalists*. New York: Simon & Schuster, 1936.

Phillips, Paul T. "One World, One Faith: The Quest for Unity in Julian Huxley's Religion of Evolutionary Humanism." *Journal of the History of Ideas* 68, no. 4 (2007): 613+.

Picard, Jacques. "Encyclopédistes Méconnus (3ème Partie): Michel Adanson et Son Projet d'Encyclopédie du Vivant." Accessed March 3, 2023. https://www.dicopathe.com/encyclopedistes-meconnus-3eme-partie-michel-adanson-et-son-projet-dencyclopedie-du-vivant/.

Pinkerton, John. *A General Collection of the Best and Most Interesting Voyages and Travels in All Parts of the World*. Longman, 1814.

Pleijel, Francine, and George Rouse. "Least-Inclusive Taxonomic Unit: A New Taxonomic Concept for Biology." *Proceedings of the Royal Society B* 267, no. 1443 (2000).

Pulteney, Richard. *A General View of the Writings of Linnaeus*. London: J. Mawman, 1805.

Raymond, Henry J. *History of the Administration of President Lincoln*. New York: J. C. Derby & N. C. Miller.

Robbins, Paula. *The Travels of Peter Kalm: Finnish-Swedish Naturalist Through Colonial North America, 1748–1751*. Fleischmanns, N.Y.: Purple Mountain Press, 2007.

Roberts, Andrew. *Napoleon and Wellington*. New York: Simon & Schuster, 2002.

Roger, Jacques. *Buffon: A Life in Natural History*. Ithaca, N.Y.: Cornell University Press, 1997.

Rosen, George. *A History of Public Health*. Baltimore: Johns Hopkins University Press, 1993.

Ross, Michael. *Bougainville*. London: Gordon & Cremonesi, 1978.

Saint-Beuve, Charles Augustin. *Portraits of the Eighteenth Century, Historic and Literary*. New York: G. P. Putnam's Sons, 1905.

———. *Causeries du Lundi*. London: George Routledge & Sons, 1909.

Schiebinger, Londa. *Nature's Body: Gender in the Making of Modern Science*. New Brunswick, N.J.: Rutgers University Press, 2004a.

———. *Plants and Empire: Colonial Bioprospecting in the Atlantic World*. Cambridge, Mass.: Harvard University Press, 2004b.

Schwann, Theodor. *Microscopical Researches into the Accordance in the Structure and Growth of Animals and Plants*. London: Sydenham Society, 1847.

Scott, John. "On the Burning Mirrors of Archimedes, and on the Concentration of Light Produced by Reflectors." *Proceedings of the Royal Society of Edinburgh* 6 (1869).

Shaw, George. *Vivarium Naturae, or the Naturalist's Miscellany*. London: Royal Society of London (1790–1813).

Shaw, George Bernard. *Back to Methuselah: A Metabiological Pentateuch*. New York: Brentano's, 1921.

Sloan, Phillip R. "The Buffon-Linnaeus Controversy." *Isis* 67, no. 3 (1976): 356–75.

Smith, James Edward. *Translation of Linnæus's Dissertation on the Sexes of Plants*. Dublin: Luke White, 1786.

———. *A Selection of the Correspondence of Linnaeus and Other Naturalists, from the Original Manuscripts*. London: Longman, Hurst, Rees, Orme and Brown, 1821.

———. *Memoir and Correspondence*. London: Longman, Rees, Orme, Brown, Green and Longman, 1832.

Smith, Lewis. "Seaside Snail Most Misidentified Creature in the World." *The Guardian*, March 12, 2015. https://www.theguardian.com/environment/2015/mar/12/seaside-snail-most-misidentified-creature-in-the-world.

Society of Gentlemen in Scotland. *Encyclopaedia Britannica*. 1771.

Spaary, E.C. *Utopia's Garden: French Natural History from Old Regime to Revolution*. Chicago: University of Chicago Press, 2000.

Stafleu, Frans A. *Linnaeus and Linnaeans: The Spreading of Their Ideas in Systematic Botany, 1735–1789*. Utrecht, The Netherlands: A. Oosthoek's Uitgeversmaataschappij, N.V. First edition, 1971.

Stearn, W. T. "The Background of Linnaeus's Contributions to the Nomenclature and Methods of Systematic Biology." *Systematic Zoology* 8, no. 1 (1959): 4–22.

Stephens, Tim. "Astronomers Use Slime Mold Model to Reveal Dark Threads of the Cosmic Web." UC Santa Cruz, March 10, 2020. https://news.ucsc.edu/2020/03/cosmic-web.html.

Stillingfleet, Benjamin. *Literary Life and Select Works of Benjamin Stillingfleet*. London: Longman, 1811.

Stoever, Dietrich Heinrich. *The Life of Sir Charles Linnaeus*. London: E. Hobson, 1794.

Terall, Mary. *Catching Nature in the Act: Réaumur and the Practice of Natural History in the Eighteenth Century*. Chicago: University of Chicago Press, 2014.

Thiéry, Maurice. *Bougainville: Soldier and Sailor*. London: Grayson & Grayson, 1932.

Thornton, John. *The Temple of Flora*. London: Thornton, 1807.

Thunberg, Carl Peter. *Travels in Europe, Africa, and Asia*. London: F. and C. Rivington, 1796.

Uncredited. "Necrologe des Artistes et des Curieux." *Revue universelle des arts* 13 (1861).

Voltaire. *A Philosophical Dictionary*. London: John and Henry L. Hunt, 1824.

Wells, Herbert George, Julian Huxley, and George Phillip Wells. *The Science of Life*. London: Cassell & Company Ltd., 1934.

White, Charles. *An Account of the Regular Gradation in Man*. London: C. Dilly, 1799.

Wright, John. *The Naming of the Shrew: A Curious History of Latin Names*. London: Blooms-bury, 2014.

Zimmer, Carl. "DNA Analysis Reveals Ancient Secrets." *New York Times,* February 11, 2020.

Index

ABOUT THE AUTHOR

JASON ROBERTS is a writer of fiction and nonfiction. His previous book, *A Sense of the World: How a Blind Man Became History's Greatest Traveler*, was a national bestseller and finalist for the National Book Critics Circle Award. A contributor to *McSweeney's, The Believer*, and other publications, he lives in Northern California.

www.jasonroberts.net

This book was set in Requiem, a typeface designed by the Hoefler Type Foundry. It is a modern typeface inspired by inscriptional capitals in Ludovico Vicentino degli Arrighi's 1523 writing manual, *Il modo de temperare le penne.* An original lowercase, a set of figures, and an italic in the chancery style that Arrighi (fl. 1522) helped popularize were created to make this adaptation of a classical design into a complete font family.